Particle Technology Forum 2013

Core Programming Area at the 2013 AIChE Annual Meeting: Global Challenges for Engineering a Sustainable Future

San Francisco, California, USA
3 – 8 November 2013

ISBN: 978-1-63439-050-7

Printed from e-media with permission by:

Curran Associates, Inc.
57 Morehouse Lane
Red Hook, NY 12571

Some format issues inherent in the e-media version may also appear in this print version.

Printed by Curran Associates, Inc. (2014)

AIChE
3 Park Avenue
New York, NY 10016-5991

Phone: (203) 702-7660
Fax: (203) 775-5177

www.aiche.org

Additional copies of this publication are available from:

Curran Associates, Inc.
57 Morehouse Lane
Red Hook, NY 12571 USA
Phone: 845-758-0400
Fax: 845-758-2634
Email: curran@proceedings.com
Web: www.proceedings.com

TABLE OF CONTENTS

(21a) Panel Discussion - Validation of Computational Codes

Cocco, R. A., Particulate Solid Research, Inc.
Curtis, J. S., University of Florida
Dave, R., New Jersey Institute of Technology
Hrenya, C. M., University of Colorado at Boulder
Lakshmanan, A., Aspen Technology, Inc.
Muzzio, F. J., Rutgers, The State University of New Jersey
Originally presented on: 11/3/2013 15:30:00 - 17:30:00

Appropriate validation of computational codes is a bane of process industry. The usefulness of results obtained from computational codes relies not only on systematic validation of basic assumptions, calculation methods and comparison with valid experimental data, but also on an understanding of how these validation efforts may be extrapolated to lend confidence in the modeling of new scenarios. Our invited panelists are experts from both industry and academia who have either led or participated in such efforts, and will be sharing their expertise with the audience

(52a) Discrete Element Simulation of the Electrostatic Effect On Particle Mixing in Vibrating Beds

Cheng, Y., National University of Singapore
Lee, L. Q.,
Zhang, W.,
Wang, C. H., National University of Singapore
Originally presented on: 11/4/2013 8:30:00 - 8:48:00

Due to particle-particle and particle-wall interactions, it is inevitable to generate electrostatic charges in a vibrating bed of granular materials. Because the electrostatic phenomenon can influence the flow dynamics in the bed, it is important to analyze the mechanism of electrostatic charging in such granular flow systems. In this study, a numerical model based on Discrete Element Method (DEM) is introduced to investigate the effects of vibrating frequency and amplitude on the electrostatic charging. The simulation model can be used to validate experimental data in the literature. With the increase of vibrating frequency and amplitude, the saturated charge on the particles will increase and the time required for the particles to reach the electrostatic equilibrium state will decrease. With the inclusion of electrostatic charge generation and transfer using the above DEM model, the particle mixing in the vibrating bed can be investigated. A new mixing index is introduced to quantify the mixing degree of different particles. It is concluded from the simulation results that the electrostatic effect can improve the particle mixing and particles in the vibrating bed can take less time to reach the well mixed state.

(52b) The Investigation of Cohesive Particle Mixing At Sub-Agglomerate Scale Using Discrete Element Method

Deng, X.,
Dave, R. N., New Jersey Institute of Technology
Scicolone, J. V., New Jersey Institute of Technology
Originally presented on: 11/4/2013 8:48:00 - 9:06:00

Recent experimental results show the magnetically assisted impaction mixing (MAIM), a high-shear mixing device, is capable of mixing nanoparticles at sub-agglomerate scale. Likewise, other high-shear and/or high-strain devices may be able to mix very cohesive powders due to their ability to break down agglomerates. In this work, MAIM system was used a model high-shear mixing device for cohesive particle mixing at sub-agglomerate scale. In order to better understand the dynamic process of the MAIM system, discrete element method (DEM) based modeling was employed. In a typical DEM mixing simulation, the initial state of the particles is in deagglomerated form. However, in order to truly capture the high cohesion of very fine powders that are micron or smaller in size, it would be desirable to examine them in an agglomerated form as the initial condition. Accordingly, a DEM analysis was performed to investigate the relationship between homogeneity of mixing (HoM) and various parameters such as magnets size, surface energy of non-magnet particles, and magnets-to-sample mass ratio. The cohesive force between non-magnets is based on the JKR model. Agglomerates were formed before mixing, thus better capturing the effect of cohesion on the initial state. Simulation results indicate that the mixing will be faster for decreasing surface energy, increasing mass ratio, and decreasing magnet size at fixed mass ratio. The results related to collision frequency and collision energy between magnets and non-magnets show that homogeneous mixing required longer processing times for higher surface energy since higher collision numbers and collision energies were necessary to deagglomerate the particles. Moreover, when the collision energy between magnets and non-magnets exceeds the cohesive energy, the HoM would achieve the steady value within a shorter processing period. The simulation results are in agreement with previously published experimental results, indicating the system model involving the evolution of agglomerates is applicable to cohesive power mixing.

(52c) Modeling of Heat Transfer in Rotary Calciners

Anderson, K. V., Rutgers University
Emady, H. N., Purdue University
Glasser, B. J., Rutgers University
Borghard, W. G., ExxonMobil
Cuitino, A., Rutgers University
Muzzio, F. J., Rutgers, The State University of New Jersey
Originally presented on: 11/4/2013 9:06:00 - 9:24:00

Rotary calciners are commonly used in industry to achieve high-temperature chemical reactions or a phase change for granular systems. Many industries use rotary calciners for applications such as sintering limestone to make cement, or calcining petroleum coke in refining. Though operational cost of the calciner is high, the design of calcination units is often conservative because many fundamental physics about the calcination process are widely unknown. If a fundamental physical understanding of both solids flow and heat transfer within the calciner were attained, rotary calciners could be optimized for higher output and/or reduced operational cost. The objective of this presentation is to provide a fundamental understanding of heat transfer within the rotary calciner. Discrete element method (DEM) modeling is used with a heat transfer code to study the heat transfer regimes within a rotary calciner heated at the walls. Additionally, an analytical model is proposed for one heat transfer regime. The analytical model describes the temperature profile within the granular bed of a rotary calciner. The model assumes that the calciner operates within the rolling mixing regime, has no flights or baffles, and heat transfer from radiation and gas convection are small. The analytical model provides a simple "back of the envelope" solution to the conduction heat transfer problem within the rotary calciner to take the place of time-consuming DEM simulations.

(52d) Experimental and Computational Investigation of Slow and Dense Granular Flow in Moving/Pebble Bed Type Reactors

Khane, V., Missouri University of Science & Technology
Al-Dahhan, M., Missouri University of Science and Technology
Originally presented on: 11/4/2013 9:24:00 - 9:42:00

Moving bed reactors are used in chemical and petrochemical industries in situations where there is a need to replace deactivated catalysts with new or regenerated catalysts. They are also under consideration as a 4th generation nuclear reactor known as Pebble Bed Reactor (PBR). The granular flow in moving/pebble bed reactors is of slow and dense type. The basic physics governing such flows is not yet fully understood and relies on experimental and computational investigations to extract useful information. It is important to study the movement of solids in moving/pebble beds, which can provide useful information about Lagrangian trajectories, residence time distributions, velocity fields, and the presence and extent of dead zones, if any. Conventional optics based velocimetry techniques are of limited use for investigation of granular flow in these reactors; as these systems are dense and opaque. Radio-isotopes based non-invasive techniques such as radioactive particle tracking (RPT) are suitable for such applications and can provide detailed information about solids flow fields, overall and local residence time distribution, stagnant zones and their sizes, and many other parameters. Discrete element method (DEM) based simulations can provide complete information in 3-D about granular flow in these reactors. DEM simulations rely on calculation of contact forces which are evaluated using phenomenological models. There is a lack of contact force models developed from the first principles and this demands assessment of contact force models with experimental bench-mark data. The design and development of a continuous pebble re-circulation experimental set-up, which simulates the flow of solids in a moving/pebble bed test reactor that measures one foot in diameter and one foot in height, is carried out as a part of this work. Advanced radioisotopes based flow visualization techniques such as RPT are implemented around this continuous pebble recirculation experimental set-up mimicking cold flow operation of moving/pebble bed reactors. RPT experiments provided crucial information about solids movement in terms of Lagrangian trajectories, velocities, residence times etc. EDEM™ (Experts of Discrete Element Method) is a commercial DEM based software from DEM Solutions Ltd., UK and is used to simulate a slow and dense granular flow. Hertz-Mindlin contact model (with no-slip) is used in these DEM based simulations. Obtained RPT results served as a benchmark data for the assessment of contact force model used in these DEM based simulations. EDEM™ based simulation results are in fair agreement with RPT experiments results.

(52e) Predicting Particle Impact Velocities in Vibrationally Fluidized Granular Flows Using the Discrete Element Method

Hashemnia, K., University of Toronto
Spelt, J. K., University of Toronto
Originally presented on: 11/4/2013 9:42:00 - 10:00:00

Vibratory finishing is widely used to deburr, polish, burnish, harden and clean metal, ceramic and plastic parts. In a tub vibratory finisher, a container filled with granular media is oscillated by an eccentric rotating shaft so that it develops a vibrationally-fluidized circulatory bulk flow of the media that is largely two-dimensional. The media have both a large-scale bulk flow velocity and a local impact velocity during each vibration cycle. Therefore, work-pieces that are entrained in the flowing media are subjected to the repetitive, high-frequency impacts of the surrounding particles.

Within a vibratory finisher, the erosive wear and plastic deformation of workpieces are largely affected by the velocity, frequency, and direction of the impacts with particles. Discrete element modeling (DEM) is a numerical method that is often used to solve problems involving transient dynamics of systems comprising a large number of moving bodies that interact with each other. Generally, DEM simulations give reasonable predictions of bulk flow velocity and volume fraction in both fluidized and non-fluidized flows. However, only a small number of studies have focused on the local behavior of the media in granular flows and some of these have investigated vibrationally-fluidized flows. Most of these studies have been about modelling, and the predicted collision scale quantities, such as impact velocity, collision frequency or impact energy, were not validated experimentally. Therefore, it is not known if the commonly used approaches and parameters in DEM simulations yield correct predictions of the impact velocity. This remains as a significant limitation since the impact forces that govern erosion, wear and fracture in granular flows are fundamentally linked to the particle impact velocity and kinetic energy.

The main objective of this research was to compare previous experimental measurements of particle impact and bulk flow velocities with those values predicted by numerical simulations using the discrete element method (DEM). The sensitivity of these predictions to the input values of the coefficients of restitution, friction and rolling resistance was also studied in a simplified model consisting of a single layer of particles.

The commercial software package, EDEM (DEM Solutions Inc. 2012), was used to perform the DEM simulations. The Hertz-Mindlin contact model was selected to calculate contact forces between the particles. In this model both the elastic and plastic deformation of particles were considered in the normal and tangential directions considering the coefficients of restitution, friction and rolling resistance.

It was of interest to study the sensitivity of the predicted bulk flow velocity and impact velocity to the contact coefficients used in the DEM. To begin, three simple case studies were considered using an analytical model of two-body collision: 1- collision of a moving disk with a fixed disk; 2- collision of a moving disk with a moving wall; 3- collision of two moving disks. In each case, the equations of the linear and angular rebound velocities of the disks were derived as functions of the contact parameters. The sensitivity of the rebound velocities to the contact parameters was determined by differentiating the velocities with respect to the coefficients. It was found that the coefficient of friction had the greatest effect on the tangential and angular velocities. However, the normal velocity was only a function of the coefficient of restitution.

The sensitivity analysis was then extended to a simplified DE model of the flow within the tub finisher consisting of a single-layer of particles moving between parallel sheets of glass. Design of experiments (Taguchi method) was used to reduce the number of required runs of the model to 25 in order to investigate the relative sensitivity of the velocity predictions to the 6 contact parameters under consideration. Analyzing the output of these 25 model runs with MINITAB™led to the conclusion that the coefficient of friction between the particles and the tub wall plays the most important role in determining the impact velocities. A pin-on-disk tribometer was used to measure the coefficients of friction between two steel balls, two porcelain balls, and between these balls and the polyurethane tub wall., he corresponding coefficients of restitution were measured using a high-speed laser displacement sensor.

The subsequent DE modelling sought to mimic the setup used in the earlier experimental study of bulk flow and impact velocities. First, the mechanical properties including shear modulus, Poisson's ratio and density of the steel balls, polyurethane wall of the tub and the glass partitions were input to the software. The glass partitions were assumed to be frictionless. The tub geometry was imported to the EDEM interface and its motion was defined according to the experimental measurements made using accelerometers. The sinusoidal motions were introduced to simulate 19 s of real-time operation. The simulation time step was set to be 25% of the Rayleigh critical time step which is defined as the time taken for a shear wave to propagate through a solid particle, and therefore was a function of the particle diameter, density and shear modulus. The simulation time step for 6.3 mm diameter steel balls was calculated to be 2.17×10^{-6}s. The data acquisition rate was 94 Hz which was twice the frequency of the tub vibration (47 Hz); i.e. the data was saved every 0.0105 s. The instantaneous velocities of particles (impact velocities) were acquired at each time step in a 9×9×6.3 mm bin in two different locations. The mean impact velocity was the average of the absolute values of the instantaneous velocities of the particles. The data acquisition was started 9 s after the tub began to vibrate in order to reach a steady flow.

Different numbers of particle layers were modeled between the glass partitions (1, 2, 4 and 8 layers). were simulated to study the differences between the simulation results of the single-layered and the multi-layered granular flows. Finally, the DE model was used to simulate the actual experimental setup, including the laser displacement sensor probe that was used to measure the impact and bulk flow velocities. In order to decrease the simulation time in this full model, only half of the geometry was modeled using a symmetry boundary made of the same material as the particles. The symmetry boundary was assumed to be frictionless to satisfy the symmetry conditions. The paper will present the comparisons of the measured impact and bulk flow velocities in a tub vibratory finisher with the measured values for two granular media – steel and porcelain balls having a diameter of 6.3 mm.

(52f) Application of DEM Simulations and Experimental Studies to Improve the Uniformity of An Active Coating Process

Toschkoff, G., RCPE GmbH

Just, S., Heinrich Heine University

Knop, K., Heinrich Heine University

Kleinebudde, P., Heinrich Heine University

Funke, A., Bayer HealthCare Pharmaceuticals

Djuric, D., L.B. BOHLE Maschinen + Verfahren GmbH

Scharrer, G., RCPE GmbH

Khinast, J. G., Graz University of Technology

Originally presented on: 11/4/2013 10:00:00 - 10:18:00

Introduction: Pan coating is commonly used to apply a functional layer to tablet cores. Tablets are placed in a rotating drum, and a coating liquid is sprayed onto the moving tablet bed. For most applications, the uniformity of coating (the equal distribution of coating over a single tablet and/or between different tablets of a batch) is of great importance. This necessitates a high degree of process understanding. A number of investigations were done both experimentally [1] and in recent years also computationally using the Discrete Element Method (DEM) [2], [3].

Objective: In this work, the application of an active coating was investigated: Gastrointestinal therapeutic systems (Adalat® GITS, Bayer Pharma AG, Germany) were coated with an aqueous suspension containing candesartan cilexetil as active pharmaceutical ingredient (API). As the variation of this API between tablets has to be small, a high inter-tablet coating mass uniformity is vital. The central aim therefore was to improve the inter-tablet coating uniformity.

Methods: To study how different process parameters influence the uniformity, both experiments and DEM simulations were performed. A lab pan coater (BFC 5, L.B. Bohle, Germany) with two or four spray nozzles (Düsen Schlick GmbH, Germany) was used. The process parameters were set according to the same statistical plan in experiments and simulations. The results of both are compared, and guidelines for optimization are extracted.

-- ***Experiments***: The factors pan load, pan speed, and spray rate were investigated in two 2^3 full factorial DoEs with three centre points. For each DoE, either two or four spray nozzles were used. The response variable was the coefficient of variation (CoV) of the API amount in the coating layer at the coating endpoint. To get the API amount, 20 active coated tablets per trial run sampled at the endpoint were analysed by HPLC (Elite La Chrom, VWR Hitachi, Germany).

-- ***Simulation*:** The Discrete Element Method (DEM) simulations were done using commercial software (EDEM 2.4, DEM Solutions, UK). The geometry of the coating apparatus was provided by the manufacturer. The material properties came from measurements [4]. Two methods for the modelling of a spray in DEM simulations were used (during run-time and in post-processing). They gave detailed information on the spray process, such as the coating mass of each tablet or the cycle and spray zone residence times.

Two 2^3 full factorial DoE (the same as in the experiments) were performed; for each run 90 seconds of process time were simulated. The response variable was the CoV calculated from the individual coating mass on the tablets. Further, additional simulations were done to provide more data points and to study the impact of e.g. axial tilt or spray nozzle setup.

Results & Conclusion: In the presented work, DEM simulations and experiments were performed to investigate active coating. The results are compared, and the insights concerning the optimization of uniformity are presented. For a deeper understanding, the outcome is also interpreted in the light of known analytic expressions for e.g. the time development of coating uniformity (cf. Fig. 1). It was possible to decrease the variation between tablets to values well below 6% as required from regulatory authorities, and to reduce the total process time.

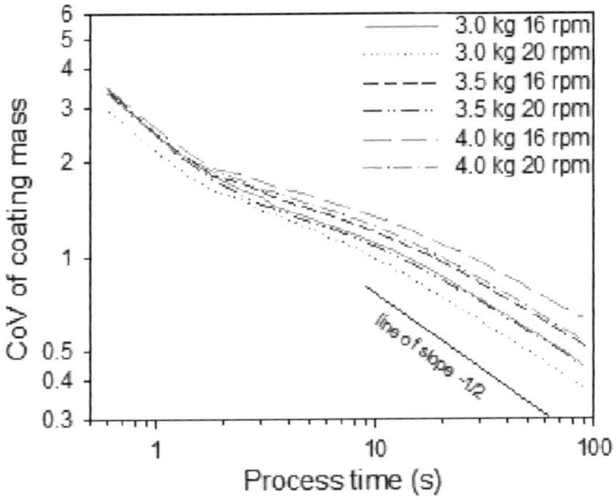

Figure 1: CoV of coating mass with increasing time for different process parameter settings from the DEM simulations. Random mixing models predict a decrease inversely proportional to the square root of coating time. (line of slope of -1/2 in the figure). After ~10 seconds, tablet movement is randomized, and the DEM results agree well with model expression.

References

[1] D. Brock, J. A. Zeitler, A. Funke, K. Knop, and P. Kleinebudde, ?A comparison of quality control methods for active coating processes.,? *International Journal of Pharmaceutics*, vol. 439, no. 1?2, pp. 289?95, Dec. 2012.

[2] B. Freireich, J. Li, J. Litster, and C. Wassgren, ?Incorporating particle flow information from discrete element simulations in population balance models of mixer-coaters,? *Chemical Engineering Science*, vol. 66, no. 16, pp. 3592?3604, 2011.

[3] A. Dubey, F. Boukouvala, G. Keyvan, R. Hsia, K. Saranteas, D. Brone, T. Misra, M. G. Ierapetritou, and F. J. Muzzio, ?Improvement of tablet coating uniformity using a quality by design approach.,? *AAPS PharmSciTech*, vol. 13, no. 1, pp. 231?246, Mar. 2012.

[4] S. Just, G. Toschkoff, A. Funke, D. Djuric, G. Scharrer, J. G. Khinast, K. Knop, and P. Kleinebudde, ?Experimental Analysis of Tablet Properties for Discrete Element Modeling of an Active Coating Process,? *AAPS PharmSciTech*, 2013.

(52g) DEM Simulation Scale-Up Study of Dry Impregnation Processes

Shen, Y., Rutgers University
Chester, A. W., Rutgers University
Tomassone, M. S., Rutgers University
Originally presented on: 11/4/2013 10:18:00 - 10:36:00

In the manufacturing of heterogeneous catalysts, the impregnation of active metals onto a porous catalyst support is a crucial preparation step that may significantly affect the activity and selectivity of the resulting catalysts particles. In a typical dry impregnation (pore filling or incipient wetness impregnation) process, metal solutions are sprayed over a particulate bed in a mixing vessel until the pore volume is reached. Current scale-up practices lead to poor fluid distribution and a wider particle size distribution. There are two issues to consider when scaling up the process: the balance between inertial and gravitational forces and the size and magnitude of the spray flow pattern as a function of system scale.

In this work, we use a Discrete Element Method (DEM) and work with two dimensionless numbers, Froude number (Fr) and Spray Flux number (Ψ), to characterize the scale up of the system. Froude number measures the ratio of inertial force to gravitational force. A good balance between the inertial and gravitational forces is important to obtain a given particle motion. The Spray Flux number measures the density of spray droplet (or flux) in the spray zone. The calculation of each dimensionless quantity converts specific operation parameters into a term value.

DEM simulations are performed in a simple cylindrical geometry whose width covers the entire spraying zone. We vary the vessel size, the rotation speed and the spray flow rate and the location of the nozzles. Our results show that for smaller Froude numbers the powder bed exhibits a flat surface profile, thus facilitating a better control of the size of the wetted area. However if the Froude number is too low, the system does not show good mixing. On the other hand, if the Fr number is too large the surface profile is two-angled and particle collision and attrition are more likely to occur. We found an optimal value for Fr number that gives the best mixing profile and particle dispersion, and also displays a flat surface profile. When the optimal Fr number is considered across various scales, an optimal Spray Flux number (Ψ) is also determined based on a given Fr number. We observe that the location and area of the spraying zone can significantly affect the overall content uniformity. Smaller Spray Flux numbers tend to give more homogeneous liquid distribution during and after the total liquid is depleted, while larger Spray Flux numbers result in greater inhomogeneities of the fluid content and poorer particle mixing in the powder bed. Hence, a homogeneous liquid distribution with optimal particle mixing is obtained by minimizing the spray rate and Ψ for an optimal Fr number. In larger scale blenders, multiple nozzles can be used and placed evenly along the axis to increase the spraying area and to keep the Spray Flux number low.

(52h) CFD Modeling of Proppant Distribution in a Multiply-Perforated Horizontal Pipe for Applications in Hydraulic Fracturing

Tsai, K., Dow Chemical
Originally presented on: 11/4/2013 10:36:00 - 10:54:00

Proppant distribution in a hydraulically stimulated horizontal well has important implications since it can significantly influence the production efficiency of an unconventional well. It has been known that a fractured reservoir with more evenly distributed proppants can have higher production over time, therefore it is important to understand the interplay between fluid and proppant distributions, and how to better promote the desired distribution of proppants.

In this work the proppant flow distribution is studied using computational fluid dynamics (CFD) with proppant particles individually traced to model the interactions between proppants and the carrier fluid (Snider, 2001; Tsai *et al.*, 2012). The flow field is solved using large eddy simulation (LES) with the interactions between proppants and carrier fluid resolved by adopting a well-established drag law (Wen and Yu, 1966) and a solid pressure model (Harris and Crighton, 1994) to account for the fluid-particle and inter-particle forces, respectively. By simulating known flow and proppant distributions in a multiply-perforated horizontal pipe (Crespo *et. al.,*2013), the driving mechanism for proppant distribution can be studied in detail.

Since proppants and carrier fluid are modeled interactively, investigation of proppant distribution's sensitivity on carrier fluid viscosity and proppant density can be carried out with high certainty. Issues regarding the validity of using acoustic data to interpret proppant distribution will also be discussed.

This is also a typical slurry flow problem often encountered in many chemical and refining processes, including the transport of polymer pellets, pneumatic transport of powdery particles, fluidized bed, *etc.* However, there are surprisingly few CFD investigations done for slurry flows in horizontal pipes. A literature survey shows that credible CFD models capable of predicting the pressure drop and solid distribution across a wide range of flow regimes are non-existent. This work attempts to tackle some of these questions by starting with proppant flows where monetary impact can be significant and hopefully can be extended to other areas of interests.

References,

Crespo, F., N. K. Aven, J. Cortez, M. Y. Soliman, A. Bokane, S. Jain and Y. Desphande, "Proppant distribution in multistage hydraulic fractured wells: A large scale inside-casing investigation," SPE Paper #163856, 2013.

Harris, S. E. and D. G. Crighton, "Solitons, solitary waves, and voidage disturbances in gas-fluidized beds," J. Fluid Mech. **266**, 243-276, 1994.

Snider, D., "An incompressible three-dimensional multiphase particle-in-cell model for dense particle flows," J. of Comp. Phys. **170**, 523-549, 2001.

Tsai, K., E. Fonseca, S. Degaleesan and E. Lake, "Advanced computational modeling of proppant settling in water fractures for shale gas production," SPE J. **18**, 50-56, 2013.

Wen, C. Y. and Y. H. Yu, "Mechanics of fluidization," Chem. Eng. Prog. Symp Ser. **62**, 100-110, 1966.

(53a) Eulerian Simulations of the Bubbling Dynamics of Binary Mixtures in Pseudo-2D and 3D Gas-Solid Fluidized Beds

Bates, R. B., Massachusetts Institute of Technology
Altantzis, C., M.I.T.
Ghoniem, A. F., Massachusetts Institute of Technology
Originally presented on: 11/4/2013 8:30:00 - 8:49:00

Fluidized bed reactors are utilized in a variety of industrial processes including fluid catalytic cracking (FCC), coal/biomass gasification and combustion, and chemical looping combustion (CLC). Most existing experimental and modeling studies of the bubbling behavior of gas solid fludized beds deal with monodispersed particles; however, in industrial practice cases this is rarely the case. During the fluidization of poly or bidisperse powders with size and/or density differences, mixing induced by bubble motion counteracts the segregation of lighter, smaller particles (flotsam) above the heavier, larger particles (jetsam). Thus the size, velocity and distribution of bubbles within the bed are crucial to achieving satisfactory gas-solid and solid-solid mixing[1]. While several experimental works from the literature examine the effect of bidispersity on global hydrodynamic characteristics[3] and mixing/segregation[4], few have analyzed the effect of bidispersity on the bubble behavior.

In this work, 2D and 3D Eulerian simulations of the fluidization of bidisperse powder mixtures was performed. Powder mixtures of varying size, density, and composition were simulated at a range of inlet superficial gas velocities. Subsequently, digital image analysis (DIA) techniques were then used to extract relevant bubble statistics. These statistics are compared with a systematic experimental study of bidisperse powder bubble dynamics by Busciglio et al. 2005[5]. By performing both 2D and 3D fine grid simulations, the effect of the walls on the bubble coalescence and splitting is clearly identified.

Sources cited:

1. Shen, L., Zhang, M. & Xu, Y. Solids mixing in fluidized beds. *Powder Technol.* **84,**207–212 (1995).

2. Rao, T. & Bheemarasetti, J. . Minimum fluidization velocities of mixtures of biomass and sands. *Energy* **26,**633–644 (2001).

3. Noda, K., Uchida, S., Makino, T. & Kamo, H. Minimum fluidization velocity of binary mixture of particles with large size ratio. *Powder Technol.* **46,**149–154 (1986).

4. Goldschmidt, M. J. V., Link, J. M., Mellema, S. & Kuipers, J. A. M. Digital image analysis measurements of bed expansion and segregation dynamics in dense gas-fluidised beds. *Powder Technol.* **138,**135–159 (2003).

5. Busciglio, A., Vella, G. & Micale, G. On the bubbling dynamics of binary mixtures of powders in 2D gas-solid fluidized beds. *Powder Technol.* **231,** 21–34 (2012).

(53b) Numerical Modeling of Sorption-Enhanced Water-Gas-Shift Reaction in a Circulating Fluidized Bed Reactor

Abbasi, E., Illinois Institute of Technology
Abbasian, J., Illinois Institute of Technology
Arastoopour, H., Illinois Institute of Technology
Originally presented on: 11/4/2013 8:49:00 - 9:08:00

One of the promising pre-combustion technologies for carbon capture in power plants is the Sorption-Enhanced Water-Gas-Shift (SEWGS). SEWGS is a high-pressure and high-temperature regenerative dry sorbent process that combines CO_2 capture with the WGS reaction. WGS is an equilibrium controlled reaction in which carbon monoxide reacts with water vapor to form carbon dioxide and hydrogen. When the amount of CO_2 production is high, as it is the case in power plant applications, multiple reactors operating at progressively lower temperatures are required to reach a high level of CO_2 removal. However, if one of the syngas products, either H_2 or CO_2, can be continuously removed from the reactor, i.e. by CO_2 capture, equilibrium will be shifted and the conversion of CO to CO_2 can be accomplished in a single reactor. Therefore, a SEWGS process for continuous separation of CO_2 or H_2 would be desirable. This sorbent augments or replaces the CO conversion catalysts currently used in WGS reactors and improve overall WGS thermal efficiency. The major advantages of this high temperature sorbent include eliminating or reducing the amount of WGS catalyst required to fully shift the syngas to CO_2 and H_2 and eliminating syngas cooling/reheating that is necessary for current CO_2 separation systems. We have developed a highly durable Potassium-promoted MgO-based sorbent for SEWGS[1,2] which has been tested successfully in packed and dispersed bed reactors at lab scales. Furthermore, to describe the behavior of the sorbent at various operating conditions an explicit reaction model suitable for CFD calculations called two-zone variable diffusivity shrinking core model with expanding product layer has been developed[3]. However, in order to utilize this sorbent continuously at large scale processes a circulating fluidized beds (CFB) is needed which ensures continuous CO_2 removal processes based on the dry sorbent concepts[4,5]. To design a CFB system and afterward to bridge the gap between lab/bench scale and large scales needed for demonstration a powerful tool, i.e. Computational fluid dynamics (CFD), is needed.

In this work, we used an Eulerian-Granular model in combination with our reaction model, to evaluate the performance of the proposed SEWGS process in a bench scale circulation fluidized bed based on the design of NETL C2U experiment[6]. The model is an extension of our previously published 2D model[5,7] which was employed in the simulation of riser part of a CFB. We have extended the model to a full-loop 3D model of CFB including riser, cyclone, down-comer, regenerator, and seal pots. A major limitation of riser section simulations is that the solid inventory of the riser or the solid circulation rate should be known a priori, which was eliminated by considering the whole system. Initially, NETL C2U experimental data on the reactor's hydrodynamics were used to validate the numerical model. Next, the numerical model will be used to evaluate the sorbent performance in the CFB reactor. This work facilitates implementation of new technologies for CO_2 capture in fossil fuel power plants and eventually paves the path toward production of H_2 as a clean fuel for electricity generation.

References

[1] Hassanzadeh, A., "A regenerative process for carbon dioxide removal and hydrogen production in IGCC", Thesis (Ph.D.), *Illinois Institute of Technology*, 2007.

[2] Hassanzadeh, A., & Abbasian, J. (2010). Regenerable MgO-based sorbents for high-temperature CO_2 removal from syngas: 1. Sorbent development, evaluation, and reaction modeling. *Fuel*, *89*(6), 1287-1297.

[3] Abbasi, E., Hassanzadeh, A., & Abbasian, J. (2013). Regenerable MgO-based sorbent for high temperature CO 2 removal from syngas: 2. Two-zone variable diffusivity shrinking core model with expanding product layer. *Fuel*, *105*, 128-134.

[4] Yi, C., K., Jo, S., J., Seo, Y., Lee, J., B., Ryu, C. K., "Continuous operation of the potassium-based dry sorbent CO_2 capture process with two fluidized-bed reactors", *International journal of greenhouse gas control*, 1(2007): 31-36.

[5] Abbasi, E., & Arastoopour, H. (2011). CFD Simulation of CO_2 Sorption in a Circulating Fluidized Bed Using Deactivation Kinetic Model. In *Proceeding of the Tenth International Conference on Circulating Fluidized Beds and Fluidization Technology, CFB-10, edited by TM Knowlton, ECI, New York* (pp. 736-743).

[6] Clark, S., Snider, D. M., & Spenik, J. (2013). CO_2 Adsorption loop experiment with Eulerian–Lagrangian simulation. *Powder Technology*.

[7] Abbasi, E., & Arastoopour, H. (2011). Numerical Simulation of CO2 Removal Process Using Solid Sorbent In a Fluidized Bed: A CFD-PBE Model. In *2011 AIChE Annual Meeting*.

(53c) Simulation of the Semi-Dry Flue Gas Desulfurization in a CFB Using the Two-Fluid Model With a Heterogeneous Drag Model

Wang, X., Institute of Process Engineering, Chinese Academy of Sciences,China
Li, Y. J.,
Zhu, T., Institute of Processing Engineering, Chinese Academy of Sciences
Jing, P. F., Institute of Process Engineering, CAS
Originally presented on: 11/4/2013 9:08:00 - 9:27:00

Semi-dry flue gas desulfurization (FGD) has been developed for low temperature in a circulating fluidized bed (CFB) reactor. The particle volume in the semi-dry FGD reactors (<5kg/m^2•s) is much more dilute than the particle volume in the classical CFB reactors, such as the fluid catalytic cracking (FCC) reactors and the CFB combustion reactors. The sorbent for the semi-dry FGD is lime (CaO) which has unique fluidizing characteristics. However, previous studies of the numerical simulation of the CFB reactor mainly focused on the process of high particle volume, the reports on CFD modeling of the semi-dry FGD process are much rare and the controlling step is controversial currently.

The semi-dry FGD was carried out in a pilot scale CFB reactor with a venturi distributor for the gas acceleration. Experimental measurements of both fluidizing and FGD had been compared with the simulation results to verify the model accuracy. A comprehensive 3D model was developed based on the two-fluid model and involved reactants mass transfer, water evaporation and desulfurization.

Since the flow pattern in the CFB reactor is heterogeneous and has significant influence on the FGD process, the flow modeling of the cold state in CFB is necessary to get clearly comprehension. Compared to the homogeneous drag models, heterogeneous drag models could better describe the particle aggregation. In all heterogeneous drag models, the O-S model has an empirical factor representing the particle clustering effect with short computation time. Therefore, the O-S model was selected in this work. The O-S model was analyzed in terms of the pressure drop, the distribution of particle volume and the distribution of the velocity in the CFB by comparing with the experimental data and the computing results of the classical Gidaspow drag model. Results show that the O-S model is more accurate than the Gidaspow drag model, but there are still some differences from experimental data due to an inaccurate description of the particle volume in clusters.

The SO_2 absorption process includes humidifying, evaporation, production of $Ca(OH)_2$and neutralization reactions. The FGD simulation is based on the O-S drag model verified above. However, there are a lot of difficulties in the simulation process. First of all, the process of heat transfer relates to the humidifying and evaporation, which involves mass transfer and phase change of water, as well as the convective heat transfer near the wall. Previous studies considering heat transfer as an adiabatic process cannot simulate accurate results according to the different conditions in the real application. Besides, it is difficult and time-consuming to simulate such complicated process involving evaporation, ionization, dissolution, neutralization and deposition in the turbulent gas-solid flow. This method of microcosmic modeling is overloaded with details and might be inefficient for engineering prediction. Moreover, doubts on the controlling step still exist between the physical process such as mass and heat transfer and the chemical process such as dissolution and ionization.

The aim of this work is to develop a simplified model that can simulate the flow behavior coupling with the heat transfer and desulfurization reaction in the CFB reactor more efficiently. Simulation results of the desulfurization efficiency and reaction rate indicate that higher particle volume and more uniform particle distribution lead to higher desulfurization efficiencies. The controlling step is predicted to be either the dissolution of CaO or the mass transfer of SO_2, corresponding to the different water spraying volume.

These considerations greatly improve the consistence of theoretical predictions with experiments and better explain the phenomena in the semi-dry FGD in the CFB. The model can also be used for design and optimization of the semi-dry FGD processes and guide industrial scale-up.

(53d) Effect of Oxygen Carrier On Particulate Emissions From Chemical Looping Combustion

He, F., North Carolina State University
Li, F., North Carolina State University
Originally presented on: 11/4/2013 9:27:00 - 9:46:00

Carbon dioxide emitted from fossil energy conversion accounts for roughly 20% of the greenhouse effect. Chemical looping processes, which include chemical looping combustion (CLC) and chemical looping gasification (CLG), utilize a novel, indirect strategy for fossil fuel conversion and in-situ CO_2 capture. The CLC process uses an oxygen carrier particle to oxidize the carbonaceous fuel into concentrated CO_2. The reduced oxygen carrier from the fuel oxidation step is subsequently combusted with air to generate heat/electricity. Although CLC is a promising approach for CO_2 capture, it can cause undesired particulate emission from the circulation of oxygen carrier particles. Small particulates, especially those with less than 10 micron diameter, are particularly harmful to respiratory systems. Despite of the various ongoing pilot scale demonstrations, little research has been conducted on characterizing particulate emissions from CLC processes, to date.

This study represents an attempt to comprehensively evaluate particulate emissions from an experimental CLC apparatus. With methane as the fuel, two Cu based oxygen carrier particles were tested in a laboratory fluidized bed apparatus. One (CuO-Al_2O_3) is prepared by an impregnation method and the other (CuO-$MgAl_2O_4$) by a solid state method. A lab scale fluidized bed coupled with an automated gas panel is constructed to conduct multi-cycle redox reactions. Emitted particulates are collected and fractionated into seven particle size ranges using two sets of MRI cascade impactors as well as the filters. The effects of operating conditions including oxygen carrier composition, redox cycles, degree of reduction, etc. on particulates properties are investigated. Our research indicates that the majority of the small particulates from both the reduction and oxidation steps have aerodynamic diameters ranging between 1 and 6 μm. Notable amount of sub-micro (< 1 μm) sized particulates are also identified. Based on our experimental results, particulate control devices besides high efficiency cyclones are likely to be necessary for CLC even for the combustion of methane.

(53e) Numerical Simulation and Experimental Study On Fouling in the FCC Flue Gas Turbine

Du, Y., China University of Petroleum, College of Chemical Engineering, Statekeylab of heavyoil processing

Zhao, H., State Key Laboratory of Heavy Oil Processing

Yang, C., State Key Laboratory of Heavy Oil Processing

Hu, R., China University of Petroleum, College of Chemical Engineering

Originally presented on: 11/4/2013 9:46:00 - 10:05:00

Numerical Simulation and Experimental Study on Fouling in the FCC flue gas turbine

Yupeng Du, Hui Zhao, Chaohe Yang, Renbo Hu

State key laboratory of heavy oil processing, China University of Petroleum, Qingdao, China

Introduction

The fluid catalytic cracking (FCC) flue gas turbine is a typical turbo-machinery and an essential device of the power recovery unit in FCC processing. It functions as converting the high-temperature heat energy and high pressure energy of exhaust gases generating from carbon burning on catalysts in FCC regenerators to mechanical energy, to drive the main air blower or the generator to recover energy contained in flue gas. Nevertheless, over recent years an urgent issue has been put on the agenda: catalyst scaling on the inner parts of the flue gas turbine, which has a considerable influence on the long-run of FCCU. We have surveyed 87 refineries in China, out of which 59 refineries confronted with this trouble [1]. Although some researches on this issue have been published [2-4], the fundamental mechanism and solution to this problem are still remained as a problem.

Numerical Methodology and Experimental analysis

The Euler-Lagrange multiphase approach was adopted for this investigation. The conservation equations of mass, momentum, and energy are solved for gaseous phase by using Euler's model, while particle trajectory equations of solid phase are solved by employing Discrete Phase Model (DPM) with Discrete Random Walk model (DRW). Besides, SST k-¦Ø model was used as the turbulence model for gaseous phase because of the SST k-¦Ø model is more accurate and reliable for a wider class of flows (e.g., adverse pressure gradient flows, airfoils, transonic shock waves) [5]. The simulation cases were carried out on an equal-radius quasi-three-dimensional plane of blading cut from a single-stage axial flow turbine (Figure 1). Typical dimensions for main structure could be described as: Blade pitch was 0.05m, Chord length was 0.05m, and Gap distance is 0.02m Four gaseous species (O_2, H_2O(g), CO_2, and N_2) and one solid specie (catalyst fines) had been defined as the model. Based on this model, gaseous phase fluid fields were numerically studied and the trajectories of entrained particles with various diameters were calculated; particle depositions on the rotor blades were investigated and analyzed numerically as well.

With the simulation results to be validated, the physical and chemical characteristics of catalysts and fouling samples were analysed by means of analytical instruments, including X-ray diffraction (XRD) for analyzing mineral compositions and scanning electron microscope (SEM) for analyzing the micro-morphology of samples.

Results and Discussion

Typical flow field between stator and rotor was shown as Figure 2. It could be found that the velocity on the pressure side of both stator and rotor was lower than that on other places. While, there were higher moisture content and temperature field on the pressure-side of the rotor than other places. Much longer particle-wall interaction time (e.g. Van der Waals' force and electrostatic force) could be expected from lower flow velocity, and so more readily particles migrate towards the wall as well. Under a circumstance of high moisture content and temperature, certain substrates, such as Ca, K, Ni, V and $CaCl_2$, on the surface of catalyst are apt to form eutectic mixture, resulting in the catalyst particles become stickier [6]. The formed eutectic mixtures could also function as one kind of binders, which enhance particles adhering junction and meantime absorb follow-up catalyst agents flowing nearby, leading to more particles conglutination and melting there.

The predicted fouling positions induced by particles with two different diameters (1μm, 3μm) were shown in Figure 3. In fact, the particle concentration in flow passage of flue gas turbine was extremely low. However, in the current study, the stack concentration of catalyst particles has been increased multifold for clarifying the different behaviors of different particles. Obviously, particles with 3μm form a particle concentrated area on the pressure-side of rotors, while particles with 1μm are well-distributed. However, 1μm particles closed to the rotor surface also readily accumulated there. The predicted particle deposition locations induced by catalyst particles were shown in Figure 4(B) and 4(C), while Figure 4(A) showed practical fouling locations on rotors in a practical flue gas turbine. Basically good agreement could be found from comparison between each other.

Comparisons between two XRD result were shown in Figure 5, remarkable difference could be observed: the peak that indicated the of crystalline for equilibrant catalyst in Figure 5(A) disappeared, instead of which an amorphous peak of fouling sample occurred in Figure 5(B). The main reason for this transition could be concluded as the deposited particles on the walls melt or sinter at the high-temperature circumstances, the same conclusion can be inferred as shown in Figure 6. From SEM spectrograms of fouling samples in figure 5, catalyst accumulation zones and compact melting zones with a large scale can be observed. We also found the melting zones on the flow-side of fouling were much larger than that on the other side due to the influence of much higher temperature on the flow-side of fouling.

Conclusions

According to the simulation results and instrumental analyses mentioned above, detailed fouling procedure inside the FCC flue gas turbine could be proposed: Firstly, some gaseous melted components condensed when they collide with walls (such as gaseous vanadic fusant), and wetting the vicinity, so the original layer comes into being; Secondly, small particles (3μm or even smaller) concentrate on the pressure-side of rotors, and thicken the deposition layer. Meanwhile, heat resistance increases, which may lead to the temperature of fouling more approximated to that of the gaseous phase. Finally, under a high temperature, more melted matters strengthen their stickiness. Besides, with the help of the condensation of cooling steam from the turbine bearing block, more catalyst fragments could be captured. Likewise, these three steps take place sequent. As a consequence, catalyst fouling in turbine.

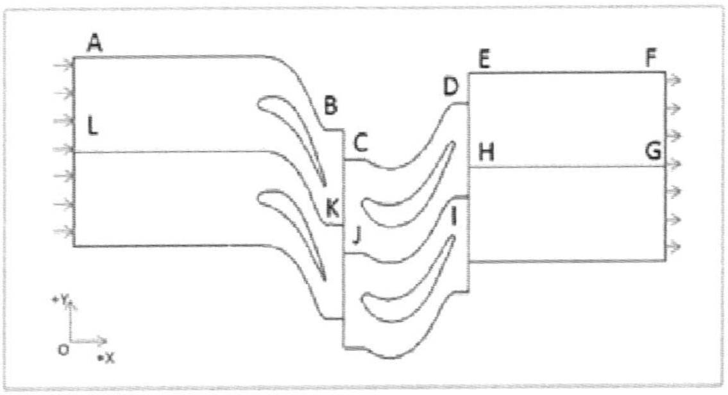

Fig.1 Geometric model and simulation domain

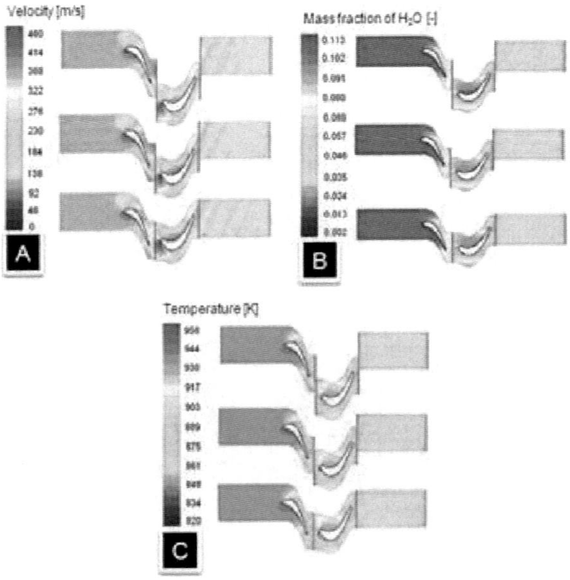

Fig.2 Flow field and profile of (A) Velocity, (B) Molar concentration of H2O, and (C) Temperature btween stator and rotor

SHAPE * MERGEFORMAT

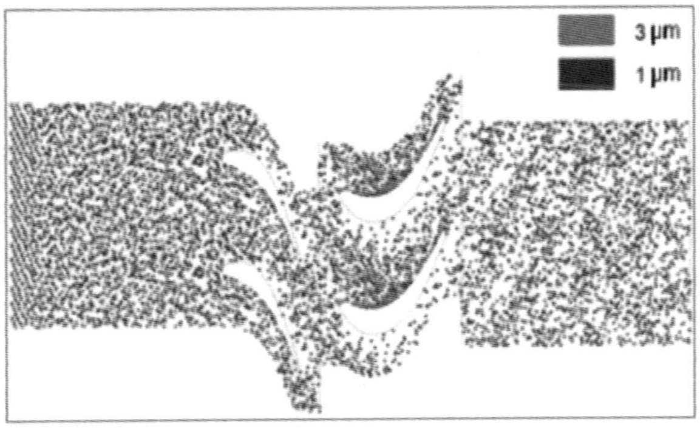

Fig.3 Distribution of particles with different diameters among blade rows

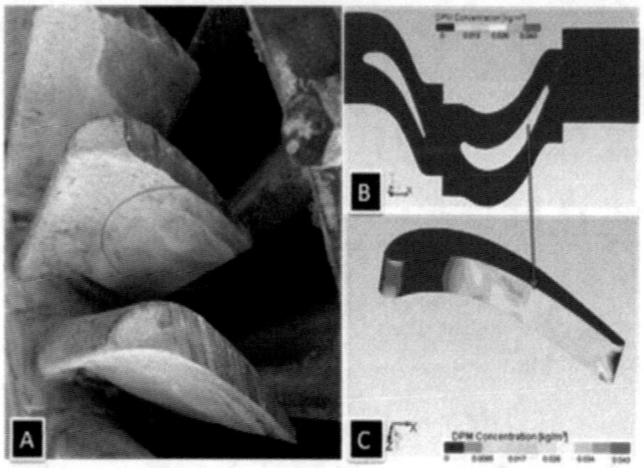

Fig.4 (A) Practical fouling locations, and (B),(C) predicted deposition locations on rotors.

SHAPE * MERGEFORMAT

A

B

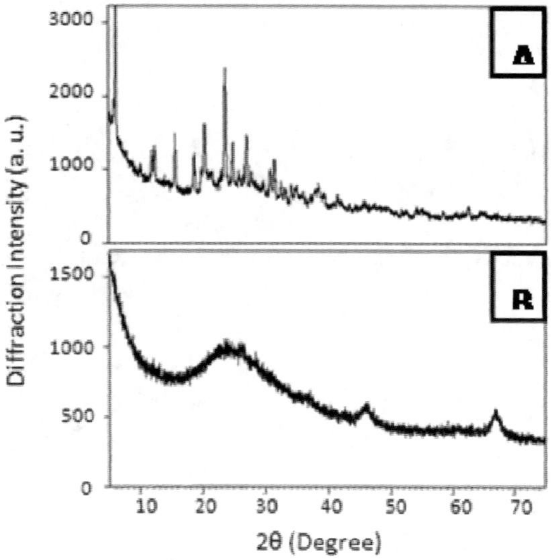

Fig.5 XRD Patterns of (A) Equilibrium catalyst and (C) Fouling sample

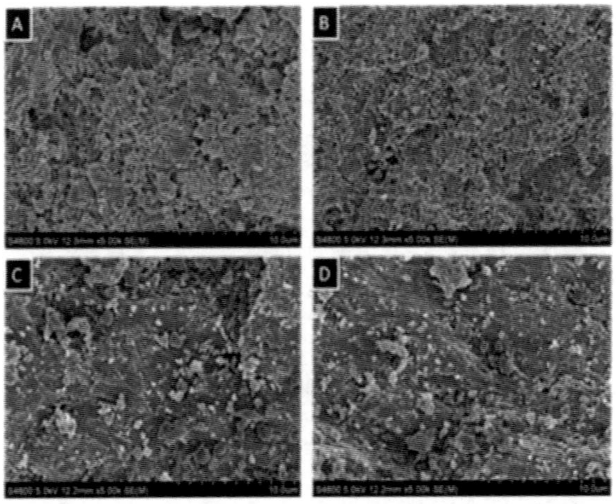

Fig.6 SEM spectrograms of (A and B) Underside of fouling and (C and D) Upside of fouling

References

1. Renbo Hu, Rui Bai, and Ming GAO (2012), "Fouling Status and Reason Analysis of FCC Expander," *40th annual conference of National Technology Center of Chemical Engineering Design*, pp. 282-287.

2. Dianguo Hou (2011), "Analysis on Causes of Scaling in Flue Gas Turbine of FCC and Countermeasures," *China Petroleum Processing and Petrochemical Technology*, 13(1), pp. 66-74.

3. Pengfei Lu, J. Ji, and L.W. Yang (2008), "A study into deposition, erosion and corrosion properties of flue gas turbines," *Sino-Global Energy*. (s1).

4. Yun Xu, Yuxia Zhu, and Huiping Tian (2010). "A Probe into Causes and Remedies for Flue Gas Turbine Fouling in FCCU," *Industrial Catalysis*,18(3), pp.56-60.

5. Menter Florian R. (1994), "Two-equation Eddy-viscosity Turbulence Models for Engineering Applications," *AIAA Journal*, 32(8), pp.1598-1605.

6. Atul Sharma, V.V. Tyagi, C.R. Chen (2009). "Review on Thermal Energy Storage with Phase Change Materials and Applications," *Renewable and Sustainable Energy Reviews*. 13(2), pp.318¨C345.

(53f) Segregation of Char and Wood in a Cold Bubbling Fluidised Bed of Sand for Biomass Gasification Applications

Cluet, B., LERMAB
Mauviel, G., LRGP, ENSIC, CNRS-Nancy Universite
Rogaume, Y., LERMAB
Authier, O., LRGP, ENSIC, CNRS-Nancy Universite
Delebarre, A., LEMTA ESSTIN
Originally presented on: 11/4/2013 10:05:00 - 10:24:00

Among the biomass gasification technologies, bubbling fluidised bed is well adapted for combined heat and power production in the 1-10 MWe range. However this technology is not yet mature.

One key point for its development is a better understanding of the biomass particles hydrodynamics inside the bed. The quality of the syngas and the efficiency of the reactor directly depend on the location of the particles at each phase of their conversion (drying, pyrolysis, char gasification). Bruni et al. [1] showed that endogenous bubbles generated by devolatilisation carry the biomass particles to the bed surface. Since Bruni et al.[1] restricted their study to incipient fluidization, this observation should be assessed for turbulent fluidization which is typical for biomass gasifiers.

The hydrodynamics of bubbling fluidized bed has already been investigated experimentally in hot device through high frequency differential pressure measurements [2].This method is used in this study to analyse segregation of biomass and char particles.

This present study achieves segregation experiments with different materiel in cold bubbling fluidized bed (20-50°C). Three types of materials are used to mimic characteristics of biomass during devolatilisation: wood particles with a solid density of 750 kg/m^3, wood char particles with a solid density of 400 $kg/m^3$and a new empirical material that could mimic biomass characteristics: density, shape, size, but also volatile matter release (same volatile content and volatile production rate).

Fluidization experiments are achieved in cold device (25 cm ID). The bed consisted of olivine particles with an average particle size of 350 μm and a solid density of 3000 kg/m^3(Geldart B classification). The mass ratio of biomass in bed is between 0 and 10%. Particles are injected over the bed or in the bed just above the air distributor. The comparison allows understanding and assessing the role of the characteristics of biomass property (density, shape) on particles hydrodynamics. The aim of this study is to produce correlations for segregation of biomass in bubbling fluidized bed.

The final goal of this work is to take into account the coupling between particles hydrodynamics and particles conversion in a global model representing a hot gasifier bubbling fluidized bed.

REFERENCES

[1] G. Bruni, R. Solimene, A. Marzocchella, P. Salatino, J. Yates, P. Lettieri, et al., Self-segregation of high-volatile fuel particles during devolatilization in a fluidized bed reactor, 128 (2002) 11–21.

[2] A. Rao, J.S. Curtis, B.C. Hancock, C. Wassgren, Classifying the fluidization and segregation behavior of binary mixtures using particle size and density ratios, 57 (2011) 1446–1458.

(53g) Computational Fluid Dynamic Simulation of a Chemical Looping Process for Methane Combustion

Breault, R. W., National Energy Technology Laboratory
Liu, Y., National Energy Technology Laboratory, U.S. Department of Energy
Huckaby, E. D., National Energy technology Laboratory
Konan, N. A., National Energy Technology Laboratory
Originally presented on: 11/4/2013 10:24:00 - 10:43:00

Chemical Looping Combustion is an emerging technology for the "carbon-capture ready" production of energy from fossil fuels. Two chemical looping units have been recently constructed at the National Energy Technology Laboratory to specifically investigate and develop this technology: a high-temperature chemical looping reactor and cold flow system with a similar geometry and optically accessible. Computational fluid dynamic simulations were performed of the cold flow system and compared to experimental pressure measurements and high-speed video at several operating conditions. The short-term goals of the experiments and simulations of this cold flow unit are to help plan and analyze the future experiments in high-temperature reactor.

The chemical looping reactor consists of two interconnected fluidized beds. The first bed, the air reactor, consists of a lower mixing zone with a diameter of 6 inches and 30 inches high connected to a riser 2.5 inches in diameter and 115 inches high. Two additional ports are provided to add secondary fluidization air to help transport solids through the riser. The second bed, the fuel reactor, is 8 inches in diameter and 64 inches high. The exit of the air reactor is connected to the fuel reactor via a cyclone, loopseal and dip leg. An L-valve connects the exit of the fuel reactor to the air reactor inlet. The pressure is measured at 19 locations around the system at 5 Hz. The circulation rate is estimated from rate of change in pressure drop across the fuel reactor bed after the solid flow to the fuel reactor has been stopped. This procedure assumes that the circulation rate is regulated by the L-valve. In the tests which were simulated glass beads are circulated. In addition to circulating experiments, the data set includes fluidization tests for the air reactor, fuel reactor and loop seal.

Predictions for methane conversion in a NETL experimental facility will be presented. The performance predictions will be assessed for a variety of operating conditions including the effects of fuel gas composition, fuel gas flow rate as well as oxygen carrier type and particle size.

(53h) Tracking of Simulated Biomass Particles in Bubbling Fluidized Beds

Daw, C. S., Oak Ridge National Laboratories
Halow, J. S., Separation Design
Finney, C. E. A., Oak Ridge National Laboratory
Originally presented on: 11/4/2013 10:43:00 - 11:02:00

Tracking of Simulated Biomass Particles in Bubbling Fluidized Beds

Stuart Daw* (Oak Ridge National Laboratory, Knoxville, TN 37932)

J.S. Halow (Separation Design Group, Waynesburg, PA 15370),

*dawcs@ornl.gov Introduction

Fluidized beds are extensively used for implementing fluid-solid reactions, and, in many cases, more than one type of solid particle is present in the reactor. Interest in the behavior of multi-particle bubbling bed reactors is specifically of interest in the context of biomass pyrolysis for liquid fuels production. Bench-scale bubbling beds are often used in laboratory experiments to study the effects of pyrolysis reactor operation and to characterize the conversion behavior of different biomass feedstocks. However, it is frequently very difficult to separate the effects of biomass particle mixing and segregation in the reactor from the intrinsic biomass particle kinetics. In this presentation we summarize results from experimental measurements of simulated biomass particle trajectories in an ambient temperature bench-scale bubbling bed of sand for ranges of fluidization intensity, bed height, and particle properties that we expect to be relevant to laboratory bubbling bed pyrolysis reactors. Our measurement approach is based on a previously published magnetic tracking technique that allows very rapid, real-time recording of 3D position and velocity for simulated biomass particles of different size and density as they mix and/or segregate in the laboratory bubbling bed. Based on our observations, it appears that we are able to estimate the parameters needed to replicate the observed particle trajectory statistics with a Langevin-type model. The observed trends in the estimated parameters with fluidization intensity and particle properties indicate potentially useful directions for constructing general correlations and models for biomass particle mixing and segregation in bubbling bed reactors.

Experimental Materials and Methods

In the present study we utilized a previously reported (1,2) particle tracking method for fluidized beds. Simulated biomass particles of varying size and density were tagged with inexpensive neodymium magnets and multi-probe magnetic detectors were used to track and record the spatial locations of the tagged particles at 0.01s intervals as they migrated through the beds for periods of several minutes under different fluidization conditions. Based on these measurements, we are able to reconstruct the detailed long-time-scale trajectory of each different simulated biomass particle, including its position, velocity, and acceleration. From this information, we derived statistical distributions for each of these quantities as well as developed a preliminary dynamic model to match the observed behavior.

For the results reported here, the experimental bubbling fluidized bed was a 55 mm in diameter glass tube fluidized with air at ambient conditions. The bed solids consisted of 200 micron glass beads and the simulated biomass particles ranged in size from about 2 mm to 12 mm with densities ranging from 0.5 g/cc to 1.0 g/cc. The fluidization air flows investigated ranged from 1 to 6 times the minimum fluidization and the bed height varied from 1.0 to 2.0 times the bed diameter. Improvements have been made in the sensitivity of the magnetic tracking system and the algorithm used to calculate position from magnetic readings from the previously reported studies. These will be discussed in the presentation.

Results and Discussion

We compare the time-average axial segregation patterns for the tracked particles in these experiments with our previously reported observations of Weibull-type axial distributions for a similar laboratory bed. We also summarize the spatially-averaged and spatially localized particle velocity distributions as functions of particle properties and bed operating conditions. Preliminary results appear to indicate that a Langevin-type model may be able to replicate the observed tracked particle velocity distributions with appropriate values for the key parameters and distribution of the stochastic perturbations. If the accuracy of the Langevin-type model can be confirmed, it should provide a useful benchmark for validating and interpreting the results from computational fluid dynamics simulations. Likewise, we conjecture that a Langevin approximation of biomass particle mixing may provide a rapid method to account for particle mixing effects in lower-order pyrolysis reactor simulations, provided that the variation in particle properties with conversion are adequately accounted for. This could provide a low-computational-overhead approach for evaluating general trends in pyrolysis reactor performance with variations operating conditions and feedstock properties.

1. Patterson et.al., An Innovative Method Using Magnetic Particle Tracking to measure Solids Circulation in a Fluidized Bed, Ind. Eng. Chem. Res. 2010, 49, 5037-5043.

2. Halow et. Al. Observed Mixing Behavior of Single Particles in a Bubbling Fluidized bed, Ind. Eng. Chem. Res. 2012, **51**, 14566-14576.

Abstract submitted for presentation at the 2013 Annual A.I.Ch.E. meeting in San Francisco,CA

(55a) Modeling Gas--Solid Heat Transfer Using Particle-Resolved Direct Numerical Simulation

Sun, B., Iowa State University
Tenneti, S., Iowa State University
Subramaniam, S., Iowa State University
Originally presented on: 11/4/2013 8:30:00 - 8:52:00

Gas–solid heat transfer is important in many emerging technologies such as carbon-neutral energy generation using biomass, chemical looping combustion, and CO_2 capture. Computational Fluid Dynamics (CFD) simulations of multiphase flow are increasingly being used as an efficient alternative to experiments for process and design optimization, because experiments are often costly and time-consuming. The averaged equations governing mass, momentum, and energy that are solved in multiphase CFD simulations require models for the average interphase transfer of momentum and energy between different phases. Specifically, the averaged fluid temperature equation from two–fluid theory contains the three unclosed terms: (i) average gas–solid heat transfer, (ii) divergence of average heat flux in the fluid phase, which includes axial conduction in the fluid phase, and (iii) transport of temperature-velocity covariance. In this work, we use Particle-resolved Direct Numerical Simulations (PR–DNS) to develop models for gas–solid heat transfer, axial conduction in the fluid phase, and transport of temperature-velocity covariance.

Preliminary results for gas--solid heat transfer using PR—DNS have been obtained using the Particle-resolved Uncontaminated-fluid Reconcilable Immersed Boundary Method (PUReIBM) [1]. Tenneti *et al.* studied a canonical problem of heat transfer in a steady flow through a homogeneous fixed assembly of spherical particles and showed that mean fluid temperature changes along the streamwise direction due to fluid heating or cooling by particles. Therefore, although the hydrodynamic problem is statistically homogeneous, the directional nature of the mean flow and fluid heating renders the thermal problem statistically *inhomogeneous*. It has been shown [2] that obtaining surface statistics from PR—DNS of gas-solid flow in inhomogeneous problems is a significant challenge. The difficulty is that averaged interphase exchange terms calculated from spatially varying surface statistics converge much more slowly with the number of realizations than in the statistically homogeneous case [2]. To alternate this difficulty, Tenneti *et al.* formulated a *thermally fully–developed* gas–solid heat transfer problem in the same canonical setup using a thermal similarity condition. The thermal similarity condition enabled simulation of the statistically inhomogeneous thermal transport problem in steady flow past a statistically homogeneous assembly of fixed particles in periodic domains. Based on the PR--DNS results, Tenneti *et al.* showed that fluid heating can change the average fluid temperature on very small length scales for high volume fraction and low Reynolds number.

Building on this previous work of Tenneti *et al.,* we verify fundamental assumptions in the averaged fluid temperature equation and develop models for gas–solid heat transfer. Local statistical homogeneity of the average fluid temperature that is assumed in the two–fluid averaged equations is investigated by developing an exponentially decaying model for the bulk fluid temperature from PR–DNS data. We develop a length scale criterion that determines the validity of the locally homogeneous average fluid temperature assumption as a function of solid volume fraction and Reynolds number.

In the range of validity of the two–fluid averaged equation formulation, the unclosed terms in the averaged fluid temperature equation are quantified and modeled using PR–DNS results. We demonstrate that using the bulk fluid temperature is the correct choice to compute average interphase heat transfer in gas–solid flow rather than using the average fluid temperature. PR–DNS results show that axial conduction in the fluid phase that is often neglected in existing one-dimensional models cannot be neglected for $Re_m < 10$. Finally, the transport of temperature-velocity covariance that is often neglected in CFD simulations is computed and modeled from PR–DNS data. It is found that this term is non-negligible since the average transport of temperature-velocity covariance is about 30%-70% of the average convective term in the averaged temperature equation. This study shows how PR–DNS can be used to evaluate fundamental assumptions underlying averaged continuum models for heat and mass transfer in gas--solid flow, and that models can be developed for interphase heat transfer, axial conduction in the fluid phase, and the transport of temperature-velocity covariance.

References

1. Tenneti S, Sun B, Garg R, Subramaniam S. 2013. Role of fluid heating in dense gas–solid flow as revealed by particle–resolved direct numerical simulation. Intl. J. Heat Mass Transfer 58:471–479
2. Xu Y, Subramaniam S. 2010. Effect of particle clusters on carrier flow turbulence: A direct numerical simulation study. Flow, Turbulence and Combustion 85:735–761. 10.1007/s10494-010-9298-8

(55b) A Quadrature-Based Uncertainty Quantification Approach With Reconstruction of the Probability Distribution Function of the System Response in Bubbling Fluidized Beds

Hu, X., Iowa State University
Passalacqua, A., Iowa State University
Fox, R. O., Iowa State University
Originally presented on: 11/4/2013 8:52:00 - 9:14:00

In simulations of industrial systems, it is important to have an estimate of the distributions of errors due to uncertainty in the model parameters and input data. This can be accomplished by developing uncertainty quantification tools that can be combined with available CFD codes. Here a non-intrusive, quadrature-based, uncertainty quantification (QBUQ) method is presented. The approach relies on Gaussian quadrature formulae to generate the set of samples of the distribution of the uncertain parameters of the model. A numerical simulation is performed for each sample, and the moments of the system response are directly computed from the simulation results by means of quadrature formulae. These moments are then used to determine the reconstructed distribution function of the values of the system response by means of the extended quadrature method of moments [1].

The QBUQ procedure is demonstrated by considering a bubbling fluidized bed as an example application. The initial bed height, the particle restitution coefficient and the mean particle size are assumed to be the independent and uniformly distributed uncertain input parameters. The system is simulated with a standard two-fluid model with kinetic theory closures for the particulate phase [2] implemented into MFIX. The effect of uncertainty on the disperse-phase volume fraction, on the phase velocities and on the pressure drop inside the fluidized bed are examined, and the reconstructed probability density function of the system response is provided for the three quantities studied.

[1] C. Yuan, F. Laurent, R.O. Fox, An extended quadrature method of moments for population balance equations, J. Aerosol Sci. 51 (2012) 1–23.

[2] D. Gidaspow, Multiphase flow and fluidization, Academic Press, 1994.

(55c) Verification of Sub-Grid Drag Models for Gas-Particle Fluidized Beds With Immersed Horizontal Cylinder Arrays

Sarkar, A., Purdue University
Sun, X., Pacific Northwest National Laboratory
Sundaresan, S., Princeton University
Originally presented on: 11/4/2013 9:14:00 - 9:36:00

Cylinder arrays are commonly used in fluidized beds to achieve heat transfer—hundreds of immersed cylinders, typically a few centimeters in diameter, can be present in large commercial devices that are many meters tall. Grid-independent CFD simulations of commercial-scale fluidized beds, where fine particle clusters and flow around individual cylinders must be explicitly resolved, are computationally infeasible. An alternate strategy is to use faster coarse-grid simulations where the cylinder array is viewed as a stationary, uniform porous medium. The influence of the cylinders on the suspension is modeled as a filtered cylinder-suspension drag force, and on the cluster/bubble formation is captured via sub-grid corrections to the gas-solid drag. Filtered models for the cylinder-suspension drag and gas-solid drag for beds with cylinder arrays have been developed in a previous study[1]. This work addresses verification of these sub-grid drag models: the fidelity of the filtered sub-grid model predictions to highly-resolved simulation results is evaluated.

Results from highly-resolved simulations of a large bed with an immersed cylinder array are generated for two operating regimes—bubbling and turbulent fluidization. These computationally expensive, grid-independent results are considered to be the 'exact' solutions. Two approaches are followed to perform model-on-model verification. First, the large bed CFD results are analyzed in an attempt to reconstruct the filtered drag models. The reconstructed models are then compared against the previously developed filtered relationships reported in Sarkar et al.[1] Second, the filtered constitutive models are implemented as sub-grid corrections in coarse-grid simulations of the large fluidized bed. Predictions for the macroscopic particle distributions and velocity profiles are compared against the exact solutions (i.e., the highly-resolved simulations). Addition of sub-grid corrections is found to improve the predictive capabilities of the coarse-grid simulations, which otherwise over-estimate the bed height. Differences observed between the coarse-grid CFD predictions and the exact solutions may be resolved by further refinement of the filtered constitutive models.

[1]Sarkar, A., Sun, X., Sundaresan, S. *A Sub-grid Model for an Array of Immersed Cylinders in Coarse-grid Multiphase Flow Simulations of a Carbon Capture Device.* AIChE Annual Meeting, Pittsburgh, October-November 2012.

(55d) Numerical Study On Discrete Element Modeling in Industrial Powder Systems

Sakai, M., School of Engineering, the University of Tokyo
Shigeto, Y., The University of Tokyo
Sun, X., The University of Tokyo
Mizutani, S., The University of Tokyo
Originally presented on: 11/4/2013 9:36:00 - 9:58:00

The Discrete Element Method (DEM) is often employed to perform a simulation of powder systems such as mixing, milling and conveying. Although the DEM was widely applied to various systems, there are some problems such as substantial limitation of number of the calculated particles and modeling of the free surface in solid-liquid flows. These problems prevent us to apply the DEM to industrial systems. To solve these problems, we develop a coarse grain model of the DEM and Lagrangian-Lagrangian approaches.

In the coarse grain model, the large-size modeled particle which is called the coarse grain particle represents the group of the original particles. Namely, when the size of coarse grain particle is l times larger than that of the original particle, l^3 original particles are included in the coarse grain particle. When a binary collision is occurred between the coarse grain particles, the l^3pairs of a binary collision are assumed to occur simultaneously. Besides, the total energy is assumed to be agreed between the coarse grain particles and original ones in the binary collision. The coarse grain model can simulate the adhesive force such as van der Waals and liquid bridge forces. In the past studies, the coarse grain model was verified by an Eulerian-Lagrangian model in gas-solid flows, where the coarse grain ratio was set to be at most 3.0. The coarse grain ratio could not be set to be large value in verification tests. This is because the calculation cost becomes huge in the original system. In the present study, we performed a validation test to show the availability of the coarse grain model with high coarse grain ratio. The calculation results are shown to be in good agreement with the experimental results in fluidized bed systems.

Two kinds of Lagrangian-Lagrangian approaches were developed to perform the solid-liquid flow in industrial systems. The Lagrangian-Lagrangian approaches indicate the DEM-MPS method and DEM-SPH method, where the DEM was coupled with the continuum Lagrangian model such as the MPS or SPH method. These Lagrangian approaches had an advantage that the free surface can be simulated accurately. The Lagrangian-Lagrangian approaches were simulated by macroscopic modeling based on the local volume average technique. The DEM-MPS and DEM-SPH method was validated in a ball mill syetem. In this study, applicability of the Lagrangian-Lagrangian approaches to a wet ball mill is shown thorough validation tests.

Consequently, we developed new technologies for the Eulerian-Lagrangian and Lagrangian-Lagrangian to perform the DEM simulations in industrial systems. These models are going to be applied to actual systems in the near future.

(55e) DEM Simulations of Fluidized Bed Using a Scaled Particle Approach

Baran, O.,
Kodl, P.,
Aglave, R. H., CD-adapco
Originally presented on: 11/4/2013 9:58:00 - 10:20:00

Fluidized beds are widely used in chemical process and refining industry due to various advantages they offer over fixed bed reactors. Nonetheless, their design is a difficult task and modeling is increasingly deployed to understand and scale-up these equipment. This paper reports the results for large-scale parallel simulations of fluidized bed using Discrete Element Method (DEM) fully coupled with fluid flow model using STAR-CCM+ software developed by CD-adapco. The simulation model uses the geometry and set-up as described in reference [1] and the simulation results are being compared with corresponding experimental results from the same reference.Â In the numerical model, the particle's drag depends on local solid fraction with Ergun equation based drag used for dense regions and Wen-Yu drag model otherwise. A novel approach of scaling up the DEM particle's size with simultaneous scaling of drag coefficient is introduced and used. For each particle size scaling factor the drag scaling factor is obtained through parametric study by reproducing the fluidization conditions obtained in experiment. The scaling of drag law as a function of particle size scaling is discussed as well as the effect of scaling approach on simulation time and accuracy of numerical model.Â Authors believe that this approach will be able to achieve a right balance between accuracy (a feature of DEM methods) and speed (a feature of computation particle approaches) to enable practical design and scale-up workflow.

[1] J. X. Bouillard, R. W. Lyczkowski and D. Gidaspow, "Porosity Distributions in a Fluidized Bed with an Immersed Obstacle," *AIChE Journal,* vol. 35, no. 6, pp. 908-921, 1989.

(55f) The Effect of Psd and Particle Load On the Grade Particle Separation Efficiency of Large Cyclone

Liu, Z., IHI Corporation
Suda, T., IHI Corporation
Originally presented on: 11/4/2013 10:20:00 - 10:42:00

Fluidized bed incinerators are used to burn various fuels, such as low- quality coal, biomass and trash. Cyclones are used to separate particles from exhaust gas. Due to the difference of fuel burned, the Particle-size Distribution (PSD) in the bed is different from the PSD of original bed material. The particle load at the inlet of the cyclone is different for different fuel, because the calorific value is different for different fuel. The gross separation efficiency of cyclone was conventionally predicted by integration of grade separation efficiency, on the assumption that the grade separation efficiency was independent of PSD and particle load at the inlet of cyclone. This paper is to investigate the precision of the above assumption when the PSD and the load of particle have a large diversification.

The numerical study is made by using Barracuda, a commercial code, in which Eulerian–Lagrangian procedure is used. In order to evaluate the code, the calculated results have been compared with measurement data taken from the literature. The comparisons showed that the numerical simulation is reasonable. A large centrifugal cyclone with the diameter of 5 m has been numerically studied. Two kinds of particles, one is sand, the other is ash, are employed in this study. The weight percentage of ash changes from 0% to 10%, and particle load changes from 16kg/(m^2s) to 64 kg/(m^2s), with a constant velocity at the inlet of the cyclone. The numerical results showed that the grade particle separation efficiency is not constant when PSD and particle load are changed. The method of predicting the gross particle separation efficiency and grade particle separation efficiency of large cyclones have been proposed based on the numerical results.

(55g) Gas-Solid Flow Pattern Analysis in the Riser of Rfcc Process Using CPFD Simulation

Cho, H.,

Cha, B., Yonsei University

Kim, S., SK Innovation

Ryu, J.,

Kim, J.,

Moon, I., Yonsei University

Originally presented on: 11/4/2013 10:42:00 - 11:04:00

Gas-solid flow pattern analysis in the riser of RFCC process using CPFD simulation

Hyungtae Cho[a], Bumjoon Cha[a],Sungwon Kim[b], Jaewook Ryu[b], Junghwan Kim[b], Il Moon[a,*]

[a]Department of Chemical and Biomolecular Engineering, Yonsei University, 50 Yonsei-ro, Seodaemun-gu, Seoul 120-749, Korea

[b]SK Innovation Global Technology, 325 Exporo, Yuseong-gu, Daejeon 305-712, Korea

One of the main problems associated with reactors in commercial Residue Fluidized Catalytic Cracking (RFCC) process is that each of six outlets of the riser has different particle mass flow rates due to uneven distribution particle flow. This problem makes catalyst loss. High conversion of cracking reaction requires uniform catalyst distribution in the riser. The riser termination device (RTD) is installed to achieve rapid separation between catalysts and product gas. RTD is composed of six close-coupled cyclone sets at the top of riser. This study focuses on the analysis of flow pattern of catalysts from the riser to the cyclone sets. The cracking reaction is taken into account in the simulation scheme and its conversion is evaluated in the riser by employing multiphase-particle in cell (MP-PIC) method, one of computational particle fluid dynamics (CPFD) methodologies. It applies both a stochastic particle model and Lagrangian method for particle phase, and Eulerian method for fluid phase, respectively. Figures 1 represents, for 30 seconds, particle flow patterns of the riser comprising six outlets. Macroscopically steady state is reached at 1.5 seconds as shown in Figure 1. Figure 2 shows six different outlet particle mass flow rates over a range between 117.7 kg/s and 122.4 kg/s. A high particle mass flow rate represents a high solid-gas ratio loaded to cyclones; Figure 2 reveals significant deviations of the ratios from its even distribution. This has an adverse effect on the cyclone efficiency, resulting in severe loss of catalysts. So, this article confirm industrial problem such as uneven distribution of particle from the riser to the cyclone.

Figure SEQ Figure * ARABIC 1. The particle flow pattern in the riser (a) velocity magnitude and (b) volume fraction

Figure SEQ Figure * ARABIC 2. Mass flow rate of particle vs. the riser outlet location.

REFERENCES (¾ç½Ä È®ÀÎ ÈÄ ¼öÁ¤z¹Á¤)

[1] Sung Won Kim, Ju Wook Lee, Cheol Joong Kim, Jae Suk Koh, Gyung Rok Kim, and Sun Choi, Characteristics of Deposits Formed in Cyclones in Commercial RFCC Reactor. Industrial and Engineering

Chemistry Research, 51, 10238-10246 (2012).

[2] Sung Won Kim, Ju Wook Lee, Jae Suk Koh, Gyung Rok Kim, Sun Choi, and Ik Sang Yoo, Formation and Characterization of Deposits in Cyclone Dipleg of a Commercial Residue Fluid Catalytic Cracking Reactor, Industrial and Engineering Chemistry Research, 51, 14279-14288 (2012).

[3] M.J. Andrews and P.J. O'Rourke, The Multiphase Particle-In-Cell (MP-PIC) method for dense particulate flows, International Journal of Multiphase Flow, 22, 379-402 (1996)

[4] V. A. Danilov, Jongkoo Lim, and Il Moon, Three-dimensional two-phase CFD model for DMFC design, Journal of Power Source, 162(2), 992-1002 (2006).

[5] Xinglong Zhou, Leming Cheng, , Qinhui Wang, Zhongyang Luo and Kefa Cen, Non-uniform distribution of gas–solid flow through six parallel cyclones in a CFB system: An experimental study, Particuology, 10(2), 170– 175 (2012)

(120a) The Kinematics of Non-Cohesive, Sphero-Cylindrical Particles in a Low-Speed, Vertical Axis Mixer

Hua, X., Purdue University
Wassgren, C. R., Purdue University
Curtis, J. S., University of Florida
Hancock, B. C., Pfizer Worldwide Research and Development
Ketterhagen, W. R., Pfizer Inc.
Originally presented on: 11/4/2013 12:30:00 - 12:48:00

A discrete element method model is used to examine the velocity, solid fraction, and particle orientation fields of non-cohesive, sphero-cylindrical particles agitated in a vertical axis mixer for a range of particle aspect ratios and bed depths. The model is validated against experimental measurements of the rotating shaft torque. The particle trajectories within the bed are similar to those that have been reported previously for spheres, with a vortex circulating in the direction opposite of the blade rotation on horizontal planes of the bed. Increasing the particle aspect ratio generally decreases the particle velocities relative to the blade, implying reduced mixing. In general, larger particle aspect ratios decrease the overall bed solid fraction as well as the solid fraction uniformity. The degree of alignment between particles increases near boundary regions. Particles with a larger aspect ratio tend to increase the degree of alignment. Particles with an aspect ratio larger than one have major axes that are offset between 10 – 15 degrees from the flow streamlines. In addition to the strong correlation between the particle principal orientation and velocity vectors, regions of larger velocity gradient magnitude result in smaller solid fractions and smaller degrees of three-dimensional alignment between particles.

(120b) Granular Shear Flows of Flexible Fibers

Guo, Y., University of Florida
Wassgren, C. R., Purdue University
Hancock, B. C., Pfizer Worldwide Research and Development
Ketterhagen, W. R., Pfizer Inc.
Curtis, J. S., University of Florida
Originally presented on: 11/4/2013 12:48:00 - 13:06:00

Flexible fibrous particles are widely encountered in nature and industry, e.g., plant stover, glass fiber, and paper pulp. Previous studies indicate that the deformability of the fibers has a significant impact on the properties of the bulk materials and the fiber suspensions. Here, the shear flows of flexible fibers are simulated using the Discrete Element Method (DEM) to systematically explore the effect of particle flexibility on the flow behavior and constitutive laws. In the simulations, a fiber is formed by connecting a number of spheres in a straight line using deformable and elastic bonds. The forces/moments induced by the bond deformation resist the relative normal, tangential, bending, and torsional movements between two bonded spheres. The bond stiffness determines the difficulty of particle deformation, and the bond damping accounts for the energy dissipation in the particle vibration process. The simulation results show that the elastically-bonded fiber particles have smaller coefficient of restitution compared to the rigidly connected particles. In addition, the coefficient of restitution decreases as the bond stiffness decreases and the bond damping coefficient increases. As a result, smaller stresses are obtained for the granular flows of the fibers with lower bond stiffness and larger bond damping coefficient. For the flexible fibers, besides the kinetic energy, the potential energy is also generated due to bond deformation. It is found that during shear flows at a specified shear rate, the average potential energy per fiber increases as the bond stiffness decreases and the bond damping coefficient decreases.

(120c) Relating Gas Pressure Gradient to Flow Profiles in Process Equipment

Johanson, K. D., Material Flow Solutions Inc.

Originally presented on: 11/4/2013 13:06:00 - 13:24:00

The influence of gas pressure gradients on the behavior of fine powders in process equipment is critical to proper process operation. Sometimes these gas pressure gradients induce preferential flow channels in process equipment. Sometimes these gradients induce a more uniform flow in the bulk material. Gas injection is often required to overcome slow or limiting flow rates. However, injecting too much gas can result in temporary arch formation and hang-ups. Clearly, the gas within a bulk material has a significant influence on the flow of the mass. This paper looks at the governing equations that relate gas flow profiles to solids velocity profiles in process equipment. The relationship between flow along process vessel walls and the gas pressure gradient is presented. This analysis allows the prediction of preferred flow channels in process equipment. The new methodology also predicts new mass flow criteria based on the influence of gas pressures in the process vessel. Finally, the analysis suggests that the controlled injection of gas into a bulk material could be the means of controlling the blending or anti-segregation behavior of the unit operation.

(120d) DEM Studies of the Ordering and Force Transmission in Packings of Non-Spherical Particles

Lu, G., ETH Zurich

Third, J., ETH Zürich

Müller, C. R., ETH Zurich

Originally presented on: 11/4/2013 13:24:00 - 13:42:00

DEM studies
of the ordering and force transmission in packings of non-spherical particles

So far most studies concerning the packing of granular materials have only considered systems composed of spherical particles, irrespective of the fact that granular materials are commonly of non-spherical shape and show highly anisotropic characteristics [1]. However, very recently, Hidalgo et al. [2, 3] and Lu et al. [4] reported that in packings of 2D, elongated particles a more uniform stress distribution than in packings of spheres can be observed. Due to the preferred horizontal alignment of the particles, the force in the vertical direction of the bed does not saturate with height. Instead, the vertical stress profile becomes similar to the linear hydrostatic pressure profile in a liquid.

However, despite the observations summarized above, the influence of the particle surface roughness on the packing characteristics of non-spherical grains remains an open question. In this study, non-spherical particles of different aspect ratios were modeled in a discrete element method (DEM) framework by either gluing together of spheres or using the super-quadric equation [5, 6]. The surface roughness of the non-spherical particles was adjusted by the number of spheres used to construct their non-spherical shape. On the other hand, a super-quadric equation-based particle possesses a smooth surface. In total 4500 elongated particles of aspect ratio five were simulated. It was found that the surface roughness of the particles crucially influenced both the orientational ordering of the particles and the force transmission within the bed of grains. Super-quadrics-based particles showed horizontal alignments (Fig. 1a), with local clusters of elongated particles forming (Fig. 1b). In addition, a smooth particle surface favored the direct transmission of the bed weight to the bottom of the container (Fig. 1c). On the other hand, the horizontal alignment of grains was less pronounced in packings of elongated particles constructed by gluing together spheres with, surprisingly, coarser particle surfaces resulting in a more pronounced horizontal alignment. Finally, the stresses measured at the bottom of a container filled with particles constructed by gluing together spheres were much smaller than for the case that the container was filled with smooth super-quadric particles.

References

[1] Aranson, I. S., & Tsimring, L. S. (2006). Patterns and collective behavior in granular media: Theoretical concepts. *Reviews of modern physics*, 78(2), 641.

[2] Hidalgo, R. C., Zuriguel, I., Maza, D., & Pagonabarraga, I. (2009). Role of particle shape on the stress propagation in granular packings. *Physical review letters*, 103(11), 118001.

[3] Hidalgo, R. C., Zuriguel, I., Maza, D., & Pagonabarraga, I. (2010). Granular packings of elongated faceted particles deposited under gravity. *Journal of Statistical Mechanics: Theory and Experiment*, 2010(06), P06025.

[4] Lu, G., Third, J. R., & M¨ller, C. R. (2012). Packing and discharging of non-spherical particles in hoppers. In: *Proceedings of the 7th International Conference for Conveying and Handling of Particulate Solids,* Friedrichshafen, Germany, 10-13 September 2012.

[5] Third, J. R., Scott, D. M., Scott, S. A., & M¨'ller, C. R. (2010). Tangential velocity profiles of granular material within horizontal rotating cylinders modelled using the DEM. *Granular Matter*, 12(6), 587-595.

[6] Lu, G., Third, J. R., & M¨'ller, C. R. (2012). Critical assessment of two approaches for evaluating contacts between super-quadric shaped particles in DEM simulations. *Chemical Engineering Science*, 78, 226-235.

(c)

(a) (b)

+

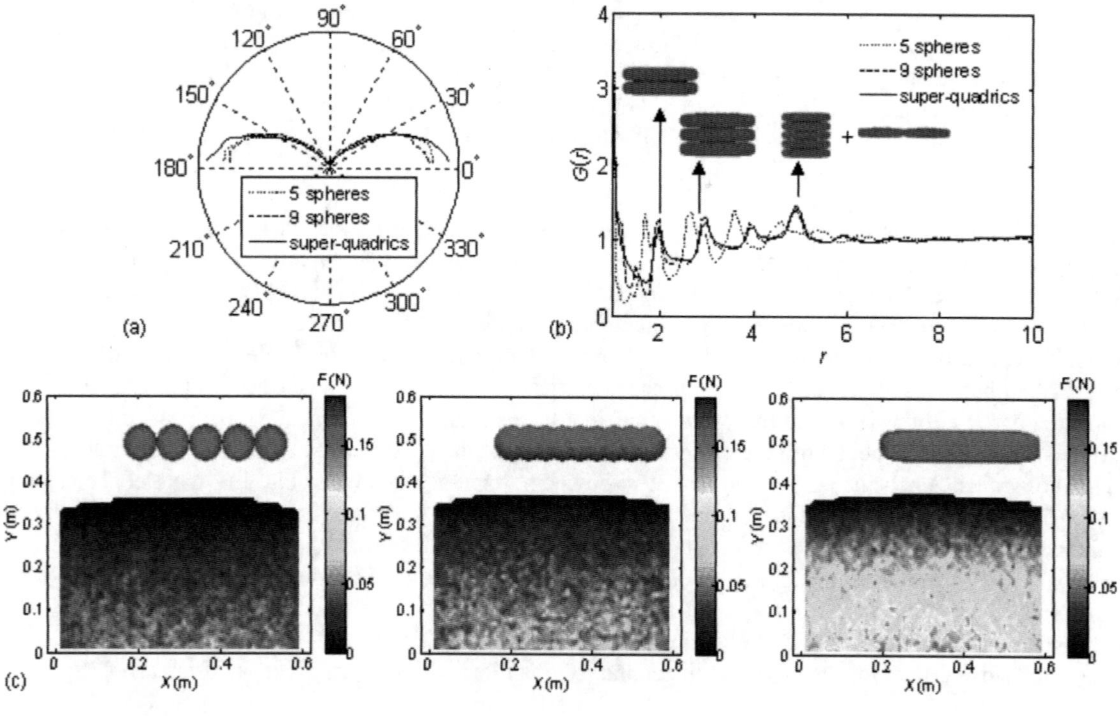

Fig. 1 DEM simulations of granular packings of elongated particles with different surface roughnesses in a 2D, flat bottom container: (a) Particle orientation distribution, (b) radial distribution function ($G(r)$ is given as $G(r)=\langle N(r+\delta r)/2\pi r\delta r\rho\rangle$, where $N(r+\delta r)$ is the number of particles in the differential area $\delta S=2\pi r\delta r$ at a distance r relative to the central particle and ρ is the average number of particles per unit area in the packing region analyzed) and (c) transmission of contact forces within the beds of grains.

(120e) Packing of Elongated Particles: DEM Simulations and Experiments

Tangri, H., University of Florida
Guo, Y.,
Curtis, J. S., University of Florida
Originally presented on: 11/4/2013 13:42:00 - 14:00:00

Packing of Elongated Particles: DEM Simulations and Experiments

Henna Tangri[a, 1], Yu Guo [a] and Prof. Jennifer S. Curtis [a]

[a] Department of Chemical Engineering, University of Florida, Gainesville, FL, 32611 USA

Abstract:

For the achievement of an efficient packing, minimizing the space of occupation and maximizing the strength of packing, it is important to understand the underlying mechanisms. The behavior of identical elongated cylindrical particles is studied via Discrete Element Method (DEM) simulations and complementary experiments. Two types of particles are employed in the DEM simulations: glued sphere elongated cylindrical particles, which are formed by rigidly connecting identical spheres, and true cylindrical particles. The effects of coefficient of friction, coefficient of restitution, drop height, fill height, packing method, deposition intensity, particle size to container size ratio, and the number of glued spheres forming the cylindrical particle are explored. In addition, two novel and accurate methods of packing density analysis - (i) a center of particle method and (ii) a top grid analysis method - are also introduced. It is found that the packing density increases with coefficient of restitution, drop height and number of constituent spheres and decreases with coefficient of friction and Young's modulus. In addition, the coordination number increases with coefficient of restitution and drop height and decreases with coefficient of friction. The most efficient packing scheme is pouring particles over the whole cross-section although deposition intensity and wall effects also play a role in choosing the most efficient packing scheme.

[1] Corresponding author: *hennatangri@gmail.com*

(120f) Dynamic Reduced-Order Modeling of Particulate Systems for Pharmaceutical Unit Operations

Rogers, A., Bristol-Myers Squibb Co.
Ierapetritou, M. G., Rutgers, The State University of New Jersey
Originally presented on: 11/4/2013 14:00:00 - 14:18:00

Particulate and granular processes are of great interest within the pharmaceutical industry, as solid-based dosage forms like tablets remain among the most commonly manufactured drug products. (1) The creation of predictive process models can play an important role in developing the requisite understanding of these processes. Process models can be used to reduce experimental requirements, enhance process understanding, and develop control strategies. Modeling is also an important component of the Quality by Design (QbD) paradigm described by the ICH Q8 guidance for pharmaceutical development. (3)

Powder flow, compaction and other particulate phenomena have been effectively modeled with a high degree of accuracy using discrete element method (DEM) simulations. (2) However discrete and finite element models are computationally expensive to evaluate and therefore cannot be practically implemented in the context of integrated manufacturing process simulation. Reduced order models can be used to incorporate information from these high-fidelity simulations into computationally efficient process models for simulation and optimization purposes. Previously the use of reduced-order models based on principal component analysis has been demonstrated for the incorporation of distributed state information from DEM or CFD simulations into lower order process models for systems at steady state. **(4),** (5) These reduced order modeling methods rely on the reduction of the dimensionality of state data using principal component analysis (PCA) or partial orthogonal decomposition (POD). (6) State data include velocity trajectories or energy and force information for particles in the case of DEM or fluid elements in the case of CFD. Once the state data has been projected into a lower dimensional space, a mapping between process inputs and the reduced state data is developed. This mapping can be based on fitting a response surface, neural networks, kriging, or any other appropriate technique. Once this mapping has been developed the state space can be reconstructed for any relevant set of input parameters, facilitating predictive modeling of distributed parameters in a computationally efficient manner.

Prior work has demonstrated the implementation of reduced order models for systems at steady state. (4)However, dynamic models are needed to study process response to disturbances and to implement effective control strategies for disturbance rejection. The current work thus focuses on the implementation of reduced order models for dynamic systems, specifically those pertaining to particulate processes for the manufacture of solids-based drugs. In this case, time is considered one of the inputs of interest from which the state data can ultimately be predicted. These dynamic models assume inherent time dependence for the underlying system. A variety of order reduction techniques, including those based on principal component analysis are examined.

A case study based on a continuous convective mixer is presented to demonstrate the methods discussed. State data including average particle velocities, kinetic energies and force trajectories within the continuous mixer are readily obtained from DEM simulations. This unit operation provides an effective case study as it has been used in prior work to demonstrate the efficacy of a discrete elements reduced order modeling method for particulate systems at steady state. (4)

1. *CONTINUOUS PROCESSING IN THE PHARMACEUTICAL INDUSTRY Changing the Mind Set.* . **Plumb, K.** A6, Chemical Engineering Research and Design, Vol. 83, pp. 730–738.

2. *Discrete particle simulation of particulate systems: A review of major applications and findings.* **Zhu, H.P., Zhou,Z.Y., Yang, R.Y., Yu,A.B.** 2008, Chemical Engineering Science, Vol. 63, pp. 5728 -- 5770.

3. *ICH Harmonised Tripartite GuidelinePHARMACEUTICAL DEVELOPMENT Q8(R2) Current Step 4 version.* s.l. : International Conference on Harmonisation of Technical Requirements for Registration of pharmaceuticals for human use, August 2009.

4. *Reduced-order discrete element method modeling.* **Boukouvala, F., Gao, Y., Muzzio, F., Ierapetritou, M.G.** Chemical Engineering Science, p. Article in Press.

5. *Optimization of IGCC processes with reduced order CFD models.* **Lang, Y., Zitney, S.E., Biegler, L.T.** 2011, Computers & Chemical Engineering, Vols. In Press, Corrected Proof.

6. *Reduced orde rmodelling method via proper orthogonal decomposition (POD) for flow around an oscillating cylinder.* **Liberge, E., Hamdouni, A.** 2010, Journal of Fluids and Structures, Vol. 26, pp. 292-311.

7. *Periodic Section Modeling of Convective Continuous Powder Mixing Processes.* **Gao, Y., Ierapetritou, M., Muzzio, F.** 1, 2012, AIChE Journal, Vol. 58, pp. 69-78.

8. *Continuum model of mixing and size segregation in a rotating cylinder: concentration-flow coupling and streak formation.* **Khakhar, D.V., Orpe, A.V., Ottino, J.M.** 2-3, 2001, Powder Technology, Vol. 116, pp. 232-245.

9. *Characterization of continuous convective powder mixing processes.* **Portillo, P.M., Ierapetritou, M.G., Muzzio, F.J.** 2008, Powder Technology, Vol. 182, pp. 368–378.

10. *Areviewon the continuous blending of powders.* **Pernenkil, L., Cooney, C.L.** 2006, Chemical Engineering Science, Vol. 61, pp. 720-742.

11. *Effect of operating conditions and design parameters in a continuous powder mixer.* **Vanarase, A.U., Muzzio, F.J.** Powder Technology , 2011, Vol. 208, pp. 26-36.

(120g) A Markov Chain Model for a Tablet Coating Process

Kumar, R., Purdue University
Wassgren, C. R., Purdue University
Originally presented on: 11/4/2013 14:18:00 - 14:36:00

A Markov chain is a mathematical model for a system undergoing random transitions from one state to another. The transition between states is described by a transition matrix and it is assumed that the transition depends only on the current state. In this work, a stationary Markov chain model is developed to model tablet coating in a rotating drum. Coating data from short Discrete Element Method (DEM) simulations are used to construct the transition matrix for the Markov process. The transition matrix is subsequently used to predict the inter-tablet coating variability at longer times. The predictions from the Markov chain model are then compared to those obtained directly from DEM. The effect of the time between transition states, number of states, and the learning time (time required to construct the transition matrix) for the Markov process are studied. It is found that the time between the transition states has a strong influence on the predicted inter-tablet coating variability. Good agreement with DEM is obtained when the transition time is chosen to be the time required for successive appearances of tablets in the spray zone, referred to as cycle time. The learning time should be large enough so that the transition matrix converges (as measured using an l_1-norm) and a transition into a state has at least one transition out of that state. The predictions get better as the number of states increases but also require a larger learning time.

(120h) Predicting Bulk Scale Properties of Surface Modified Fine Powders From Particle Scale Interactions

Capece, M., New Jersey Institute of Technology
Huang, Z.,
Dave, R. N., New Jersey Institute of Technology
Originally presented on: 11/4/2013 14:36:00 - 14:54:00

Packing or porosity is a fundamental property of solid particulates systems and is of prime importance to many industries which handle or process such material. In the case of fine dry particles, attractive van der Waals forces are dominant and can result in high cohesion among particles and poor bulk scale powder properties such as high porosity. This study investigates the effect of van der Waals force on porosity of powder beds consisting of fine cohesive particles (5 μm – 223 μm). Nano-dry coating is utilized to modify the surface roughness of particles in order to further elucidate the effect of the van der Waals force on porosity. Results conclude that nano-dry coating significantly reduces the van der Waals force resulting in improved porosity. The granular bond number, defined as the ratio of the van der Waals force to particle weight was also shown to accurately predict porosity. A subsequent theoretical investigation determined the effect of surface modification, specifically the size, surface energy, and surface area coverage of asperities on the porosity. Further investigation showed that the surface roughness of non-surface modified particles may be poorly defined and may lead to erroneous calculation of inter-particle forces and ultimately the erroneous determination of bulk scale properties such as porosity. Therefore, in addition to dry coating being a useful process to improve powder properties such as porosity, dry coating can be used to define surface properties to accurately predict bulk level powder properties from particle scale interactions defined by the granular bond number.

(122a) Co-Gasification of Sewage Sludge and Biomass in a Fixed Bed Downdraft Gasifier

Jiang, X.,
Gao, W., National University of Singapore
Cheng, Y., National University of Singapore
Neoh, K. G., National University of Singapore
Wang, C. H., National University of Singapore
Originally presented on: 11/4/2013 12:30:00 - 12:49:00

Sewage sludge, an unavoidable product of municipal waste-water treatment plant, is one of the wastes that can be converted into useful energy. The amount of sewage sludge will increase considerably due to municipal development and increasing populations in cities. Currently, sewage sludge is mainly being handled via landfill, incineration or anaerobic digestion. This study investigates the co-gasification of sewage sludge with wood chips in a fixed bed downdraft gasifier. The produced syngas is to be used to generate heat and power in an internal-combustion engine. The characterization of sewage sludge and wood chips was carried out via thermogravimetric analysis (TGA), elemental analysis and Inductively Coupled Plasma (ICP) analysis. The effects of different sewage sludge fractions in the mixture with woodchips on product yields, gas composition, gross heating value and gasification performance are investigated. These results can help us better understand co-gasification of sewage sludge and biomass in a fixed-bed downdraft gasifier.

(122b) Sintering Inhibition of Fe-Based Fluidized Beds for Use in High Temperature Redox Reactions

Greek, B., University of Florida
AuYeung, N., University of Florida
Imponenti, L., University of Florida
Klausner, J., University of Florida
Originally presented on: 11/4/2013 12:49:00 - 13:08:00

In the design of high-temperature thermochemical reactors, iron has been considered as a reactant for hydrogen production via a two-step water-splitting cycle. Use of particles as sintering inhibitors to allow fluidized and fixed bed operation of Fe-based reactions at temperatures exceeding 550 °C was investigated. Various mass percentages of elemental iron (Fe) were mixed with various nonreactive particles, including ZiO_2, SiO_2, and Al_2O_3. In addition, use of reactive CeO_2 particles was investigated. Various particle sizes were used, ranging from 75-180μm. Oxidation was carried out under a stream of H_2O at temperatures from 550-950 °C, while reduction was carried out under a stream of CO at temperatures ranging from 800-950 °C. During preliminary investigations, sintering inhibition was achieved at temperatures up to 950 °C with mixtures consisting of approximately 20 % Fe, 53% ZiO_2, and 27% SiO_2 by mass.

(122c) Greenhouse-Gas (GHG) Emissions From Denitrifying Fluidized Bed Bioreactors (FBBR)

Eldyasti, A. K., University of Western Ontario
Nakhla, G., The University of Western Ontario
Zhu, J., Western University
Originally presented on: 11/4/2013 13:08:00 - 13:27:00

Anthropogenic greenhouse gases (AnGHGs) emissions are globally recognized by the United Nation framework Convention on Climate Change (UNFCCC) (UNFCCC, 2007). GHGs include carbon dioxide (CO_2), methane (CH_4), and nitrous oxide (N_2O). N_2O, the dominant ozone-depleting substance, is the third most important GHGs with a global warming potential (GWP) of 310 times that of CO_2. Nitrous oxide (N_2O) is a significant anthropogenic greenhouse gases (AnGHGs) emitted from biological nutrient removal (BNR) processes. This study tries to fill the gap of knowledge and get a deeper insight into N_2O emissions from denitrifying fluidized bed bioreactors (DFBBRs). The DFBBR was operated on a synthetic municipal wastewater at organic and nitrogen loading rates of 5.9 kg COD/(m^3.d) and 1.2 kg N/(m^3.d), respectively in phase I for 150 days. In phase II, the DFBBR columns were tested for 50 days at higher organic and nitrogen loading rates of 7 kg COD/(m^3.d) and 2 kg N/(m^3.d), respectively. The aforementioned organic and nitrogen loading rates represent two different chemical oxygen demand (COD)-to-nitrogen (COD/N) ratios of 5 and 3.5 for phases I and II, respectively.

After the limited carbon phase, the DFBBR was tested at COD/N of 5 again for 50 days to investigate the dynamics of N_2O emissions for the particulate biofilm. Results showed that the N_2O conversion rate from the DFBBR was 0.53% of the total nitrogen loading in phase I while in phase II at the low COD/N ratio of 3.5, the N_2O conversion rate increased by 196% to 1.57% of the influent nitrogen loading with a 7 fold increase in liquid nitrite concentration. Subsequent increase to a COD/N ratio of 5, affected a 78% increase in emissions of the influent nitrogen compared to the first phase. Comparing the N_2O emission of the DFBBR system with other systems (up to 14.6%), showed that the N_2O emission was in a lower range (0.53% to 1.57%).

(122d) Chemical Looping Simulations Using Aspen Plus

Ocone, R., Heriot-Watt University
Porrazzo, R.,
White, G.,
Originally presented on: 11/4/2013 13:27:00 - 13:46:00

Carbon capture from power plants is recognised as one of the primary remits of governments worldwide as part of their duty to mitigate climate change. Chemical looping combustion (CLC) is potentially the technology best suited for the efficient, low cost and low energy capture of CO_2 by separating the CO_2 from flue gases. The process consists of cyclic reduction and oxidation of a metal oxygen carrier (in the form of particulates) which is exchanged between two reactors. The oxidation reactor is usually a circulating fluidised bed and the combustor is often a bubbling bed reactor. The oxygen carrier transfers oxygen from the air to the fuel, hence direct contact between air and fuel is avoided. Consequently, the outlet gas from the combustor contains only CO_2 and H_2O; the latter is easily removed by condensation and the CO_2 is readily captured for storage and/or utilisation. The main attractiveness of the process is that it does not involve any penalty for extra highly intensive separation of CO_2 from the other combustion products. Indeed, separation always involves high energy consumption with the possible consequent production of additional CO_2.

A number of challenges exists for the optimal operation of the CLC set-up and those involve the characterisation of the oxidation reactor, the optimal circulation rates, the thermal output, the reactors' residence time.

The present paper presents the process integration analysis of the CLC set-up and the implementation of a novel method to simulate fluidised beds using Aspen Plus software. The fuel reactor is modelled as PFR and CSTR reactors in series: the PFR reactor is employed to describe the bubble phase whereas the CSTR describes the emulsion phase. The air reactor is modelled via four CSTRs in series which take into account the variation of the void inside the reactor. Two different models are considered: the uniform conversion model (UCM) and the shrinking core model (SCM). In addition, various controlling steps of reaction are investigated and comparisons between the fluidised bed models proposed here and the Gibbs reactor model (usually implemented in Aspen Plus to simulate fluid beds) are carried out.

Results are shown from the two models and the net heat duty of the whole system is verified.

(122e) The Effect of Pressure On Bubble Parameters in Jetting Fluidized Bed At High Fluidizing Gas Velocity

Li, J., Institution of Coal Chemistry, Chinese Academy of Sciences

Cheng, Z., Institute of Coal Chemistry, Chinese Academy of Sciences

Fang, Y., Institute of Coal Chemisty, Chinese Academy of Sciences

Originally presented on: 11/4/2013 13:46:00 - 14:05:00

The effect of pressure on bubble parameters in jetting fluidized bed at high fluidizing gas velocity

Junguo Li, Zhonghu Cheng, Yitian Fang

State Key Laboratory of Coal Conversion, Institute of Coal Chemistry, Chinese Academy of Sciences, Taiyuan, 030001, P. R. China

Introduction

Lignite is abundant and makes up approximately 40% of coal reserves in the world. Commercial use of lignite is forecast to grow substantially owing to the occurrence of deposits in such countries as India, China, and Southeast Asia, which are experiencing the strongest growth in electricity demand.[1] Also inferior anthracite and non-coking bituminous coal is abundant in China. What is worse, more than 50 percent of coal has a high ash melting point. Commercial use of the above mentioned coal is forecast to grow substantially owing to the lack of oil in China, which is experiencing the strongest growth in the development of coal chemical industry.[2] However, inferior anthracite usually has high ash and sulfur content and high ash fusion temperature. In addition, lignite has high water and oxygen content, low energy density, being prone to weathering and spontaneous combustion. All these characteristics limit their high efficient utilization. Fortunately, fluidized bed is considered as a promising gas producing technology due to its better adaptability to the poor coal and effectiveness in the control of pollutant emissions. However, it has the disadvantages of difficult scale-up, and relative lower capacity. Considering the developed large-scale gasification technology is all on the base of increasing of pressure, the behaviour of gas-fluidized beds at high pressures is of interest from theoretical and practical points of view.

Jetting fluidized bed (JFB), another type of spouted bed, where the jetting phenomena is an important considerations, is the spouted fluid bed, especially when the jet does not penetrate through the bed like that in a spouted bed.[3] One of the important applications of the JFB is the ash-agglomerating fluidized-bed coal gasifier (U-gas, KRW, and ICC). Several demonstration projects of this coal gasification process have been established in China and USA.[4] Many researchers pay attention to the jet penetration depth; however, too few papers report on the bubble dynamic of the JFB. After all, bubble characteristics have been well recognized as one of the most important parameters of conventional gas-solid fluidized bed.

The present work aimed to perform experimental investigations on the bubble dynamic in a two dimensional JFB. It focused on systematically examining the effects of pressure, fluidizing gas velocity, jetting gas velocity and static bed height on the bubble parameters and its new correlation.

Experimental Setup

Apparatus and materials

Experiments were conducted utilizing a high-pressure fluidization cold model facility. Briefly the facility comprises a pressure vessel, a nitrogen supply system, CCD video cameras, a video imaging system, an optic fiber system and peripheral equipment. The pressure vessel, capable of operating at pressures up to 3.3MPa, is 1.6 m high with an inside diameter of 1.2 m. The interior of the vessel is lighted by internal LED lamps. The nitrogen supply system uses a high-pressure reciprocating pump. No nitrogen recirculation system is used. Supply pressure in the reciprocating pump is maintained by a central liquid nitrogen system. Flow control is accomplished by use of a parallel mass flow meter setup. The control valve is used for fluidized bed distributor as the fluidized gas, and the control valve is used for jetting tube as the jetting gas. An upstream pressure control valve and a back pressure control valve are used to maintain the system pressure. Experiments were carried out in a 2D Plexiglas jetting fluidized bed which has a cross section of a = 300 mm, b = 30 mm and total height of 1450 mm (see Fig. 1). The area of jetting nozzle is c = 50 mm, b = 30 mm and height l_3 = 400 mm. A V type gas distributor which has a b = 90 ° conical angle and height l_2 = 100 mm was located at the bottom of the bed and the total area of all orifices is 1.2%.

The bed material was polystyrene with the sphericity close to 1. The mean particle diameter was 1.32 mm and the particle density was 1020 Kg/m^3, which is close to the group B/group D boundary using Geldart's definition. The minimum fluidization velocity, u_{mf}, was then determined from the plot of pressure drop vs. velocity of nitrogen in the traditional way at 0.7, 1.1, 1.6 and 2.1 MPa. The terminal velocity, u_t, was then determined from the terminal velocity of a falling particle under different pressure. A summary of experimental conditions and the particles fluidization properties studied in this work were listed in Tables 1 and 2, respectively.

Video imaging system and image processing

The low-cost CCD cameras which are able to work under pressure are customized and the image is clear under lower pressure. However, as the pressure increases, the density and refractive index of the gas are constantly larger and camera image becomes increasingly blurred. Fortunately, changing the focal length appropriately, the image becomes clear again under higher pressure. In the experiment three groups of CCD cameras which have different focal length are installed to observe bubbles and solids motion form in the high-pressure fluidization cold model facility. The image of the first group cameras is clear only under lower pressure (0.1 - 0.7 MPa), the second is clear under middle pressure (0.8 - 1.4 MPa), and the last is clear only under higher pressure (1.5 - 2.1 MPa). The cable, sealed in the flange, is capable of transmitting imaging signal to an external video imaging system. And then the films of digital video image from the video camera are transferred to PC computer. With the image processing toolbox of Matlab, the software is developed to automate the procedures for picture acquisition, data processing and analysis for frame by frame. Then the data about bubble diameter in the bed is acquired.

Table 1 Experimental conditions

P (MPa)	H_0 (mm)	u_f-u_{mf} (m/s)	u_j / u_t
0-2.1	240-300	0.2-04	0.8-1.5

Table 2 Minimum and terminal velocities in fluidization of polystyrene under pressure

P (MPa)	u_{mf} (m/s)	u_t (m/s)

0.7	0.19	2.10
1.1	0.16	1.68
1.6	0.13	1.39
2.1	0.11	1.21

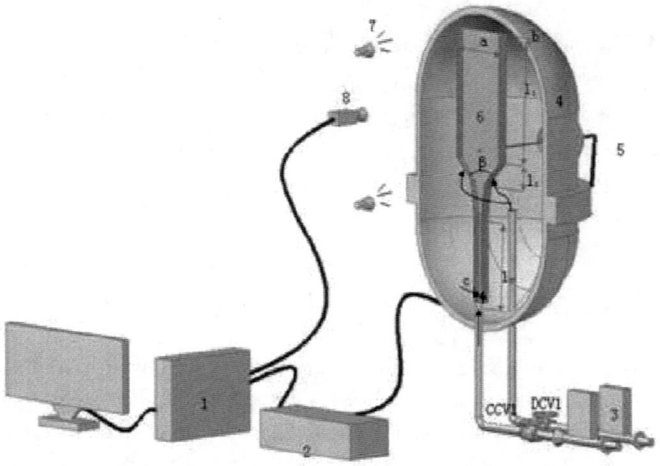

Fig. 1. Scheme of the experimental set-up. 1-computer; 2-data processing system; 3-mass flow meter/controller; 4-pressure vessel; 5-optical fiber probe; 6-two dimensional jetting fluidized bed; 7- LED lamps; 8-CCD camera.

Results and Discussions

Experimental data about bubble equivalent diameter distribution versus height above the top of air distributor at 2.1 MPa are presented in Fig. 2 for an identical static height of $H_0/D = 0.87$, jetting gas velocity $u_j = 1.25\ u_t$ and fluidizing gas velocity $u_f - u_{mf} = 0.20$ and 0.3m/s. It is shown that with the increase of bubble height or fluidizing gas velocity, bubble diameters almost keep constant. It may be due to the effect of jetting gas velocity. In the jetting fluidized bed, the jetting gas is very high and almost occupies half of the total gas.

To highlight the trend of bubble growth along bed height, the arithmetic average bubble diameters are also plotted in Fig. 2. It is reflected that the average diameters remain stable and then decrease at the top of jetting fluidized bed. The reason may that the equilibrium between the coalescence and splitting of bubbles is broken up. The bed is in the turbulent regime.

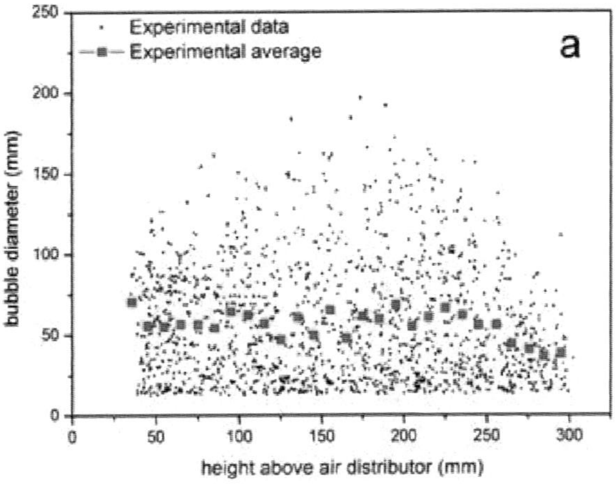

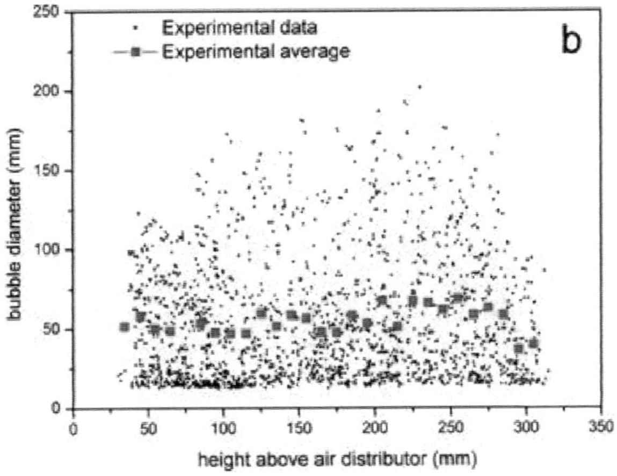

Fig .2. Bubble diameter distributions as a function of height above the distributer at 2.1 MPa: (a) $u_f - u_{mf} = 0.20$ m/s, (b) $u_f - u_{mf} = 0.30$ m/s.

The systematically effects of pressure, fluidizing gas velocity, jetting gas velocity and static bed height on the bubble parameters and its new correlation will be presented in the congress.

Notation

D	Width of 2-dimensional jetting fluidized bed, mm
H_0	Static bed height, mm
u_f	Fluidizing gas velocity, m/s
u_j	Jetting gas velocity, m/s
u_{mf}	Minimum fluidization velocity, m/s

u_t Terminal velocity of a free-fall particle, m/s

Literature Cited

1. Chadwick BL. Development and application of an on-line fouling gauge for brown coal power stations. *Industrial & Engineering Chemistry Research.* Mar 1999;38(3):1159-1162.

2. Li FH, Huang JJ, Fang YT, Wang Y. Formation Mechanism of Slag during Fluid-bed Gasification of Lignite. *Energ Fuel.* Jan 2011;25:273-280.

3. WC Y. Fluidization, Solids Handling, and Processing: Industrial Application. *Noyes Publication.* 1998.

4. Zhang K, Zhang J, Zhang B. Experimental and numerical study of fluid dynamic parameters in a jetting fluidized bed of a binary mixture. *Powder Technology.* 2003;132(1):30-38.

(122f) CFD Study of Lignocellulosic Biomass Gasification in a Fluidized Bed Gasifier: A Comparison of Eulerian and Lagrangian Representations of the Biomass Fuel

Stark, A. K., M.I.T.
Altantzis, C., M.I.T.
Ghoniem, A. F., Massachusetts Institute of Technology
Originally presented on: 11/4/2013 14:05:00 - 14:24:00

Thermochemical conversion of biomass to drop-in ready fuels and chemicals offers an attractive option to directly displace petroleum consumption and effectively utilize the energetic and chemical resources available in waste lignocellulosic materials as well as dedicated energy crops. Further, thermochemical conversion technology has the benefit of being governed by a shared set of chemical pathways, such that depending on the desired product(s) reactor conditions (temperature, pressure and availability of oxidant) and design parameters (geometry, freeboard height, etc) can be optimized for many different products from the production of hydrogen-rich syn-gas (gasification) to oxygenated bio-oil (fast pyrolysis).

In this work a Computational Fluid Dynamic (CFD) investigation of biomass gasification is undertaken utilizing the open-source code, MFiX, with a particular focus on the general applicability of the modeling framework to the multitude of reactive regimes in the conversion process. The proper representation of the pyrolysis of the virgin biomass feedstock is tantamount in predictively modeling the gasification process, furthermore it has been demonstrated that the pyrolysis process of biomass is controlled by both chemical and particle-scale transport processes. In CFD modeling of fluidized systems the gaseous species are represented in an Eulerian framework and the solid phases can be represented as either an Eulerian fluid or as a collection of individual Lagrangian particles. When the reacting particles are represented as an Eulerian continuum individual particle-scale transport phenomena are neglected, whereas in a Lagrangian representation this can be directly modeled.

In this paper an Eulerian-Eulerian-Eulerian (EEE) and an Eulerian-Eulerian-Lagrangian (EEL) representation of the gases, silica bed material and fuel (biomass/char) respectively are compared. The trade-offs between these two representations - the benefits of added model fidelity versus computational complexity and cost - are analyzed. Additionally, the mixing dynamics are considered in the context of reactor design considerations such as the location of the fuel and oxidant injection ports.

(122g) Modeling Refractory Degradation Due to Slag Penetration in An Entrained-Flow Gasifier

Pednekar, P., West Virginia University
Bhattacharyya, D., West Virginia University
Turton, R., West Virginia University
Rengaswamy, R., Texas Tech University

Originally presented on: 11/4/2013 14:24:00 - 14:43:00

Modeling Refractory Degradation due to Slag Penetration in an Entrained-Flow Gasifier

Pratik Pednekar, PhD Student, West Virginia University

Department of Chem. Eng., West Virginia University, Morgantown WVÂ 26506

ppedneke@mix.wvu.edu

Debangsu Bhattacharyya, Assoc. Professor, West Virginia University

Department of Chem. Eng., West Virginia University, Morgantown WVÂ 26506

Debangsu.Bhattacharyya@mail.wvu.edu

Tel: 3042939335, Fax: 3042934139

Richard Turton, Professor, West Virginia University

Department of Chemical Engineering, WVU, Morgantown, WV 26506

Richard.Turton@mail.wvu.edu

Tel: 3042939364, Fax 3042934139

Raghunathan Rengaswamy, Professor, Texas Tech University

Department of Chemical Engineering, Texas Tech University, Lubbock, TX 26507

Raghu.Rengasamy@ttu.edu

Tel: 8067421765, Fax 3042850903

Abstract

In slagging gasifiers, molten slag flows down the inner refractory wall. The slag can penetrate into the refractory causing thinning and spalling of the refractory lining due to the build-up of stress. Replacement of the refractory lining is required every 1-2 years and is very expensive. This also leads to significant down-time lowering the overall availability of gasifier-based power plants. Due to the harsh operating condition inside a slagging gasifier, an in-situ measurement of the refractory degradation is not possible with the current state of technology. Therefore, a mathematical model may be used to develop a better understanding of the effects of various operating conditions and physicochemical properties on the degradation characteristics. Such a model may eventually lead to development of better monitoring and prevention techniques.

With this motivation, first a slag submodel was developed to obtain the temperature profile in the wall and the thickness profile of the slag along the gasifier wall. A mechanistic model was developed by considering slag formation and detachment from the char particles followed by transport and deposition of the slag droplets onto the refractory wall. A model for the slag layer was then developed by considering mass, momentum, and energy conservation equations. Due to the temperature gradient that exists in the molten slag layer, a solid slag layer is formed between the refractory and the molten slag layer. A model of the solid slag layer was also developed using energy balance equations. Two phenomenological models were then developed for refractory degradation - one based on the tensile stress and the other based on the compressive force developed due to thermal strains and shrinkage/swelling due to slag penetration.

The slag sub-model was integrated with a 1D steady-state model of a single-stage, downward-firing, oxygen-blown, slurry-fed, entrained-flow gasifier developed previously at WVU in the Aspen Custom ModelerÂ® (ACM) environment. The gasifier model included mass, momentum and energy balance equations for solid and gas phases. The model also included a number of heterogeneous and homogeneous chemical reactions along with devolatilization and evaporation of the slurry feed. The presentation will include results that show the effects of gasifier operating conditions on slag layer thickness, wall temperature profile, and refractory degradation.

(122h) A Gasification Characteristics of Indonesian Kideco Coal in a Segregated Fluidized Bed Gasifier

Jin, G. T., Korea Institute of Energy Research
Lee, S. Y., Korea Institute of Energy Research
Park, Y. C., Korea Institute of Energy Research
Moon, J. H., Korea Institute of Energy Research
Ryu, H. J., Korea Institute of Energy Research
Originally presented on: 11/4/2013 14:43:00 - 15:02:00

In fluidized bed coal gasification reaction, coal particles were reacted with oxygen agent (in operating condition), and were being reduced the size and density during converting to char. The hydrodynamic values such as minimum fluidizing velocity and terminal velocity of particle will be reduced. With the increase of outlet gas velocity and the decrease of the hydrodynamic values of particle by gasifying, it is hard to maintain the sufficient particle residence time for complete conversion in the reactor.

If the gasifier was operated with the low gas velocity for preventing the particle elutriation, the mixing and mass transfer will become worse and apparent chemical reaction will be reduced.

To maintaining the sufficient residence time and hydrodynamic operating values in the reactor, the gasifier operate with two flow regimes. One is the turbulent regime with high superficial gas velocity, and large coal particle with high density. The other is the bubbling regime with low superficial gas velocity, small char particle with low density. Its behaviour will be segregated fluidized bed.

The tests were carried out in a fluidized bed with 0.05 m ID and 0.4 m height for turbulent regime, and 0.1 m ID diameter and 0.7 m height for bubbling regime. Coal was fed into the lower part of gasifier by screw feeder up to 2 kg/h. The gas mixture of O2 , air and N2 for adjusting the O2/C ratio and Uo/Umf is injected into bottom of the gasifier. The proximate and ultimate analysis of Kideco coal is volatile matter, fixed carbon, moisture and ash 49.26%, 38.36%, 9.30% and 3.08%, respectively. And C 62.17%, H 4.08%, N 1.32%, S 0.24% and O 19.80%, respectively.

The operating conditions are temperature were between 750 and 850 Â°C, O/C ratio were between 40 and 70%. The typical gas concentration of CO, H2, CO2 and CH4 in gas outlet was 39.72%, 28.46%, 19.91% and 2.07%, respectively.

(125a) Dynamics of a Homogeneous Bidisperse Gas-Solid Flow Using Particle-Resolved Direct Numerical Simulation

Mehrabadi, M.,
Tenneti, S., Iowa State University
Subramaniam, S., Iowa State University
Originally presented on: 11/4/2013 12:30:00 - 12:49:00

Gas-solid flows are commonly found in industrial applications such as fluidized-bed combustion, fluid catalytic cracking, coal gasification processes, and biomass energy generation. Practical applications of gas-solid flows usually involve a polydisperse solid phase and the interphase transfer of momentum and energy through gas-particle and particle-particle interactions is complex, resulting in phenomena such as segregation. Better understanding of the physics governing these interactions leads to more accurate models that can be used to improve computational fluid dynamic (CFD) approaches used in device-scale calculations of engineering problems. These interactions have been systematically studied for ideal systems of monodisperse gas-solid flows both experimentally and numerically [1-7]. However, the gas-particle and particle-particle interactions of polydisperse gas-solid flows is still a subject of active research.

In this study, we use particle-resolved direct numerical simulation (PR-DNS) to study canonical bidisperse gas-solid flow problems that capture the essential interactions in more complex polydisperse systems. We consider a homogeneous bidisperse gas-solid suspension in zero gravity where the mean pressure gradient in the fluid phase balances the average solids drag. Both fixed particle assemblies and freely evolving suspensions are simulated. When all the particles move with the same velocity (zero relative velocity between size classes), we can make a Galiliean-invariant (GI) frame transformation such that the particles are at rest in the new frame. Therefore, the fixed particle simulations are performed with zero velocity assigned to all the particles in both size classes, while the mean fluid velocity is specified. Beetstra et al. [6] expressed a drag law for each particle size class in a bidisperse suspension as the product of a quadratic function in y_α (the particle diameter of the size class α normalized by the Sauter mean diameter of the bidisperse suspension) and the drag experienced by an equivalent monodisperse suspension. Our drag results from PR-DNS of bidisperse suspensions indicate that the quadratic dependence of the mean drag on normalized particle size class diameter y_α is valid, although our simulation results indicate that the correlation proposed by Beetstra et al. [6] is not accurate because it is based on a monodisperse drag law [5, 6] that deviates considerably from the high-resolution monodisperse drag law proposed by Tenneti et al. [7]. Fixed particle simulations of bidisperse simulations as described above have no mean slip between the size classes, and are of limited value because in reality each size class experiences a different mean drag force, leading to relative acceleration and a finite mean slip between size classes.

In the case of freely evolving suspensions it is difficult to access the entire range of mean slip Reynolds numbers in sedimentation problems. This is because in sedimenting suspensions the mean slip Re_m is determined by the settling velocity, which is in turn determined by the density ratio and solid volume fraction for a fixed value of gravitational acceleration. Therefore, the simulations of freely evolving suspension undergoing elastic collisions are performed in an accelerating frame of reference [8] so that arbitrary mean slip Reynolds numbers Re_m can be accessed over a range of solid volume fractions and solid to gas density ratios. PR-DNS of freely evolving bidisperse suspensions undergoing elastic collisions show that the slip velocity between the two size classes attains a nonzero steady value corresponding to a balance between hydrodynamic and collisional forces for each size class that is determined by the physical parameters of the problem. These simulations allow us to quantify the mean drag on particle size classes as they evolve to the mean slip velocities corresponding to each size class as dictated by the natural dynamics of the problem. The attainment of finite mean slip between size classes implies that a particle flux corresponding to each size class can arise even in the absence of a number density gradient. This has implications for kinetic theories of bidisperse gas-solid flow that require a gradient in the number density of the size class in order to generate a number flux for that size class.

In addition to the mean momentum in the gas and solid phases, we extract the second moment of velocity fluctuations in both gas and solid phases. The kinetic energy associated with the velocity fluctuations in the solid phase, also called the granular temperature, reaches a steady state corresponding to a balance between the hydrodynamic source term and viscous dissipation for each size class in the freely evolving suspension. On the other hand, quantification of kinetic energy in the gas-phase velocity fluctuations, called pseudo-turbulence, reveals that the level of pseudo-turbulence in gas-solid flows of the same total solid volume fraction and mixture Reynolds number are in good agreement, even with equivalent monodisperse gas-solid suspensions reported by Tenneti et al. [9]. The simulations of bidisperse gas-solid flows considered in this study show that at steady state the interphase transfer of turbulent kinetic energy acts as the source of gas-phase velocity fluctuations. This is consistent with the model proposed for monodisperse gas-solid flows [9] modified by a correction factor due to the presence of two size classes in the particle assembly. As long as the correction factor remains close to unity, the monodisperse interphase TKE transfer is retrieved, resulting in the same level of pseudo-turbulence. Therefore, in this regime, the eddy viscosity model proposed by Tenneti et al. [9] is still valid.

In summary, this study gives insights into the physics of gas-solid and solid-solid interactions in bidisperse gas-solid flows that are crucial for developing CFD models for industrial applications.

References

1. Ergun, S., *Fluid Flow through Packed Columns.* Chemical Engineering Progress, 1952. **48**(2): p. 89-94.

2. Wen, C.Y. and Y.H. Yu, *A Generalized Method for Predicting Minimum Fluidization Velocity.* Aiche Journal, 1966. **12**(3): p. 610-&.

3. Hill, R.J., Koch, D. L., Ladd, A. J. C., *The first effects of fluid inertia on flows in ordered and random arrays of spheres.* Journal of Fluid Mechanics, 2001. **448**: p. 213-241.

4. Hill, R.J., Koch, D. L., Ladd, A. J. C., *Moderate-Reynolds-number flows in ordered and random arrays of spheres.* Journal of Fluid Mechanics, 2001. **448**: p. 243-278.

5. Van der Hoef, M.A., Beetstra, R., Kuipers, J. A. M., *Lattice-Boltzmann simulations of low-Reynolds-number flow past mono- and bidisperse arrays of spheres: results for the permeability and drag force.* Journal of Fluid Mechanics, 2005. **528**: p. 233-254.

6. Beetstra, R., van der Hoef, M. A., Kuipers, J. A. M., *Drag force of intermediate Reynolds number*

flow past mono- and bidisperse arrays of spheres. AIChE Journal, 2007. **53**(2): p. 489-501.

7. Tenneti, S., R. Garg, and S. Subramaniam, *Drag law for monodisperse gas-solid systems using particle-resolved direct numerical simulation of flow past fixed assemblies of spheres.* International Journal of Multiphase Flow, 2011. **37**(9): p. 1072-1092.

8. Tenneti, S., et al., *Direct numerical simulation of gas-solid suspensions at moderate Reynolds number: Quantifying the coupling between hydrodynamic forces and particle velocity fluctuations.* Powder Technology, 2010. **203**(1): p. 57-69.

9. Tenneti, S., Garg, R., Subramaniam, S., *Quantification of gas–phase velocity fluctuations in statistically homogeneous gas–solid flow using particle–resolved direct numerical simulation.* Submitted to AICHE Journal, 2013.

(125b) Direct Numerical Simulation of Particle Rotation Effects in Gas-Solid Flows Using An Immersed Boundary Lattice Boltzmann Method

Zhou, Q., The Ohio State University
Fan, L. S., The Ohio State University
Originally presented on: 11/4/2013 12:49:00 - 13:08:00

A new and efficient direct numerical method with second-order accuracy is presented for fully resolved simulations of incompressible viscous flows laden with rigid particles. The method combines the state-of-the-art immersed boundary method (IBM), the multi-direct forcing method and the lattice Boltzmann method (LBM). First, using the new IB-LB method, steady Stokes flows in ordered arrays of non-rotational and rotational spheres are examined extensively. The results of the non-rotational spheres show excellent agreement with existing theories. The present data are more accurate than previous computational results in the literature because of the higher accuracy of the IB-LB method. The rotational Reynolds number is employed to characterize rotational movements of spheres. The solid volume fractions are varied from 0.01 to the close-packing limit. For each solid volume fraction, rotational movements of spheres are simulated at several different rotational Reynolds numbers such as 0.1, 1, 10, 50 and 100. It is found that the lift force produced by rotation movement is directly proportional to the rotational Reynolds number and the drag force is barely affected by them. The lift force is very insignificant when rotational Reynolds number is lower than 0.1. However, it can be larger than the drag force as rotational Reynolds numbers increase especially in the presence of low solid volume fractions. It is shown that the lift force is around 43% larger than the drag force at the solid volume fraction of 0.01 and the rotational Reynolds number of 50. In the case with larger solid volume fraction, e.g., 0.5, the lift force can be as big as 39% of the drag force at the rotational Reynolds number of 100. Second, steady Stokes flows in random arrays of non-rotational and rotational spheres are also examined extensively. The lift force is also found to be very significant compared to the drag force when rotational Reynolds numbers are up to 100. It is noted that, through high speed imaging, investigators have found that about 20% of particles are rotating at high speeds in riser flow fields. The corresponding rotational Reynolds numbers of their reported mean and maximum rotations rates are 130 and 520, respectively. Therefore, simulations in the present study at the rotational Reynolds number up to 100 are not outside of the practical range. In previous drag laws, effects of the particle rotation on lift forces are not considered and thus the lift force is completely ignored. This study demonstrates that the lift force caused by the particle rotation can be very significant compared to the drag force and must be considered in practical simulations adopting drag laws, e.g., two-fluid simulations.

(125c) A Model for Two-Dimensional CFD Simulation of Pseudo-2d Gas-Solid Fluidized Beds

Li, T., URS Corp. National Energy Technology Laboratory
Shahnam, M., National Energy technology Laboratory
Guenther, C., National Energy Technology Laboratory
Originally presented on: 11/4/2013 13:08:00 - 13:27:00

Rectangular fluidized beds with limited bed thickness are often used to characterize the gas-solid flow behaviors such as bubble movement, solids mixing and clustering using the non-invasive measurement techniques. Even though experimental data from such systems are very useful for model development and validation it has been reported in the literature that the two-dimensional computational fluid dynamic (CFD) simulations usually yield poor agreement with the experimental data for solid velocity field. It has been reported by several researchers that the 2D numerical simulation significantly over-predicts the solid velocity in pseudo-2D bubbling fluidized bed [1-3]. In this paper, a new model is proposed for 2D numerical simulations of pseudo-2D gas-solid fluidized beds. The proposed model accounts for the important frictional effect of the front and back walls. The model has been applied to numerical simulations of different pseudo-2D experimental systems. Significant improvement in the numerical predictions has been observed which results in much better agreement with the available experimental data.

(125d) CFD Validation of Coal Gasification Process for Entrained Flow Gasifiers

Nakod, P., ANSYS Inc.
Orsino, S., ANSYS Inc.
Walavalkar, A., ANSYS Inc
Sami, M., ANSYS Inc.
Ozarkar, S., ANSYS Inc.
Originally presented on: 11/4/2013 13:27:00 - 13:46:00

Environmental drivers like carbon capture and other emission controls are creating a great opportunity for growth and adoption of clean coal technologies. Gasification is one of the more established and matured clean coal technologies and numerical simulations are playing crucial role in its development. Different models and approaches have been proposed and discussed in the literature.
In the present work, CFD simulations for ConocoPhillips' EGas technology based two stage oxygen blown gasifier; Mitsubishi Heavy Industries (MHI) research scale two stage air blown gasifier and Brigham Young University (BYU) laboratory scale oxygen blown gasifier are performed and validated including volatile break-up approach developed earlier in ANSYS FLUENT 14.5. The reaction rates are computed using the Finite rate/Eddy dissipation reaction model. The Discrete Ordinates radiation model with Weighted Sum of Gray Gas Model (WSGGM) for gas absorption coefficient along with particle radiation interaction is employed in these simulations. The predicted syngas composition and exit temperature are compared with the experimental results. The simulation predictions are within 10% of accuracy compared to the experimental measurements.

(125e) Numerical Analysis of the Effect of Constitutive Parameters On the Fluidization Characteristics of Particle Flow in Liquid-Solid Fluidized Bed

Liu, G., Harbin Institute of Technology
Lu, H.,
Wang, P.,
Wang, S., Harbin Institute of Technology
Originally presented on: 11/4/2013 13:46:00 - 14:05:00

The constitutive parameters are analyzed with the numerical simulation approach based on Eulerian-Eulerian two-fluid formulation of the transport equations for mass, momentum and kinetic energy. Kinetic theory of granular flow which accounts for the viscous drag influence of the interstitial liquid phase is employed in combination with two-fluid model to simulate the unsteady liquid-solid two-phase flow. Two-dimensional liquid-solid fluidized bed simulations are performed to validate the simulation results with the experimental work of Zenit et al. We focus here on the effect of constitutive parameters on local unsteady feature as predicted by the numerical models. Effective elasticity of particles is taken into account by variation of the restitution coefficients of particle-particle and particle-wall collisions. Viscosity of particles is considered with different viscosity models of granular materials. Radial distribution function influence on liquid-solid flow is investigated by applying various radial distribution function formulations of particles at different volume fractions. Virtual mass force, lift force coefficient as well as different drag models are compared in liquid-solid fluidized bed. The results in present work give some valuable insight into the internal mechanism of liquid-solid two-phase flow characteristics in liquid fluidized beds.

(125f) Numerical Study On Coal/Biomass Gasification in the Downer of a Triple-Bed Combined Circulating Fluidized Bed

Cheng, Y., National University of Singapore
Wang, C. H., National University of Singapore
Originally presented on: 11/4/2013 14:05:00 - 14:24:00

Gasification can convert the carbonaceous feed stocks into synthetic gases, which can be used in the chemical processes or electricity generation in the downstream processing. Through gasification, the energy can be used more effectively, and the emission of CO_2 can also be reduced greatly. In this study, the gasification of both coal and biomass was studied by numerical simulation. The coal gasification through partial oxidation and stream gasification was simulated in the downer of a triple-bed combined circulating fluidized bed. It was found that the steam gasification could produce more synthetic gases than partial oxidation gasification. With decreasing particle size, increasing wall temperature or inlet gas temperature, the conversion ratio of volatiles and char could be enhanced. Biomass gasification in the bubbling fluidized bed was simulated numerically, and the gasification agent was the mixture of CO_2 and air, thus the renewable biomass could not only provide the clean energy, but also help capture CO_2. CO_2 percentage in the gasifying mixture, CO_2-to-biomass ratio, moisture content and size of wood chips on the gasification performance were investigated in this study.

(125g) Coupled Lbm-DEM Simulations of Gas Fluidised Beds
Mueller, C.,
Third, J., ETH Zurich
Originally presented on: 11/4/2013 14:24:00 - 14:43:00

COUPLED LBM-DEM SIMULATIONS OF GAS FLUIDISED BEDS

J. R. Third[a]* and C. R. Mueller[a]**
[a] ETH Zuerich
Leonhardstrasse 27

Zuerich, CH-8092
* jthird@ethz.ch, **muelchri@ethz.ch

Gas-fluidised bed reactors feature in numerous industrial processes, with the fluidised catalytic cracking of hydrocarbons being perhaps the most important example. Other notable applications of fluidised beds include fertiliser manufacture, the production of high purity silicon and the gasification of coal. Despite the importance of fluidisation to industry, many aspects of the behaviour of fluidised particles remain poorly understood. Consequently, there is significant industrial and academic interest in the computational modelling of gas-fluidised beds.

Since its introduction by Tsuji *et al.* (1), the coupled computational fluid dynamics-discrete element method (CFD-DEM) technique has become one of the most widely used techniques for modelling gas-fluidised beds. In these simulations each particle is modelled as a distinct entity, whereas the fluid flow is modelled in a volume-averaged manner using a large cell size, typically 3 times the particle diameter. Due to the low spatial resolution of the fluid model, it is not possible to compute the fluid-particle interactions directly in these simulations with the result that these models require a closure relationship, often called the drag law. In recent years closure relationships have been developed for CFD-DEM simulations using the lattice Boltzmann method (LBM), *e.g.* Beetstra *et al.* (2). However, the drag law remains the greatest source of error in CFD-DEM methods since even the most sophisticated expressions describe only the average force experienced by a particle and make no allowance for particle rotation or relative motion between the particles. The aim of this work is to examine the ability of the CFD-DEM technique accurately to model gas-fluidised beds. The predictions of a CFD-DEM model are compared with those of a coupled lattice Boltzmann method-discrete element method (LBM-DEM) technique, which does not require a closure relationship for the fluid-particle interactions.

Significant differences can be observed between the results of the two modelling approached. For example, figure 1 shows the height of the centre of mass of the particles above the distributor as a function of time for CFD-DEM and LBM-DEM models. These data indicate that the LBM-DEM model predicts larger fluctuations in the average height of the particles than the CFD-DEM model. Furthermore, on a time-averaged basis the location of the centre of mass of the particles is higher for the LBM-DEM model, suggesting that this model predicts a larger bed expansion. This is consistent with a detailed analysis of the fluid-particle interaction forces, which indicates that the force acting on the particles in the LBM-DEM simulation is approximately 125 % of the force that would be predicted by the equation proposed by Beetstra *et al.* (2).

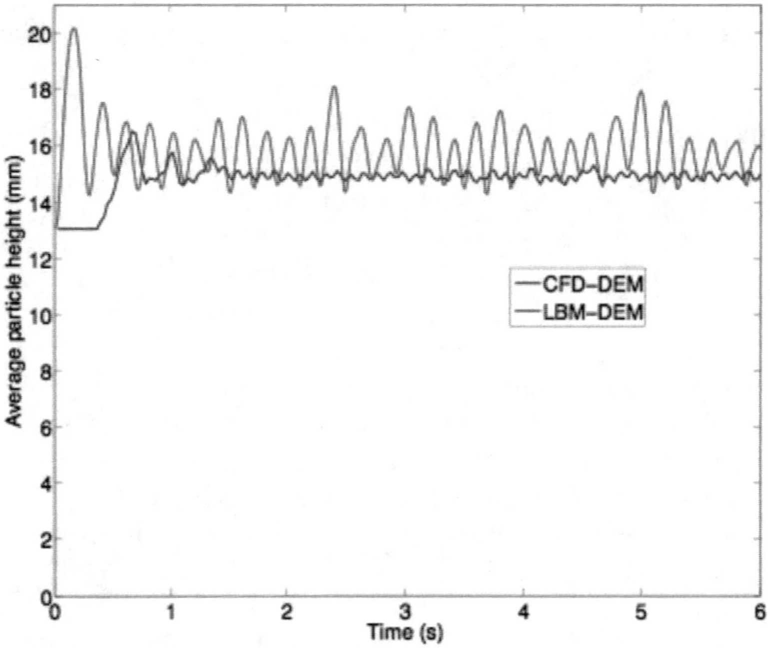

Figure 1: Height of the centre of mass of the particles above the distributor as a function of time for CFD-DEM and LBM-DEM models.

REFERENCES

1. Y. Tsuji, T. Kawaguchi and T. Tanaka, Discrete particle simulation of 2-dimensional fluidized beds. *Powder Technol.*, 77(1): 79- 87, 1993.

1. R. Beetstra, M. A. van der Hoef and J. A. M. Kuipers, Drag force from lattice Boltzmann simulations of intermediate Reynolds number of past mono and bidisperse arrays of spheres. A.I.Ch.E. Journal 53(2): 489-501, 2007.

(125h) Mineral Matter Behavior during Coal Gasification: A New Approach to Evaluate the Slag Blockage Probability in Entrained Flow Coal Gasifiers

Hosseini, S., University of Alberta
Gupta, R., University of Alberta
Originally presented on: 11/4/2013 14:43:00 - 15:02:00

In commercial IGCC applications, the predominant type of coal gasifier is the entrained-flow slagging gasifier operating under high pressures and at high temperatures. Like other processes there are some issues during gasification process. Environmental issues due to production of various emissions such as GHG emission and fine aerosol particles, issues related to injection of feedstock during gasification process and issues related to mineral matters and slag formation. Slag flow during gasification process can cause various problems such as corrosion, erosion and tapping-hole blockage which may lead to emergency shout down. All these problems affect strongly the reliability of gasification process. The phenomenon of micromechanical interactions between inorganic mineral matter and char with slag layer under various condition and at different locations on the wall of gasifier are pivotal keys in predicting and controlling slag flow. Surface tension, viscosity and mineral fusion temperature are essential parameters affecting this phenomenon. Despite the profusion of studies on viscosity and fusion temperature measurements as a function of temperature and mineral compositions, very less is known about the axial and radial variations of slag properties under realistic operating conditions. This research study tries to address these issues.

In this study, gasification experiments on two types of Canadian coal were performed under different operating conditions. The innovation of this study was using slag collector plates to take slag samples at different locations inside the gasifier at specific time intervals. At the first phase of experiments stationary non-continuous slag samples at different axial location were collected and the results showed that collected samples at the baffles of collection plate are different in color and morphology which is an indication of different composition and interaction at each elevation in the gasifier. The morphology and composition of slag at each location were determined by SEM/EDX analysis. Slag sample at higher elevation is smooth and bright but at lower elevation is granular and dark. Analysis of composition using EDX showed different mineral compositions at different axial length and also some un-reacted carbon in porous surface of some slag samples.

Proposing a method to investigate and characterize the blockage probability of slag at the bottom of gasifier is the main purpose of this study. So as the next step of experiments a combination of cylindrical and cone shape ceramic plates were used to collect flowing slag samples. This kind of configuration can be used for simulating the bottom of gasifier for modeling the blockage phenomenon. After installing the ceramic slag collector plates at specific location inside the gasifier gasification experiments were performed with feed stock with different size ranges from 28μm to 106 μm. One of the main advantages of this innovation is that the plate can be removed at desired time intervals (half an hour in these experiments), so the characteristics of minerals correspond to first slag layer are analyzed. The results showed that small iron and calcium particles are the first particles that stick to wall of gasifier. By putting the plates inside the gasifier for another time interval the interaction of new incoming particles with the previous layer were studied. SEM/EDX analysis showed that at vertical locations large alumina silicate particles can stick to particles in the first layer mostly if they have enough iron on their surface but in inclined locations the hydrodynamic of carrier gas, properties of first layer and conversion of char particle affect strongly the fate of incoming particles.

Utilizing the solid residue of each experiment as the feed stock of next gasification experiments showed that carbon conversion in coal particles plays a crucial role in interaction between char particles and slag layer. Char particles rarely stick to other particles on the wall and mostly stick to molten slag layer at high carbon conversion. The SEM/EDX analysis of char particles with different carbon conversion showed that in particles with high carbon conversion more sticky mineral particles are exposed to the surface of char particle making it very sticky which is likely to stick to wall at high temperatures.

In order to take into account the carrier gas hydrodynamic and slag layer properties two dimensionless number were used as the particle interaction indexes. Re number is used as index for gas hydrodynamic and We number is used as index for slag layer properties. The experiments were performed for different temperatures and different composition of gasifying agents and at the end of each experiment, Carbon conversion, Re and We number and slag layer thickness were calculated and all the data were gathered in a table. The results showed that when the temperature is lower than the T_{cv} (temperature of critical viscosity) only particle sintering is likely to occur and having more iron and calcium particles will increase the growth rate of sintered layer and carbon conversion, Re and We number don't have an essential role. At temperatures higher than $T_{cv,}$ size and composition of particles, carbon conversion, Re and We are important and there usually will be a critical number for each of these parameters beyond which particles behaviour will change. The results showed that for particles having high carbon conversion and at low Re, blockage is probable when We number is very low. But the role of Re number is different for vertical and inclined walls. Particles are not likely to stick to slag layer on vertical wall at high Re, but they are most likely to stick to slag layer at high Re on inclined slag surfaces. The results also showed that blockage is mostly probable working with feed stock with small size range at temperatures higher than T_{cv}, very high carbon conversion and very low We. (slag blockage occurred three times and slag bridging occurred 2 times in our experiments at the bottom of slag collector plates).

Having a data bank for slag blockage probability as a function of operating condition parameters is essentially beneficial in controlling slag flow in gasifiers which is performed in this study. It's also crucial to extend these data bank for various feed stock compositions which is the next aim of these study which is planned to be done by doing the gasification experiments with different coal blending.

(172a) Multidimensionality in Fluidized Nanopowder Agglomerates

de Martin, L., Delft University of Technology
van Ommen, J. R., Delft University of Technology
Originally presented on: 11/4/2013 15:15:00 - 15:33:00

Fluidization is a promising but challenging technique for processing of nanoparticles, e.g. to produce coated nanoparticles [1]. Due to van der Waals forces, dominant at the nano-scale, nanoparticles do not fluidize individually but form agglomerates [2]. These agglomerates are usually described with fractal geometry and by a fractal dimension around 2.5 [3, 4]. However, this dimension is typically obtained with settling experiments [3], which include multiple assumptions [5] and can only focus on the macro-scale of the agglomerates, since the smallest scales are not visible. The characterization of the whole agglomerate structure with only one fractal dimension seems in contradiction with the –also broadly accepted– fact that the agglomerates have a hierarchical structure [2, 6]. If the mechanism of formation of the different hierarchical levels is different, one would expect to find different fractal dimensions in each level, reflecting the mechanism that governed its formation.

We aim to clarify this contradiction measuring in-situ the structure of the fluidized nanoparticles agglomerates with spin-echo small-angle neutron scattering (SESANS) [7]. The SESANS outcome is polarization as a function of length scale, which can be related to the autocorrelation function of the sample.

We investigate two models to describe the autocorrelation function $\gamma(r)$ of the agglomerates structure. One of them is the classical fractal description on the mono-dimensional form $\gamma(r) \propto r^{Df-3}$. The other model includes two fractal dimensions, one describing the smallest scales $\gamma(r) \propto r^{Df1-3}$ and another one describing the largest scales $\gamma(r) \propto r^{Df2-3}$.

The mono-dimensional form fails to predict the experimental polarization. Contrary, the agreement between the experimental and the modeled polarizations is excellent for the bi-dimensional $\gamma(r)$, indicating that fluidized TiO_2 P25 agglomerates have at least two fractal dimensions. One of them is approximately 2.1 and characterizes the primary strong chain-like aggregates that the particles typically form in the production process. The second one is about 2.8 and characterizes the secondary agglomerates formed by the primary agglomerates.

The results are in line with the hierarchical levels proposed by Yao *et al.* [2]. The fractal dimension changes between these levels as a result of the different mechanisms that govern their formation. This is opposite to the classical description that uses only one fractal dimension to characterize the scaling in the whole structure. The bi-fractal characterization can help to unravel the mechanisms that form each hierarchical level, and to explain diffusion processes inside the agglomerates.

1. Ommen, J.R., J. Valverde, and R. Pfeffer, Journal of Nanoparticle Research, 2012. **14**(3): p. 1-29.

2. Yao, W., G. Guangsheng, W. Fei, and W. Jun, Powder Technology, 2002. **124**(1–2): p. 152-159.

3. Castellanos, A., J.M. Valverde, and M.A.S. Quintanilla, Physical Review E, 2001. **64**(4): p. 041304.

4. Nam, C.H., R. Pfeffer, R.N. Dave, and S. Sundaresan, AIChE Journal, 2004. **50**(8): p. 1776-1785.

5. Tang, P. and J.A. Raper, Powder Technology, 2002. **123**(2–3): p. 114-125.

6. Valverde, J.M. and A. Castellanos, AIChE Journal, 2006. **52**(2): p. 838-842.

7. Andersson, R., L.F. van Heijkamp, I.M. de Schepper, and W.G. Bouwman, Journal of Applied Crystallography, 2008. **41**(5): p. 868-885.

(172b) Novel Fluid Grid and Voidage Calculation Techniques for a Discrete Element Model of a 3D Cylindrical Fluidized Bed

Boyce, C. M., University of Cambridge

Originally presented on: 11/4/2013 15:33:00 - 15:51:00

Fluidized beds are used widely in industry, but despite this, the fundamental physics underlying fluidized beds is still not well understood. Computational modelling of fluidized beds has grown significantly recently, due in part to the drastic increase in computational power. An accurate model is advantageous because it can provide detailed time-averaged and instantaneous information on essentially all aspects of fluid and particle motion in fluidized beds. This paper describes the development and validation of a 3D cylindrical model incorporating novel fluid grid and voidage calculation methods to ensure accurate and stable simulations of fluidized beds.

Discrete element modelling (DEM) to simulate individual particle motion combined with computational fluid dynamics (CFD) to model fluid motion has proven an effective way to model laboratory size fluidized beds. Since it model particles individually, governed by contact and Newtonian mechanics, DEM-CFD provides a more accurate and insightful description of the particles in fluidized beds than do two-fluid models (TFMs) which treat the particles as a continuous phase. The main imperfection in DEM-CFD is that it requires a drag law to be used to describe the force of interaction between fluid and particles; however, more detailed models which do not require a drag law, such as direct numerical simulation (DNS), are too computationally expensive to model laboratory-scale fluidized beds.

Recently, researchers have been able to develop DEM-CFD models which go beyond simulating 2D rectangular beds and now model fully 3D cylindrical and more complex geometries. These models have generally used one of two techniques to simulate more complex geometries: (1) using a rectangular fluid grid with the immersed boundary method [1] or (2) using an unstructured fluid grid generated by a commercial CFD package, *e.g.* [2]. One difficulty associated with both methods is that it is very difficult to ensure that the fluid cells retain relatively equal volumes as well as reasonable shapes. In order to satisfy the assumptions made in deriving the volume-averaged fluid equations [3] used in DEM-CFD, it is necessary to ensure that the fluid cells are significantly larger in volume than the particles and do not have oblong shapes. If these criteria or not satisfied, the equations can break down, corrupting the solution.

An additional issue raised by using unstructured grids in DEM-CFD stems from trying to measure voidage in arbitrarily shaped CFD cells. Voidage is the volumetric fraction of a CFD cell free of particles, ideally defined as:

$\varepsilon_{cell} = 1$? $(\Sigma V_{particles})/V_{cell}$

where ε_{cell} is the voidage of the cell, $\Sigma V_{particles}$ is the total volume of all the particles in the cell and V_{cell} is the volume of the cell. This calculation needs to be accurate and stable in DEM-CFD because terms in the fluid equations and the drag law depend heavily on voidage. The main difficulty in calculating voidage stems from the fact that individual particles often lie in multiple fluid cells, and their volume must be divided between those cells. A crude approximation, here forward referred to as the direct method, would be to assume that the entire volume of a particle lies in a fluid cell if its centre lies in the cell. A method originally proposed for rectangular grids [4], here forward referred to as the grouping method, involves calculating voidage for each cell via the direct method and then reassessing the voidage for each cell by averaging its voidage with the voidage in all of its surrounding cells. This methodology was developed to add stability to the voidage calculation by spatially smearing the results and can be directly translated for use in arbitrarily shaped fluid cells. Another methodology for calculating voidage in rectangular grids [5], here forward known as the cuboid approximation, treats particles as being encapsulated by a cube, with the fraction of the volume of a particle divided into fluid cells being equal to the fraction of the volume of the cube located in the cells. This methodology provides stability and accuracy to voidage calculation in rectangular grids, yet it is not readily applicable to arbitrarily shaped fluid cells, since it is computationally expensive to divide the volume of a cube amongst arbitrarily shaped cells.

Thus a 3D cylindrical DEM-CFD model was developed here, using novel methods to address the issues of fluid cell volumes and voidage calculation associated with complex geometry models. A structured CFD grid in cylindrical coordinates was used, in which the length of cells in the radial direction remained constant, angle subtended by cells decreased with distance from the centre of the cylinder. This CFD grid conformation allowed the fluid cells to be kept with constant volume and reasonable shapes, to ensure the volume averaged fluid equations would be satisfied, while still fitting a 3D cylindrical geometry. Additionally, a novel voidage calculation technique, referred to as the ?square grid? technique, was developed to stably and accurately calculate voidage in the cylindrical fluid cells. In this technique, voidage was first calculated on a square grid, using the cuboid approximation [5] to divide particle volume between square cells. Then, based on the fraction of each cylindrical fluid cell which lay in each square cell, calculated during initialisation, the voidage in each square cell was translated to calculate voidage in each cylindrical fluid cell. Figure 1 shows the cylindrical fluid grid setup with the square grid for calculating voidage overlaid.

These novel techniques were built on top of the rectangular DEM-CFD code described in Müller *et al.[6]*. Particles were still modelled in rectangular coordinates, yet contained by cylindrical walls. Fluid particle interaction was modelled using the Beetstra *et al.* [7] drag law. Fluid velocities were converted into rectangular coordinates at the position of the particle of interest, in order to calculate the relative velocity between the fluid and the particle for calculating drag on the particles. The drag force was converted back into cylindrical coordinates to calculate the cumulative drag on the fluid.

Figure 2 shows instantaneous voidage and vertical fluid velocity maps of steady state flow through a packed bed using the square grid, grouping and direct voidage calculation methods. While the voidage calculations are very similar, instabilities in the direct method caused even by minor movement in particles between cells in a packed bed, make its simulated fluid velocity unphysical.

Figure 3 shows instantaneous and time-averaged voidage and vertical fluid velocity maps in a short bubbling fluidized bed using the square grid and grouping methods for calculating voidage. While both have similar voidage results and time-averaged fluid velocity results, only the square grid method shows the stability to physically simulate instantaneous fluid velocity.

The 3D cylindrical DEM-CFD model was validated by comparison to MR experiments. Figure 4 shows a comparison of time-averaged particle velocity in a short bubbling bed; this comparison is discussed further elsewhere [8]. Additionally, DEM-CFD results for slug rise velocity in a tall slugging bed were compared to MR measurements [9] and theoretical values. Simulated slugs had rise velocities of 0.23 m/s and 0.25 m/s at locations 85 mm and 50 mm above the distributor in a 24 mm diameter bed with a settled height of 350 mm with an excess gas velocity $U\text{-}U_{mf}$ = 0.1 m/s. These results match those from MR experiments of the same bed [9], which determined that at these heights in this range of excess gas velocity, slug rise velocities should lie in between the theoretical values for wall and axisymmetric slugs of 0.19 m/s and 0.30 m/s respectively.

The validation of the 3D cylindrical DEM-CFD model by experimental results in vastly different fluidized bed systems lends confidence to its use in predicting behaviour in a wide variety of fluidized bed systems.

References:

[1] Guo, Y., Wu, C.Y., Thornton, C. (2012). A.I.Ch.E. J., 59, 1075-1087.

[2] Chu, K.W., Yu, A.B. (2008a). Powder Technology, 179, 104-114.

[3] Anderson, T. B., & Jackson, R. (1967). Industrial & Engineering Chemistry Fundamentals, 6, 527?539.

[4] Yu, A.B. (2006), A.I.Ch.E. J., 52, 496-509.

[5] Khawaja, H.A., Scott, S.A., Virk, M.S., & Moatamedi, M. (2012). The Journal of Computational Multiphase Flows, 4, 183-192.

[6] Müller, C.R., Scott, S.A., Holland, D.J., Clarke, B.C., Sederman, A.J., Dennis, J.S., Gladden, L.F. (2009). Particuology, 7, 297-306.

[7] Beetstra, R., van der Hoef, M. A., & Kuipers, J. A. M. (2007). AIChE J., 53, 489?501.

[8] Boyce, C.M., Holland, D.J., Scott, S.A., Dennis, J.S. (submitted 2013). Phys Rev E.

[9] Müller, C.R., Davidson, J.F., Dennis, J.S., Fennell, P.S., Gladden, L.F, Hayhurst, A.N., Mantle, M.D., Rees, A.C., Sederman, A.J. (2007). Chemical Engineering Science, 62, 82-93.

Figures:

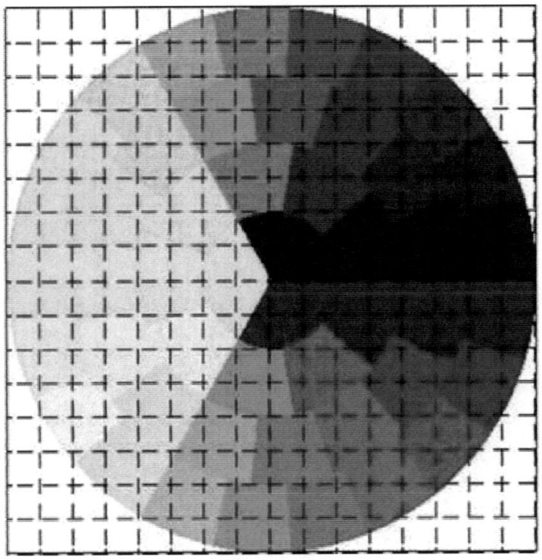

Figure 1 Novel cylindrical CFD grid, each colour representing a different fluid cell, with square grid for voidage calculation overlaid.

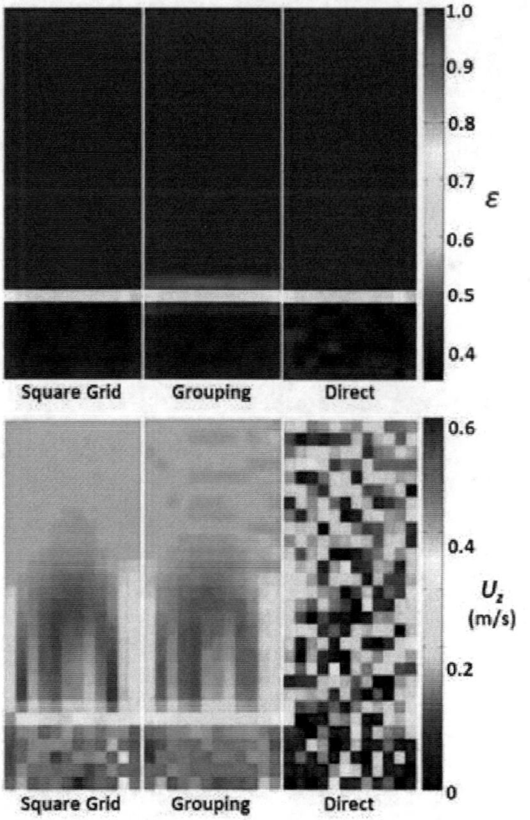

Figure 2. Comparison of voidage calculation methods for packed bed flow.

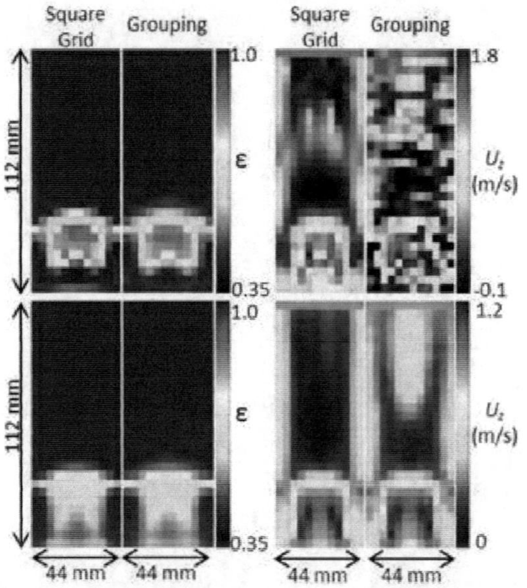

Figure 3. Comparison of voidage calculation methods for instantaneous (left) and time-averaged (right) fluidized bed flow.

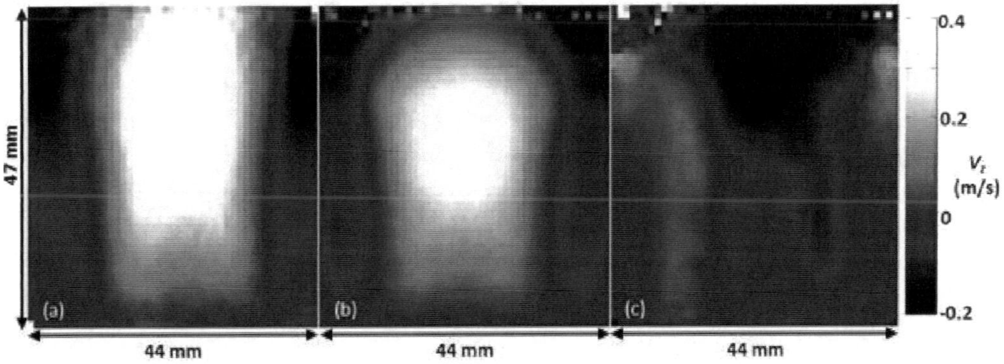

Figure 4 Comparison of time-average axial particle velocity images from: (a) experimental MR imaging and (b) DEM-CFD simulation of a bubbling fluidized bed. A difference map is shown in (c) (Reproduced from Boyce *et al.* [8])

(172c) The Effect of a Binary Size Distribution On Lunar Regolith Ejection

Berger, K. J., University of Colorado Boulder
Brown, B. K.,
Metzger, P. M., NASA
Hrenya, C. M., University of Colorado at Boulder
Originally presented on: 11/4/2013 15:51:00 - 16:09:00

Upon landing of a spacecraft on the Moon (or other extraterrestrial body), the rocket exhaust plume causes regolith (soil-like material) to be blown in all directions. The lack of atmosphere and corresponding drag on the regolith motion causes the ejection to be extremely hazardous to objects both near and far. Work was previously performed on this system to determine the role of collisions in the erosion and ejection of the regolith by investigating a monodisperse system using the discrete element method (DEM), which incorporates collisions explicitly. Following up on this study, we seek to perform a similar investigation using a binary-sized system. In order to do this, we utilize two version of DEM. Namely, one which resolves collisions using a soft-sphere model and the other which ignores collisions and allows for the (unrealistic) crossing of particle trajectories. The system simulated is located 6m from the impingement point of the rocket and lift and drag correlations are used to simulate the forces on the particles. A one-way particle-gas coupling is used such that the particles are affected by the gas but the particles do not influence the gas. The binary mixtures are chosen such that we vary the size ratio and composition. The results show that the initial erosion rate tends to drop significantly during the onset of collisions above the surface when smaller sizes are present, but the steady-state erosion rate appears to differ only slightly with composition and size ratio for the systems investigated. In addition, the character of the collisions (small particle-small particle, small-large, large-large) as they change with size ratio and composition is investigated.

(172d) Effective Wall-Boundary Conditions for Dense Flows of Granular Materials

Chialvo, S., University of Florida
Sundaresan, S., Princeton University
Originally presented on: 11/4/2013 16:09:00 - 16:27:00

A longstanding question in continuum modeling of wall-bounded granular flows and gas-particle flows is the choice of appropriate boundary conditions (BCs) for the solid phase. Among the most common formulations are the simple no-slip condition and the Johnson-Jackson model [1], the latter of which prescribes a slip velocity dependent on the fluctuation energy of the flowing material and on particle properties. Despite their common use, however, very little work has been done to validate these BCs at the micro- or mesoscale, either by experiment or by discrete particle simulations. Furthermore, even if the slip velocity can be predicted accurately, this accuracy may be lost when performing coarse-grid continuum-model simulations, as the boundary layer in a wall-bounded granular flow is typically very small compared to the simulation grid size. In the present work, we aim to address these problems by 1) defining an 'effective' slip velocity that subsumes the boundary layer into the BC and 2) constituting this effective slip velocity in terms of flow and particle properties.

To this end, we perform discrete element method (DEM) simulations of frictional particles in simple-shear flow between two flat, frictional walls. Two distinct regions are observed in the flow: a dense 'core' region, near the center and away from the walls, exhibiting a linear velocity profile; and the boundary layers, which exhibit exponential velocity profiles and lower particle concentrations. The pressure and shear stress in the core region closely follow the local kinetic-theory (KT) model of Chialvo and Sundaresan [2], while the near-wall regions do not because of nonlocal conduction of fluctuation energy. To characterize the slip behavior, we can ignore these nonlocal effects by extrapolating the core velocity profile to the wall and defining the effective slip velocity as the difference between this value and the wall velocity. A linearized version of the KT model, with near-wall behavior denoted as a perturbation from the core behavior, is then combined with Johnson-Jackson-type expressions for the energy and momentum balances at the wall to produce a suggested functional form for the dimensionless slip velocity in terms of the bulk friction coefficient. The DEM data support a model of this form, the coefficients of which we constitute in terms of interparticle and particle-wall friction coefficients. The resulting BC expression is advantageous for its applicability to coarse-grid simulations, as it eliminates the need to resolve the boundary layer. Additionally, its pairing with an algebraic version of the KT model removes the need to track near-wall fluctuation energy or its spatial gradients.

[1] P.C. Johnson and R. Jackson, J. Fluid Mech. 176, 67 (1987).
[2] S. Chialvo and S. Sundaresan, Phys. Fluids, (accepted).

(172e) Large-Scale Coupled CFD-DEM Simulations of Fluidized Granular Systems

Radeke, C., Research Center Pharmaceutical Engineering GmbH
Jajcevic, D., Research Center Pharmaceutical Engineering GmbH
Siegmann, E.,
Scharrer, G., RCPE GmbH
Khinast, J. G., Graz University of Technology
Toschkoff, G., Research Center Pharmaceutical Engineering (RCPE)
Originally presented on: 11/4/2013 16:27:00 - 16:45:00

Granular flows are characteristic to numerous natural phenomena and are part of many processes of the pharmaceutical and chemical industries. In these industrial processes granular flows have a critical impact on the product's performance and safety. For example, in continuous powder blending of active pharmaceutical ingredients (APIs) and excipients prior to tableting, the granular flow properties of the components determine the content uniformity of the tablets made later in the process. Only a detailed understanding of the influence of particle size, shape, surface texture and other properties on the mean and the fluctuating flow fields in granular flows allows a precise design and optimization of this manufacturing operation.

In recent years Discrete Element Model (DEM) simulations have increasingly been used to study flows of granular systems. In the pharmaceutical and other industries, fine granular materials are encountered (typically in the order of ten to several hundred microns), where both particle-particle interactions and fluid drag have significant contributions. Examples include fluid-bed drying, hopper filling/discharge, chute flows, die filling or coating processes. Thus, accounting for fluid-particle interaction it is necessary to correctly describe such processes, and to design and to scale-up industrial devices.

Combining Computational Fluid Dynamics (CFD) and Discrete Element Method (DEM) allows simulating fluidic-granular systems. The main advantage of CFD–DEM is that detailed particle-scale information is obtained, including particle trajectories and forces acting on individual particles. The coupling of CFD and DEM is realized through a fully two-way particle–fluid interaction forces, where the high-performance GPU-based DEM code and the CFD code (AVL-Fire®) are coupled. The advantage of our CFD–DEM hybrid CPU/GPU simulation method is, that the codes runs inside a single workstation but on separate computing platforms. We are able, to do simulations for up to 25 million particles.

The numerical results of fluidized beds obtained in this work were validated against experimental data. The experimental and CFD–DEM simulation data include profiles of the time averaged vertical particle velocity and time-averaged particle velocity fields. The results of our work show excellent agreement with the experimental data.

Our application of the validated code is the simulation of a bottom-spray Wurster-Coater. Air streams into the device through a diffusor plate at the bottom, which consists of regions with different porosities. The plate has larger orifices below the Wurster draft tube and therefore the fluidization gas enters at higher velocity below the tube. Particle coating evolution and residence time distribution is studied for a coating process of several million particles.

(172f) On the Filling Rate of a Liquid Bridge Between Wet Particles

Radl, S., Graz University of Technology

Sundaresan, S., Princeton University

Khinast, J. G., Graz University of Technology

Originally presented on: 11/4/2013 16:45:00 - 17:03:00

While previous research in the area of wet granular flows mainly focused on forces connected to liquid bridges, there is much less theory concerned with the process of liquid transfer upon particle-particle or particle-wall collision. For example, to study the liquid transfer upon bridge rupture, solution of the Navier-Stokes equation (i.e., direct numerical simulations, DNS), or solution based on a quasi-static approximation of the bridge shape can be used [1,2]. For this second stage of the liquid transfer (i.e., liquid bridge rupture), models are already available in literature [3]. However, for the initial fast process of bridge filling, little is known about the rate with which liquid drains into the meniscus.

We study the drainage process of free liquid at the surface of two wet particles using (i) DNS (see Figure), as well as (ii) solution of the film height equation. The latter approach neglects the fluid's inertia, and is based on a fixed shape of the velocity profile across the film height. By scanning a large parameter space using DNS, we parameterize a dynamic model for the bridge volume during filling. This model assumes that the particles' relative motion has no effect on the filling rate. Finally, we present results of DNS of moving liquid coated particles to illustrate the effect of a finite Stokes number on the filling process.

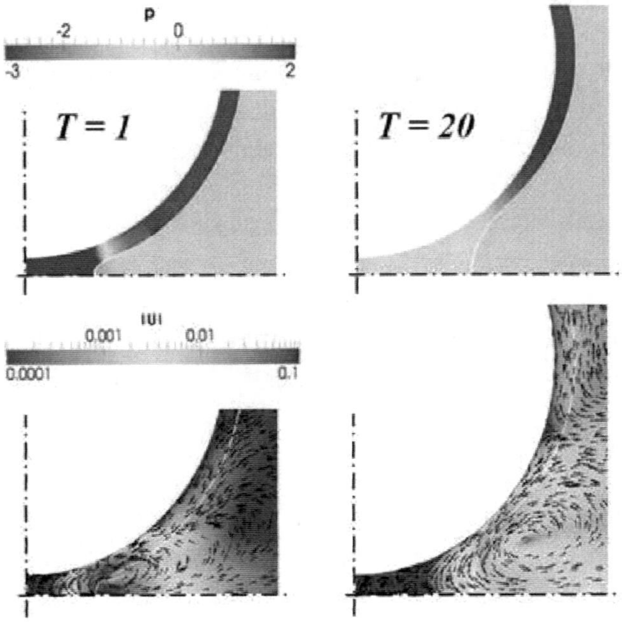

References

[1] P. Darabi, T. Li, K. Pougatch, M. Salcudean, D. Grecov, Modeling the evolution and rupture of stretching pendular liquid bridges, Chemical Engineering Science. 65 (2010) 4472?4483.

[2] S. Dodds, M. Carvalho, S. Kumar, Stretching liquid bridges with moving contact lines: The role of inertia, Physics of Fluids. 23 (2011) 092101.

[3] D. Shi, J.J. McCarthy, Numerical simulation of liquid transfer between particles, Powder Technology. 184 (2008) 64?75.

(172g) Settling of a Swarm of Suspended Particles At Low and Moderate Reynolds Numbers

Ayeni, O.,
Wu, C.,
Nandakumar, K., Louisiana State University
Originally presented on: 11/4/2013 17:03:00 - 17:21:00

The sedimentation of a swarm (cloud) of particles in a viscous fluid at low and moderate Reynolds numbers has been studied using an Eulerian-Lagrangian multiphase flow approach.

We looked at the volume fraction dependence of the settling cloud and find a similar dependence in the simulations as in the theoretical predictions of Nitsche and Batchelor 1997. The average cloud settling velocity and the velocity fluctuations around this average are found to have a linear dependence on $\varphi^{1/3}$ at negligible Reynolds number. The velocity fluctuations display strong anisotropy with the magnitude of the vertical component almost three times the magnitude of the horizontal component.

Particle leakage at low Reynolds number was established and found to be directly related to the initial number of particles in the swarm.

At higher Reynolds numbers, the cloud of particles evolved into an open torus and subsequently loses its axi-symmetry and breaks-up into a number of secondary clouds. This process is a type of Rayleigh-Taylor instability and the number of secondary drops was found in our simulations to be dependent on the shape of the boundaries of the flow domain used rather than the nature of the boundaries.

Breakup at moderate cloud Reynolds number, Re_c is found to occur after a critical aspect ratio is reached and a scaling was proposed for the dependence of the breakup length and breakup time on Re_c.

(172h) Orifice Jamming of Fluid-Driven Granular Flow

Lafond, P. G., Colorado School Of Mines
Gilmer, M. W., Colorado School of Mines
Koh, C. A., Colorado School Of Mines
Sloan, E. D., Colorado School of Mines
Sum, A. K., Colorado School of Mines
Wu, D. T., Colorado School Of Mines
Originally presented on: 11/4/2013 17:21:00 - 17:39:00

The three-dimensional jamming of neutrally buoyant monodisperse, bidisperse, and tridisperse mixtures of particles flowing through a restriction under fluid flow has been studied. During the transient initial accumulation of particles at the restriction, a low probability of a jamming event is observed, followed by a transition to a steady-state flowing backlog of particles, where the jamming probability per particle reaches a constant. Analogous to the steady-state flow in gravity-driven jams, this results in a geometric distribution describing the number of particles that discharge prior to a jamming event. We develop new models to describe the transition from an accumulation to a steady-state flow, and the jamming probability after the transition has occurred. Predictions of the behavior of the geometric distribution see the log-probability of a jam occurring proportionally to (R2 − 1), where R2 is the ratio of opening diameter to the second moment number average particle diameter. This behavior is demonstrated to apply to more general restriction shapes, and collapses for all mixture compositions for the restriction sizes tested.

(191a) Engineering of Multilayered Nanoparticles Structures for Stretchable Conductors With Self-Organized Conductive Pathways

Kim, Y., University of Michigan, Ann Arbor
Zhu, J., University of Michigan
Yeom, B., University of Michigan
Prima, M. D.,
Su, X., University of Michigan
Kim, J. G.,
Yoo, S. J.,
Uher, C., University of Michigan
Kotov, N. A., University of Michigan
Originally presented on: 11/4/2013 15:15:00 - 15:31:00

Stretchable conductors are essential components of next generation electronic devices that can facilitate integration of human-machine interfaces. Despite the fundamental difficulties of combining stretchability and conductivity tremendous research and development efforts have been put into exploring possibilities envisioned from nanoscale building blocks with various polymeric matrixes. Carbon nanotubes (CNTs) have been chosen mainly because of its high aspect ratio that enabled low percolation threshold and good electrical properties during mechanical deformation. Typical stretchable conductors from CNTs, however, have several problems; high quality of composite is not reproducible, metallic conductivity is unlikely, and properties are anisotropic. To overcome those problems, we present a systematic approach making stretchable conductors from extreme content of nanoparticles (NPs) with polymer matrix by layer-by-layer (LBL) assembly technique. To highlight the structural influences on the properties of composites, layered composites by vacuum-assisted flocculation (VAF) method were also prepared. NPs, in general, might be an inadequate choice as fillers because they have minimal aspect ratio, however specially synthesized NPs with judiciously chosen polymer matrix are expected to have better electromechanical properties. Indeed, unexpectedly high electrical and mechanical properties with metallic charge transport mechanism were observed and these were originated from dynamic self-assembly of NPs under stress. Theoretical study about dynamic conducting percolation showed excellent match with experimental data. Also, stretchable nanoparticle conductor composites showed a new approach of electro-tuning mechanical properties. Scanning electron microscope, transmittance electron microscope, atomic force microscopy, electrical properties measurement system, dynamic mechanical analysis, small-angle X-ray scattering were performed to analyze properties of nanocomposites.

(191b) Formation of High-Aspect Ratio Helical Nanorods Via Peptide Driven Self-Assembly of Fullerodendrimers

Hilmer, A. J., Massachusetts Institute of Technology
Shimizu, S., Massachusetts Institute of Technology
McNicholas, T. P., Massachusetts Institute of Technology
Strano, M. S., Massachusetts Institute of Technology
Originally presented on: 11/4/2013 15:31:00 - 15:47:00

The ability to selectively form contiguous, high aspect-ratio electron acceptor domains could aid in the control of active layer morphologies in polymer photovoltaics. We have synthesized two novel, asymmetric methanofullerenes, which self-assemble into one-dimensional supramolecular structures in cyclohexane. We show that, through the introduction of a dipeptide sequence to one terminus of the dendritic methanofullerene, it is possible to transform the assembly behavior of these molecules from poorly formed aggregates to high aspect ratio nanorods. These nanorods have diameters of 3.76 ± 0.52 nm, and possess lengths which can exceed 10 μm. Both direct AFM observations and circular dichroism (CD) measurements indicate that these structures are comprised of helical assemblies of dendritic fullerenes. Temperature-dependent absorbance and CD measurements indicate that self-assembly is initiated by intermolecular peptide interactions, which simultaneously drive self-assembly and impart chirality to the resulting supramolecular structures. Here, we present optical and morphological characterization of these assembled structures.

(191c) Engineering Localized Surface Plasmon Resonances in Si Nanowires

Chou, L. W., Georgia Institute of Technology
Filler, M. A., Georgia Institute of Technology
Originally presented on: 11/4/2013 15:47:00 - 16:03:00

Nanoscale semiconductors are rapidly emerging as an alternate materials platform with which to engineer localized surface plasmon resonances (LSPRs). These quantized excitations, which result from the collective oscillation of free carriers at a metallodielectric interface, promise new opportunities for ultrafast information transfer and efficient energy conversion. Here, we show how phosphorus-doped regions positioned along the length of Si nanowires permit user-programmable spectral responses in the mid-infrared. Highly aligned Si nanowire arrays are synthesized via the vapor-liquid-solid (VLS) technique with a combination of Si_2H_6 and PCl_3 precursors. All phosphorus-doped sections are sandwiched between intrinsic base and tip regions to deconvolute substrate and catalyst effects, respectively. *In-situ* infrared absorption spectroscopy measurements, which mitigate the complications that arise from surface oxidation, reveal intense absorption bands between 800 and 2200 cm^{-1} that exhibit dopant concentration and shape-dependent spectral shifts consistent with longitudinal LSPRs. Discrete dipole approximation calculations confirm the observed spectral response results from resonant absorption and that free carrier densities are on the order of 10^{20} cm^{-3}. Multimodal spectral responses, achieved by introducing multiple doped regions in each Si nanowire, are also possible. Our findings open the door to surface plasmon engineering in a ubiquitous semiconductor and also highlight the utility of VLS synthesis for this purpose.

(191d) Impurities in Semiconductor Nanostructures

Holmberg, V. C., University of Texas at Austin
Sahu, A.,
Norris, D. J., Swiss Federal Institute of Technology (ETH ZÃ¼rich)
Korgel, B. A., The University of Texas at Austin
Originally presented on: 11/4/2013 16:03:00 - 16:19:00

As in bulk semiconductors, impurities play an important role in semiconductor nanostructures. Solid-state reactions such as silicide formation are used to make low-resistance electrical contacts in transistor applications, and cation-exchange reactions can be used to completely (or partially) transform a nanostructure from one material to another. Likewise, impurities are used to store electrochemical energy in devices such as Li-ion batteries. In confined semiconductor nanostructures, even single atomic impurities can have dramatic effects on optical, electrical, and magnetic properties. For example, the addition of only 2–3 Ag atoms to a CdSe nanocrystal can lead to an order-of-magnitude increase in fluorescence efficiency. Controlled impurity incorporation and characterization pose significant challenges in these systems. Here, we will discuss the effects of impurity incorporation in semiconductor nanostructures, as well as how *in situ* transmission electron microscopy (TEM) can be used to study the dynamics of solid-state reactions and impurity diffusion on the nanoscale. Dramatic morphological changes can be observed due to the volume change associated with silicide formation in Si nanowires, and in some systems, the rate of impurity diffusion can be orders of magnitude different than what is observed in the bulk.

(191e) Dense Semiconducting Nanowire Arrays Grown Directly On Graphene

Alper, J.,
Gutes, A.,
Carraro, C., University of California Berkeley
Maboudian, R., University of California, Berkeley
Originally presented on: 11/4/2013 16:35:00 - 16:51:00

Semiconducting nanowires have been demonstrated as promising in a number of devices due to novel properties which are intrinsic to their nanoscale. Silicon nanowires are promising for lithium battery electrodes, where the nanowire geometry allows for relaxation of mechanical stress associated with lithium insertion. Silicon nanowires are also attractive for thin film solar cells, where the nanowire geometry results in enhanced broadband absorption and enables thinner cells with increased efficiency. As well as silicon, the properties of silicon carbide nanowires including large specific surface area and high aspect ratio lend them to application as supercapacitor electrodes and field emitters respectively. Typically these materials are grown via vapor deposition techniques at high temperatures in harsh chemical environments. These growth conditions require refractory, rigid substrates. Yet in order to integrate these materials into devices, they must be accurately transferred and electrically contacted on the array scale. Previous efforts in the transfer step have utilized the Langmuir-Blodgett technique or polymer stamping. The arrays are then electrically contacted through metal evaporation methods. Here we demonstrate an improvement over these previous techniques by growing the nanowires directly on a conductive, flexible, and mechanically robust graphene. The arrays of wires may then be transferred in a facile manner while maintaining good electrical contact throughout the array. The wire-graphene hybrids are structurally characterized via Raman spectroscopy, scanning electron microscopy and x-ray diffraction confirming their chemical identity and crystalline nature. As well, Raman spectroscopy is used to confirm that high quality graphene is maintained during the nanowire growth process. Transfer of the hybrids is demonstrated to a number of arbitrary substrates while maintaining electrical contact. Silicon carbide nanowire – graphene hybrids are also applied as a supercapacitor electrodes with performance indicating good electrical contact throughout the array. Lifetime cycling studies on these electrodes indicate that the hybrid is a robust material with over 95% capacitance retention after 10,000 charge discharge cycles in aqueous electrolyte.

(191f) Inorganic Organic Ligand Exchange On the Surface of Zinc Sulfide Nanocrytals

Lawal, Q., Stanford University
Herron, S., Stanford University
Bent, S. F., Stanford University
Originally presented on: 11/4/2013 16:51:00 - 17:07:00

Inorganic ligand exchange on the surface of zinc sulfide nanocrystals

Qudus Lawal, Steve M. Herron and Stacey F. Bent

Department of Chemical Engineering, Stanford University, Stanford, CA 94305

Nanocrystals (NCs) have been the subject of intense interest because of their current and potential applications in optical and opto-electronic devices, sensing, and medical therapeutics. Control over the surface chemistry of NCs is crucial to the performance of the NCs in opto-electronic devices, and the NC surface properties play a major role in thin-film formation from nanocrystal inks. Specifically, small inorganic species such as ammonium sulfide can provide colloidal stability in polar solvents and may facilitate charge transport between individual nanocrystals, creating more opportunities for device integration of these materials. In these studies, zinc sulfide nanocrystals were synthesized with long-chain hydrophobic ligands including oleylamine. Subsequently, a series of ligand exchanges were performed on these nanocrystals to replace the nascent nonpolar ligands with a metal-free inorganic species, ammonium sulfide. This was done in a two-step process, in which the first step was to reduce the length of the bulky oleylamine ligand to butyl amine, to facilitate transport of ammonium sulfide to the nanocrystal surface. The NC's were dispersed in the polar solvent formamide to stabilize this matrix before finally extracting excess hydrophobic ligand with hexanes. Different mass ratios (0.1:1, 1:1, 1:10) of NC to sulfide ion were used to determine the optimal ratio for a desirable exchange. The NCs before and after ligand exchange were characterized using FTIR spectroscopy to determine the efficacy of ligand exchange by monitoring the change in surface chemical environment, and UV-Vis spectroscopy to examine the optical characteristics of the NCs before and after ligand exchange. The effect of ligand exchange on the NC properties will be discussed.

(191g) Leveraging Molecular and Quantum Confinement in Arrayed Nanostructures for Energy Technologies

Crosser, L. A., Mainstream Engineering Corporation
Nicol, K. T., Mainstream Engineering Corporation
Chester, G. E., Mainstream Engineering Corporation
Zastrow, D. J., Mainstream Engineering Corporation
Schwartz, N. R., Mainstream Engineering Corporation
Wong, J. C., University of Florida
Chew, J. O., University of Florida
Ziegler, K. J., University of Florida
Hill, J. J., Mainstream Engineering Corporation
Reeves, R., Mainstream Engineering
Originally presented on: 11/4/2013 17:07:00 - 17:23:00

Structural- and energy-based anisotropy in nanomaterials can enhance the material's ability to store, transport or convert energy. As a result, researchers have demonstrated that nanomaterials can have advantageous properties relative to their bulk material. For example, transport mechanisms can be selectively tuned by altering nanostructure geometry. Many novel lab-scale experiments have demonstrated this phenomenon and how it can be used to enhance device performance. However, assembly of nanostructures at the device-scale while maintaining the property enhancement is the dominant engineering challenge. Mainstream Engineering is currently developing several nanomaterial-based energy conversion and storage technologies with a focus on device manufacturability. We will highlight our current efforts to not only fabricate nanomaterials with unique and advantageous properties, but also assemble these nanostructures in a configuration that maintains the properties at macro-scale. Using template-directed assembly of nanoparticles and wires, we will discuss their application to thermoelectric composites, hydrogen, oxygen and thermal energy storage, photovoltaics, as well as the development of electrode materials for electrochemical double layer capacitors.

(191h) Development and Application of Tin/Indium (Sn/In) Nanoparticles As Low Melting Temperature Nanosolder Materials

Shu, Y., University of Massachusetts Lowell
Rajathurai, K., University of Massachusetts Lowell
Gao, F., University of Massachusetts Lowell
Cui, Q., University of Massachusetts Lowell
Gu, Z., University of Massachusetts Lowell
Originally presented on: 11/4/2013 17:23:00 - 17:39:00

Low melting temperature lead-free solder materials are attractive for many electronic assembly and packaging processes. There are several advantages for low temperature applications, for example, applicable for flexible electronics, environmental friendly, and higher energy efficiency. Herein we report the synthesis and development of tin/indium nanosolder particles to serve as a new nanosolder material for low temperature processing and applications. The nanosolder particles are synthesized through a surfactant-assisted chemical reduction method in an aqueous phase. Parameters that affect the size and shape of the nanosolder particles have been investigated. SEM, TEM, XRD and EDS have been used to determine the structure and composition of the nanosolders. DSC measurement has been employed to determine the thermal property of the nanosolder particles. Furthermore, the characterizations of melting behavior and intermetallic formation are provided from the reflow and soldering tests. The results indicate that the tin/indium nanosolder particles have good potential for low temperature soldering applications.

(199a) Efficient Simulation of Population Balance Models By An Improved Method of Characteristics Approach

Zhu, X., Massachusetts Institute of Technology
Zhou, L., Massachusetts Institute of Technology
Braatz, R. D., Massachusetts Institute of Technology
Originally presented on: 11/4/2013 15:15:00 - 15:35:00

Particulate processes are commonly encountered in various systems (e.g., crystallization [1], aerosols [2], cell dynamics [3], and polymerization [4, 5]), where the particle size distribution (PSD) plays a key role on the product quality[1]. Recent advancements in particle technology have allowed on-line acquisition of PSD, and have inspired increasing efforts in modeling and computation of particulate systems in order to utilize such information for improved process and quality control [6].

The application of model predictive control (MPC) to particulate systems has been limited by the computational cost of numerically solving the population balance equations (PBEs) that describe the dynamics of these systems. Discretization techniques including the finite volume/difference methods can be applied, but are typically computationally expensive and exhibit problems with numerical diffusion or dispersion [7]. Although more advanced numerical schemes, such as the parallel high-resolution finite-volume method, provide much better tradeoffs between numerical accuracy and computational cost [8-10], the cost is still prohibitive for real-time MPC applications. As an alternative, some papers [11-13] have proposed approaches that combine standard method of moments or quadrature method of moments with the method of characteristics (MOCH). Such approaches require solving the moments equations as the first step to obtain the temporal information on supersaturation and nucleation rate during crystallization for subsequent computation of PSD using MOCH.

This work proposed an approach that utilizes the method of characteristics alone to solve population balance models together with physical constraints (for example, mass conservation), to accurately simulate the PSD by transforming the PBEs into a system of ordinary differential-algebraic equations (DAE). The resulted DAE is conveniently handled by available commercial solvers such as ode15s in Matlab. Four example systems are simulated and analyzed for demonstration purposes, including (i) growth-dominated crystallization with saturation control; (ii) growth and nucleation during cooling crystallization (both size-independent and size-dependent growth); (iii) a growth and agglomeration system; and (iv) evaluation of temperature-dependent growth in crystallization.

Compared with existing PSD simulation methods, the proposed approach is more advantageous for its straightforward procedures of model formulation, efficient computation, accurate results (no numerical diffusion or dispersion, with accuracy comparable to analytical solutions for those problems that have analytical solutions), and broad applicability. The proposed approach is sufficiently computationally efficient to enable the implementation of real-time MPC to particulate systems with aggregation and multiple internal and external dimensions, as well as facilitating the use of PBEs in robust parameter estimation, design, and control algorithms, which typically involve large numbers of PBE simulations.

Acknowledgements: This research was supported by Novartis pharmaceuticals.

References:

[1] A. D. Randolph and M. A. Larson, *Theory of Particulate Processes: Analysis and Techniques of Continuous Crystallization*. New York: Academic Press, 1971.

[2] T. E. Ramabhadran, T. W. Peterson, and J. H. Seinfeld, "Dynamics of aerosol coagulation and condensation," *AIChE Journal,* vol. 22, p. 840-851, 1976.

[3] N. V. Mantzaris, J.-J. Liou, P. Daoutidis, and F. Srienc, "Numerical solution of a mass structured cell population balance model in an environment of changing substrate concentration," *Journal of Biotechnology,* vol. 71, p. 157-174, 1999.

[4] T. J. Crowley, E. S. Meadows, E. Kostoulas, and F. J. Doyle III, "Control of particle size distribution described by a population balance model of semibatch emulsion polymerization," *Journal of Process Control,* vol. 10, p. 419-432, 2000.

[5] X. X. Zhu, B. G. Li, L. B. Wu, Y. G. Zheng, S. P. Zhu, K. D. Hungenberg, S. Mussig, and B. Reinhard, "Kinetics and modeling of vinyl acetate graft polymerization from poly(ethylene glycol)," *Macromolecular Reaction Engineering,* vol. 2, p. 321-333, 2008.

[6] Z. K. Nagy and R. D. Braatz, "Advances and new directions in crystallization control," *Annual Review of Chemical and Biomolecular Engineering,* vol. 3, p. 55-75, 2012.

[7] R. Gunawan, I. Fusman, and R. D. Braatz, "Parallel high-resolution finite volume simulation of particulate processes," *AIChE Journal,* vol. 54, p. 1449-1458, 2008.

[8] R. Gunawan, I. Fusman, and R. D. Braatz, "High resolution algorithms for multidimensional population balance equations," *AIChE Journal,* vol. 50, p. 2738-2749, 2004.

[9] D. L. Ma, D. K. Tafti, and R. D. Braatz, "High-resolution simulation of multidimensional crystal growth," *Industrial & Engineering Chemistry Research,* vol. 41, p. 6217-6223, 2002.

[10] S. Qamar, M. P. Elsner, I. A. Angelov, G. Warnecke, and A. Seidel-Morgenstern, "A comparative study of high resolution schemes for solving population balances in crystallization," *Computers & Chemical Engineering,* vol. 30, p. 1119-1131, 2006.

[11] E. Aamir, Z. K. Nagy, C. D. Rielly, T. Kleinert, and B. Judat, "Combined quadrature method of moments and method of characteristics approach for efficient solution of population balance models for dynamic modeling and crystal size distribution control of crystallization processes," *Industrial & Engineering Chemistry Research,* vol. 48, p. 8575-8584, 2009.

[12] S. Qamar, S. Mukhtar, A. Seidel-Morgenstern, and M. P. Elsner, "An efficient numerical technique for solving one-dimensional batch crystallization models with size-dependent growth rates," *Chemical Engineering Science,* vol. 64, p. 3659-3667, 2009.

[13] M. J. Hounslow and G. K. Reynolds, "Product engineering for crystal size distribution," *AIChE Journal,* vol. 52, p. 2507-2517, 2006.

(199b) Consequences of Non-Linear Particle Breakage: Falsified Breakage Kinetics and Delayed Attainment of a Self-Similar Particle Size Distribution

Bilgili, E., New Jersey Institute of Technology
Capece, M., New Jersey Institute of Technology
Dave, R. N., New Jersey Institute of Technology
Originally presented on: 11/4/2013 15:35:00 - 15:55:00

Comminution is an important unit operation in the production of raw materials, specialty chemicals, and value-added products. Population balance models (PBMs) have been used as a quantitative tool to model comminution. Most PBMs used in the last 60 years assumed the linearity of the breakage rate, i.e., first-order breakage kinetics. In this presentation, we organize key results from experimental studies that exhibit strong deviations from first-order breakage kinetics and present recent theoretical developments that address the complex phenomenon of non-linear particle breakage. The traditional linear model, the time-variant model, and the non-linear functional model [1] have been critically analyzed in view of experimental data. Extensive numerical simulations of dry ball milling systems [2,3] and particle bed breakage experiments [4] suggest that the non-linear functional model can serve as a unified framework by which non-linear particle breakage in a multitude of comminution systems can be systematically studied and quantified at the process length scale. Specifically, the presentation will focus on our recent efforts in exploring *falsified breakage kinetics* [5] and *self-similarity* of the milled particle size distributions [6]. We present how multi-particle mechanical interactions neglected by the traditional linear model can affect these phenomena and why elucidation of such complex phenomena entails rational, model-based design of particle breakage experiments, while indicating future research directions.

[1] E. Bilgili, J. Yepes, B. Scarlett, "Formulation of A Non-Linear Framework for Population Balance Modeling of Batch Grinding: Beyond First-Order Kinetics," Chemical Engineering Science, Vol. 61, 2006, pp. 33–44.

[2] E. Bilgili, B. Scarlett, "Population Balance Modeling of Non-Linear Effects in Milling Processes," Powder Technology, Vol. 153, 2005, pp. 59–71.

[3] M. Capece, E. Bilgili, R. Dave, "Identification of the Breakage Rate and Distribution Parameters in a Non-Linear Population Balance Model for Batch Milling," Powder Technology, Vol. 208, 2011, pp. 195–204.

[4] E. Bilgili, M. Capece, "A Rigorous Breakage Matrix Methodology for Characterization of Multi-Particle Interactions in Dense-Phase Particle Breakage," Chemical Engineering Research and Design, Vol. 90, 2012, pp. 1177–1188.

[5] M. Capece, E. Bilgili, R. Dave, "Emergence of Falsified Kinetics as a Consequence of Multi-Particle Interactions in Dense-Phase Comminution Processes," Chemical Engineering Science, Vol. 66, 2011, pp. 5672–5683.

[6] M. Capece, R. Dave, E. Bilgili, "Influence of Non-Linear Breakage Kinetics on the Attainment of Self-Similarity for Dry Milling Processes," Chemical Engineering Science, doi: 10.1016/j.ces.2013.04.017, 2013, in press.

(199c) A General Procedure for Solving the Population Balance Equation in Flocculating Suspensions

Yeow, L. Y.,

Liow, J. L., University of New South Wales Canberra

Leong, Y. K., James Cook University

Originally presented on: 11/4/2013 15:55:00 - 16:15:00

A General Procedure for Solving the Population Balance Equation in Flocculating Suspensions

Leong Y Yeow

The University of Melbourne

Jong-Leng Liow

University of NSW Canberra @ADFA

Yee-Kwong Leong

The University of Western Australia

In a flocculating suspension large particles are continuously being formed by agglomeration of smaller ones and at the same time small particles are being generated by fragmentation of larger ones. At any time t the particle size distribution (PSD) is described by the number density function $n(t,v)$ which gives the number concentration density of size v particles.

The evolution of the PSD n(t,v) is governed by the following population balance equation (PBE)

$$\frac{\partial n(t,v)}{\partial t} = \frac{1}{2}\int_{v_s=v_{min}}^{v-v_{min}} k_A(v_s, v-v_s)\,n(t,v_s)\,n(t,v-v_s)\,dv_s - n(t,v)\int_{v_t=v_{min}}^{v_{max}-v} k_A(v,v_t)\,n(t,v_t)\,dv_t$$
$$+ \int_{v_r=v+v_{min}}^{v_{max}} k_F(v_r,v)\,n(t,v_r)\,dv_r - \frac{1}{2}n(t,v)\int_{v_p=v_{min}}^{v-v_{min}} k_F(v,v_p)\,dv_p \qquad \text{Eq (1)}$$

In this equation, $k_A(v_s,v_t)$ describes the kinetics of formation large particles [$v_s+v_t \rightarrow v_u$] by aggregation, $k_F(v_r,v_p)$ describes the kinetics of the production of small particles [$v_r \rightarrow v_p+v_q$] by fragmentation.

The first integral on the RHS of Eq. (1) gives the rate of generation of size v particles as a result of the agglomeration of particles of size v_s with size v ? v_s. The second integral is the rate of disappearance of size v particles as a result of the agglomeration with particles of size v_t. The third integral is the rate of generation of size v particles arising from the fragmentation of all size v_r particles into a size v particle and a second size v_r ? v particle. The final integral accounts for the rate of disappearance of size v particles as a result of their fragmentation into all possible combinations of two smaller particles. Together these four terms give the net rate of change of the number concentration density n of size v particles.

The PBE is an integro-differential equation that does not have a simple analytical solution and for practical flocculation problems, solving it numerically results in a major difficulty because of the large span in particle volume; usually spread over 9 or more decades of particle volume or 3 decades of particle diameter.

Uniform discretization of the independent variable v then leads to inadequate numerical representation for small particle sizes.

A popular procedure for overcoming the numerical problem is to discretize the span of particle size geometrically so that $\Delta v_{j+1} = 2\Delta v_j$. This rectifies the under representation of small particles but is less intuitive. In the present method this problem is surmounted by the introduction of the logarithmic variable $s = \log_{10} v$ as the new independent variable providing both flexibility as well as a continuous function for the particle size which can be easily be handled within the PBE. The new dependent variable is n(t,s) and the PBE is then reformulated. In the computation that follows the span of s is discretised into a uniform grid *i.e.* $\Delta s_{j+1} = \Delta s_j$.. This ensures that the PBE is adequately represented over the entire span of particle volume.

Another key development is to apply trapezoidal rule to approximate the four integrals on the RHS of the PBE. After discretization, the integro-differential equation becomes a set of first order ODE for the discretised analogue of n(t, s) *i.e.* $n_1(t)$, $n_2(t)$,? $n_j(t)$? $n_N(t)$. where N is the number of discretization steps. This set of ODE can be written in a matrix form as

$$\begin{bmatrix} dn_1/dt \\ dn_2/dt \\ \vdots \\ dn_i/dt \\ \vdots \\ dn_N/dt \end{bmatrix} = \begin{bmatrix} a_{11} & a_{12} & \cdots & a_{1j} & \cdots & a_{1N} \\ a_{21} & a_{22} & \cdots & a_{2j} & \cdots & a_{2N} \\ \vdots & \vdots & & \vdots & & \vdots \\ a_{i1} & a_{i2} & \cdots & a_{ij} & \cdots & a_{iN} \\ \vdots & \vdots & & \vdots & & \vdots \\ a_{N1} & a_{N2} & \cdots & a_{Nj} & \cdots & a_{NN} \end{bmatrix} \text{ or } \frac{dn}{dt} = a(t,n)$$

The matrix **a** depends on the aggregation and fragmentation kinetics. For a given kinetics, a_{ij} are known functions of the discretised unknowns $n_i(t)$. This is a set of nonlinear first order ODEs in time that we solved numerically with $n_i(0)$, the PSD at the start of the flocculation, as the initial condition. This method of handling the integro-differential is an extension of the methods of line (MoL) technique used to solve partial differential equations.

To cope with wide span in particle volume, the number of discretization interval (N) can be as larger than 801. Although **a** is large matrix, it is sparse and can be handled efficiently by numerical procedures found in most scientific computing software.

Results

Typical computed results for different simulated kinetics and initial PSD are presented in Figures 1 to 3 in dimensionless form, where τ, V and N are the dimensionless time, particle volume and number concentration density function respectively.

In addition, the results are also presented as volume fraction density function G(X) as a function of $X = \log_{10} V$. Experimental PSD are usually reported as G(X). The area under each G(X) curve is by definition equal to unity and provides a means to validate the results. The area under the G(x) in all the cases remained below 2% confirming the reliability of the computed results.

For all the aggregation kinetic plots each curve represents a constant value of V_t, and all fragmentation kinetic plot each curve represent a constant value of V_p. In the number concentration density plots and the particulate volume plots the discrete points represent the initial PSD. The curves show the evolving PSD with a constant dimensionless time on each curve. The results in Figure 3 for rain drops show that the method is capable of capturing the evolution of the PSD for a natural event, particularly for the small values of particles sizes.

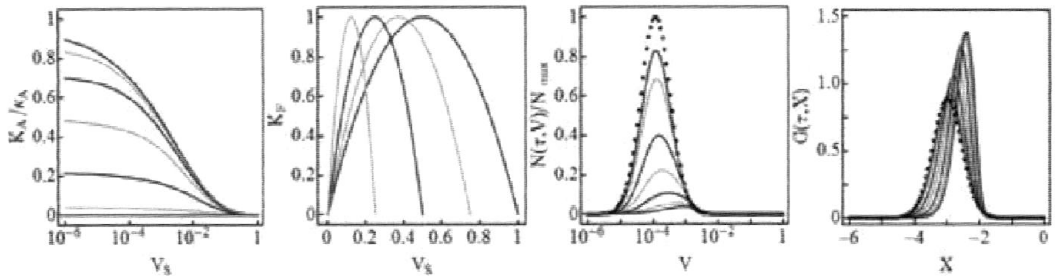

(a) (b) (c) (d)

Figure 1 (a) and (b) Aggregation and fragmentation kinetics; (c) Dimensionless number concentration density function; (d) Volume fraction density function.

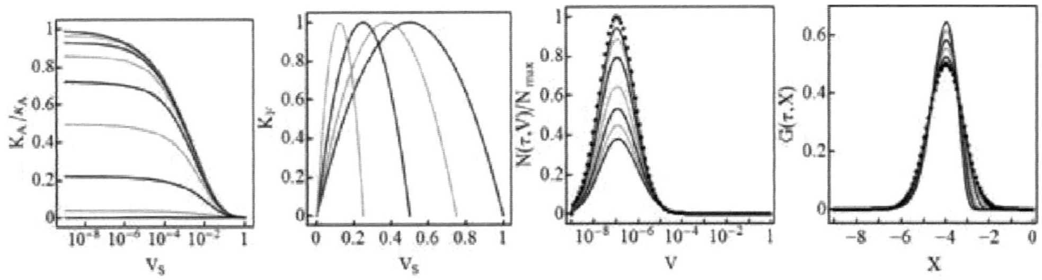

(a) (b) (c) (d)

Figure 2 (a) and (b) Aggregation and fragmentation kinetics; (c) Dimensionless number concentration density function;(d) Volume fraction density function.

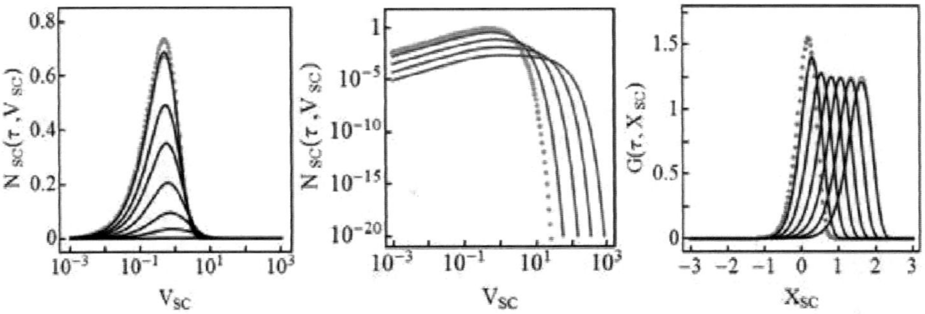

(a) (b) (c)

Figure 3 Coalescence of rain droplets under constant aggregation kinetics. (a) Dimensionless number concentration density function; (b) Replot of (a) with log scale on both axes. (c) Volume fraction density function. Dark continuous curve are the computed results of MoL and the lighter curves are exact results of Scott (1968).

Reference

Scott WT. Analytic studies of cloud droplet coalescence I. J. Atmos Sci. 1968; 25:54-65.

(199d) Exact Solutions of Population Balances Equations Via An Auxiliary Equation Method for Growth, Nucleation, Breakage and Aggregation Processes

Pinar, Z.,
Dutta, A., Groep T - Leuven Engineering College (Associatie KU Leuven)
Majumder, A., Loughborough University
Beny, G., Groep T - Leuven Engineering College (Associatie KU Leuven)
Ozis, T.,
Originally presented on: 11/4/2013 16:15:00 - 16:35:00

There has been a growing interest in the field of population balance equations (PBEs), especially for applications in processes such as granulation and crystallization (Immanuel and Doyle III, 2005; Ma et al., 2002), in aerosol science and biochemical research (Kim and Seinfeld, 1992; Roussos and Kiparissides, 2012). A continuous population balance equation (PBE) involving particulate growth and aggregation, growth, aggregation and breakage, aggregation, nucleation, breakage and growth are investigated in this study. Usually numerical methods are employed to solve these PBEs, because of the absence of an analytical solution. However, to overcome this, a new technique based on auxiliary equation method (AEM) with a sixth-degree (Pınar and Öziş, 2013a,b) even-power polynomial is proposed to obtain the exact solutions of PBEs under various conditions of growth, nucleation, aggregation and breakage. The proposed method is already applied to growth, nucleation and aggregation processes by Dutta et al. (2013). The proposed method has been found superior to standard numerical methods as it provides analytical solutions for a wide range of initial and boundary conditions. To illustrate the validity and efficiency of the proposed method, analytical solutions obtained with AEM are compared with the numerical solutions, and with the analytical solution as found in the standard literature.

References:

C.D. Immanuel, F.J. Doyle (2005). Solution technique for a multidimensional population balance model describing granulation processes. Powder Technology, 156. 213–225.
D.L. Ma, D.K. Tafti, R.D. Braatz (2002). High resolution simulation of multi-dimensional crystallization. Ind. Eng. Chem. Res. 41, 6217-6223.
Y.P. Kim, J.H. Seinfeld (1992). Simulation of multicomponent aerosol dynamics. J. Colloid Interface Science 149, 425-449.
A.I. Roussos, C. Kiparissides (2012). A bivariate population balance model for the microbial production of poly(3-hydroxybutyrate). Chemical Engineering Science, 70, 45–53.
Z.Pınar, T.Öziş, (2013a) An Observation on the Periodic Solutions to Nonlinear Physical models by means of the auxiliary equation with a sixth-degree nonlinear term, Commun Nonlinear Sci Numer Simulat18

2177–2187.
Z.Pınar, T.Öziş, (2013b) The Periodic Solutions to Kawahara Equation by means of the auxiliary equation with a sixth-degree nonlinear term, Journal of Mathematics (http://dx.doi.org/10.1155/2013/106349).
A. Dutta, Z. Pinar, Z., B. Guid., T. Öziş (2013). An auxiliary equation approach to exact solutions of population balance equations for growth, nucleation and aggregation processes. International Conference on Applied Analysis and Mathematical Modelling (ICAAMM). Istanbul, 2-5 June 2013.

(199e) Experimental and Theoretical Study of Simultaneous Agglomeration and Drying in Fluidized Bed Granulator

Ricart, B. G., Merck & Co.
Rajniak, P., Merck & Co., Inc.
Originally presented on: 11/4/2013 16:35:00 - 16:55:00

In the present pharmaceutical drug product process, fluid bed granulation is used to co-granulate two APIs (active pharmaceutical ingredients) at high drug loading using a binder solution spray. This granulation was studied in a twelve batch, half-fractional, D-optimal Design of Experiments (DOE) with factors of superficial air velocity, spray flowrate/air flowrate ratio (S/Q) , inlet air temperature, and solid components ratio. The D-optimal design allowed interrogation of large ranges for spray rate, inlet air flowrate, and inlet temperature while controlling process moisture within a proven acceptable range.

Mathematical models of increasing complexity were used to analyze the experimental data as follows:

1. Total heat and mass balance model was employed to evaluate the model's adjustable parameters (heat loss coefficient and evaporation rate coefficient) by fitting to experimental bed temperature and granule moisture (Loss on Drying = LOD). Decreasing external surface area of growing granules was calculated from experimentally measured particle size distributions (PSDs) and used as a model input. An average heat loss coefficient was evaluated before the correlation between the evaporation rate coefficient, S/Q ratio and inlet temperature was calculated (Correlation 1).
2. A population balance model was employed for the evaluation of agglomeration kernel parameters by fitting to experimental PSDs. A correlation between the experimental granule LOD and granule growth kinetics (agglomeration kernel) was observed and evaluated (Correlation 2).
3. A model combining total balances with population balances was developed and used for direct prediction of granule growth at different process conditions. The heat losses coefficient and Correlation 1 were used for "a priori" calculation of the evaporation coefficient and the granule LOD. Correlation 2 was used for calculation of agglomeration kernel and granule growth.

The combined model can be used to simulate the fluid bed granulation process of the particular formulation at the given scale. The main challenges that remain for process transfer and scalability are the physical meaning of the evaporation rate coefficient, its scalability, and relation to the granulator design (i.e. the control volume of the granule bed).

(199f) Thermodynamics of Distributions

Matsoukas, T., Penn State

Originally presented on: 11/4/2013 16:55:00 - 17:15:00

The Population Balance Equation (PBE) is a mean-field equation that tracks changes in the distribution of a population under a variety of physical mechanisms such as aggregation, breakage, nucleation, etc. Here we present an ensemble theory for generic populations that allows us to treat population balances in the statistical language of formal thermodynamics. The theory reduces to the classical population balance equation when the population is infinitely large, but has the further advantage that it (a) permits the treatment of finite populations and (b) provides the tools to study problems such as gelation and shattering as formal phase transitions. The basis of the theory is the ensemble introduced by Gibbs. We construct the ensemble of all distributions obtained when M individuals are divided into N groups. An important part of the theory is the introduction of the "selection bias" as functional that biases the selection of distributions from the ensemble. We let the system pass to the thermodynamic limit, derive the most probable distribution and formulate a rigorous analogy between the cluster ensemble and thermodynamics to obtain the temperature and pressure of the population. We predict that for certain forms of the selection bias the system undergoes explosive percolation and provide an exactly solvable model that demonstrates this behavior. Finally, we discuss how this new formulation can be applied to solve common problems in particle technology.

(199g) Effect of the Dispersed Phase Volume Fraction On Emulsion Drop Size Distributions Obtained From High Pressure Homogenization

Maindarkar, S. N., University of Massachusetts
Dubbelboer, A., University of Technology
Hoogland, H., Unilever R & D
Henson, M. A., University of Massachusetts Amherst
Originally presented on: 11/4/2013 17:15:00 - 17:35:00

Oil-in-water emulsions are ubiquitous dispersed phase systems with diverse applications including natural and processed foods. Emulsified products exhibit a wide range of physicochemical and sensory characteristics based on the system formulation, which influence dispersed and continuous phase properties such as density, viscosity, pH, electrical conductivity and interfacial tension. Process operations such as homogenization also have a substantial effect on product properties, including appearance, taste, mouthfeel, odor and safety. Emulsion system formulation and processing operations both impact the drop size distribution, a key property that influences dispersion rheology, stability, texture, appearance and food safety. In high pressure homogenization, a coarse emulsion is passed through a small orifice under very high pressure. The fluid stream passes radially through the narrow gap formed between the piston and the valve seat at high velocity, creating a local environment of high turbulence and shear stress that causes drop deformation and breakage. Under typical industrial conditions where the oil/surfactant ratio is large to reduce manufacturing costs, drop coalescence is prevalent due to insufficient surfactant to stabilize newly formed drops.

The population balance equation (PBE) modeling framework is particularly well suited for emulsification processes because functions describing single droplet events such as breakage and coalescence can be incorporated within a fundamental number balance equation to predict the evolution of the drop size distribution. While PBE models have been developed for a wide variety of emulsification processes including high pressure homogenizers, these models are typically formulated for prediction at constant and relatively low oil fractions. In our previous work (Maindarkar et al., 2012), we incorporated coalescence into breakage-only PBE model of high pressure homogenization to allow prediction of drop size distributions at high oil-to-surfactant ratios. While the model was capable of accurate prediction at a constant oil concentration, extensibility to other oil concentrations proved to be unsatisfactory.

In this presentation, we describe the development of a new PBE model that allows prediction of homogenized drop size distributions over a wide range of oil concentrations. Our model oil-in-water emulsion system consisted of 1% Pluronic F68 non-ionic surfactant and 10-50% vegetable oil. Our previous PBE model was modified in two important ways: (1) a new mechanistic function for drop breakage due to turbulent shear was developed and combined with a previous breakage function for drop collisions with turbulent eddies; and (2) the constant continuous phase viscosity used in the breakage and coalescence functions was replaced with an emulsion viscosity calculated using a model that describes the shear thinning behavior of the emulsion as a function of the oil concentration. Nonlinear optimization was used to estimate viscosity model parameters and adjustable parameters in the breakage and coalescence functions from viscosity and drop distribution data collected at 10% oil. The resulting model was shown to produce satisfactory emulsion viscosity and drop size distribution predictions at 30% and 50% oil without re-estimation of the model parameters.

Maindarkar S.N., Raikar N.B., Henson M.A., "Incorporating emulsion drop coalescence into population balance equation models of high pressure homogenzation", *Colloids and Surfaces A: Physicochemical and Engineering Aspects*, 396 (2012), 63-73.

(199h) Mechanism Identification of Population Balance Models With Limited Data

Irizarry, R., DuPont

Originally presented on: 11/4/2013 17:35:00 - 17:55:00

The formation of metal spherical particles by reaction precipitation is a very fast process (of the order of seconds), thus hindering the capability of sampling to generate kinetic data. For such systems UV-vis optical kinetic is collected using a stopped flow reactor for fast kinetics [1]. The dynamic optical response of these systems is characterized by broad bands, making their analysis difficult to interpret in terms of particles sizes. To overcome this problem, the data is analyzed in terms a new strategy called simulated dynamic optical response (SDOR) [2]. In this work the capability of combining this strategy with global optimization to identify mechanism and parameters is studied systematically. The identified model is compared with a more ideal case of having a complete particle size distribution as a function of time for the same system, but in this case, it is assumed that the data is only available in limited time intervals. In theory this data can be collected by a sequence of freeze quench experiments. The problem of limited/noisy data is studied using Bayes analysis developed for these systems including its synergy with the global optimization approach.

[1] Irizarry R., Burwell L., and M. S. Leon-Velazquez (2011) Preparation and formation mechanism of silver particles with spherical open structures, *Ind. Eng. Chem. Res.,*50, 8023–8033

[2] Irizarry R. (2010) Simulated dynamic optical response strategy for model identification of metal colloid synthesis, *Ind. Eng. Chem. Res.*, 49, 5588–5602.

(365a) Flow Regime Maps for Pneumatic Conveying

Kalman, H., Ben-Gurion University of the Negev

Originally presented on: 11/4/2013 15:15:00 - 15:40:00

In previous works we have shown that by defining various threshold velocities in terms of the modified Reynolds number as a function of the modified Archimedes number a Generalized Master-Curve can be plotted. Then, using this curve, the ratio between various threshold velocities can be easily determined. Based on the above, a generalized flow regime diagram for vertical pneumatic conveying and CFB systems was developed. We analyze dozens of experimental and theoretical studies in order to understand the influence of the various parameters including particle and gas properties, pipe diameter and particle concentration. Ee found that the four Geldart classification groups explain the appearance of different fluidization and pneumatic conveying states also. Based on the empirical and theoretical analysis it was shown that the generalized flow regime diagram for vertical systems can be presented by the relationship between the Archimedes and the Reynolds numbers modified by the solid flow rate or the volumetric concentration of circulating solids and the gas void fraction. However, it is much more complicated to develop the flow regime map for horizontal pipes. Based on the threshold velocities analysis, the border line between stationary bed and moving bed can be defined, in the case that the superficial velocity is lower than the saltation velocity.

Figure : Regimes of suspension, stationary bed and moving bed as a function of the particles concentration.

REFERENCES

[1] H. Kalman and E. Rabinovich, "Analyzing Threshold Velocities for Fluidization and Pneumatic Conveying," *Chemical Engineering Science*, Vol. 63, pp. 3466-3473, 2008.

[2] E. Rabinovich and H. Kalman, "Flow Regime Diagram for Vertical Pneumatic Conveying and CFB Systems," *Powder Technology*, Vol. 207, pp. 119-133, 2011.

(365b) Experimental Investigation of the Effects of Material Properties On By-Pass Pneumatic Conveying Systems Performance

Jones, M. G., The University of Newcastle
Williams, K. C., The University of Newcastle
Chen, B., The University of Newcastle
Originally presented on: 11/4/2013 15:40:00 - 16:05:00

Bypass pneumatic conveying systems provide a passive capability to reduce conveying velocity and therefore reduce attrition and abrasion in the process of conveying many fragile and erosive particulate solids. Because of these capabilities, bypass pneumatic conveying systems have been widely used in process industry for the last couple of decades. The pneumatic conveying system design and operation are critically dependent on the physical properties of the materials to be conveyed. The objectives of this work were to experimentally investigate the performance of pneumatic conveying of dust-like, powdery and granular materials in bypass systems and conventional pipelines, i.e. flow regimes, minimum conveying velocity and energy consumption. The bypass pneumatic experimental system was built with a main pipe of 79 mm in diameter with an internal bypass pipe with an orifice plate flute arrangement. Alumina, fly ash, sand and plastic pellets, which are located in different zones of the Geldart classification diagram, were tested in the study.

The effect of material properties on the flow regimes and the mechanism of material blockage inhibition of a bypass pneumatic transport system has been investigated. For alumina and fly ash, it was found that particulate material blockages were inhibited in bypass systems due to the air penetration into the particulate volume as a result of orifice plate airflow resistance. For the bypass pneumatic conveying of sand, a long plug being split into two smaller plugs was observed. For bypass pneumatic conveying of plastic pellets, the results showed the full bore plug could not be split into smaller plugs due to the large size and high permeability of the material.

Minimum conveying velocity is one of the key parameters for pneumatic conveying system design. It was found that, compared to the conventional pipeline, the minimum conveying velocity of alumina in the bypass system was much lower as bypass pneumatic conveying systems provide a low velocity dense phase capability to non-dense phase capable bulk materials. However, for plastic pellets, due to the formation of long full bore plugs, the minimum conveying velocity in the bypass system was higher than that in conventional system. Therefore, bypass systems may not be applicable to the transport plastic pellets due to high minimum conveying velocity and the poor performance for blockage inhibition.

When conveying different materials, the specific energy consumption in conventional and bypass systems is also discussed in this paper. For all the materials tested in this study, it was found that the bypass system consumed more energy than a conventional system when using the same air mass flow rate due to the increase in the surface friction. The results also showed that specific energy consumption increased when decreasing the diameter of orifice plates and the spacing of bypass pipe flutes. Conversely, the specific energy consumption increased with increasing the bypass diameter because of the reduction in cross-sectional area of the main pipe and the increase in wall friction due to larger wall surface area

(365c) A New Explanation to the Shear Stress of Friction During Plug Flow Pneumatic Conveying

Sommer, K.,

Originally presented on: 11/4/2013 16:05:00 - 16:30:00

Pneumatic conveying is commonly used for industrial loading, unloading and

for the internal transport of bulk solids. Depending on mass flow rate and

the velocity of the conveying gas different conveying phases appear.

Reducing the speed of the conveying gas will result in two phase flow

setting, i.e. dense phase conveying. Plug conveying is reached at very low

gas velocities and high pressure drop. During the last few years pneumatic plug

conveying has become more important due to lower energy consumption,

lower particle attrition and lower wear of pipelines. Therefore, for many

industrial applications dense phase pneumatic conveying is the preferred

transport mode for granular media.

A lot of research has been carried out concerning the prediction of pressure

drop of dense phase conveying. Pressure drop models for dense flow

conveying are derived corresponding to Janssen's silo theory under

consideration of bulk solid mechanics and constant bulk solid properties along

the plug. In this context the relationship between the radial to the axial stresses within a single plug is particularly important to understand. It reveals the strengths acting inside the plug and the forces exerted on the wall by the plug. So far plugs are regarded as compact bulk solid columns. Unfortunately, these models provide strongly different results concerning the pressure loss of single plugs and determination of pressure drop for the entire conveying pipeline is not feasible. Furthermore most of the models are based on bulk solid mechanics or a fixed bed column with constant properties along the plug.

A sensor was constructed to measure directly and concurrently normal stress, wall shear stress and pressure drop for a conveyed plug. The experimental results for vertical pneumatic conveying show a change of porosity along the plugs. This requires a new approach independent of bulk solid mechanics to explain wall shear stress.

Thanks to a new measuring device including both stress and pressure sensors, also horizontal single slugs were investigated to determine porosity and stress states within slugs of cohesionless granular material. Moreover, a special slug-catcher was developed to investigate the porosity profile over the slug height. Independent of conveying velocity, slugs were found to be fluidized over their whole length. Nevertheless, high wall shear stresses and normal stresses were detected within each slug, which could not be explained by bulk solids mechanics.

The astonishing result was that the bulk porosity never exists in a slug. The essential assumption for the existing theories could no longer be maintained.

The existing theories lost their physical bases and degenerated into pure fitting models. Because slugs are not compact structures, no axial stress exists within slugs. The existing wall friction can be explained by postulating a change of momentum similar to what happens with gas friction. The molecules are replaced by the single particles. With some simplifications of the kinetic gas theory, it was possible to estimate the wall friction and the pressure loss.

Experimental investigations on moving slugs of granular materials with respect to slug aspect, particle velocity, pressure gradient, slug porosity, and stress states revealed that the wall shear stress induced by slugs can be accurately described by applying the kinetic theory.

To improve the assumptions a twin-plane electrical capacitance tomography (ECT) was used to investigate the dynamic porosity changes during horizontal plug flow conveying of polypropylene granules. The results obtained from ECT measurements were compared to the porosity calculated from pressure drop measurements according to the Ergun equation. Using the aforementioned measuring device, wall shear and normal stresses caused by plugs were investigated. The calculated wall shear stresses were in good agreement with the measured wall shear stresses. With these results, predictions of pressure loss could be made on the basis of useful physical assumptions and the new model.

(365d) Controlling Conveying Velocity In Pneumatic Conveying Systems

Dhodapkar, S., The Dow Chemical Company

Originally presented on: 11/4/2013 16:30:00 - 16:55:00

Key performance attirbutes, namely attrition and energy consumption, are related to the conveying velocity during pneumatic transport. This paper explore various ways of controlling and manipulating the conveying velocity in a system. It is provides an overview along with specific control schemes to achieve velocity control. Both dense and dilute phase conveying is covered. Discussion of particle and gas velocities is covered as well.

(365e) Empirical Erosion Tests: Uniflow Cyclones As An Alternative to Reverse Flow Cyclones for Longer Life Time?

Kraxner, M., MCI

Originally presented on: 11/4/2013 16:55:00 - 17:20:00

Empirical Erosion Tests: Uniflow Cyclones as an Alternative to Reverse Flow Cyclones for longer Life Time?

M. Kraxner[1], M. Portenkirchner[1], M. Pillei[1], T. Kofler[1], U. Muschelknautz[1]

[1]MCI – The Entrepreneurial School, Environmental, Process & Energy Engineering, Maximilianstraße 2, A-6020 Innsbruck, AUSTRIA

Erosion, as main cause for the lifetime of cyclones, is in some cases more interesting for plant operators than high separation efficiency. The lifetime of cyclones depends mainly on the solids loading and the solids properties, as well as the operating parameters (e.g. gas temperature, flow rate) and the geometric design parameters. *Y.-M. Chen* identifies the erosion of cyclones in the FCC-process as the most expensive part of the working expenses in petroleum refineries [1]. To increase the lifetime of cyclones *S.B. Karri* et al. investigated the influence of solids loading, gas velocity and geometric design of cyclones to reduce the erosion [2]. The most erosion-sensitive region in a standard reverse flow cyclone is the lower cone and in second stage cyclones the upper dipleg.

An alternative cyclone type where the gas passes through in one direction is the uniflow cyclone. Uniflow cyclones used for the separation of particles from gases are characterized, compared to standard reverse flow cyclones, by its compactness and its low energy consumption. The compact construction gives plant operators the possibility of easy implementation in piping systems. Therefore emissions of fine dust can be reduced by a low fabrication effort. The simple design of this cyclone type leads to a low erosion potential compared to the standard reverse flow cyclone.

Erosion tests with quartz dust show a relative abrasion of the uniflow cyclone compared to the erosion in standard cyclones. Additionally the most critical area of abrasion can be located.

[1] Chen, Y.-M., Evolution of FCC – Past Present and Future – and the Challenges of Operating a High-Temperature CFB-System, CFB10 – 10th International Conference on Circulating Fluidized Beds and Fluidized Bed Technology, Sunriver/OR, 2011.

[2] Karri, S.B., Cocco, R., Knowlton, T.M., Erosion in Second Stage Cyclones: Effects of Cyclone Length and Outlet Gas Velocity, CFB10 – 10th International Conference on Circulating Fluidized Beds and Fluidized Bed Technology, Sunriver/OR, 2011.

(365f) Effect of the Solids Loading On the Uniflow Cyclone Performance At Low Solids Concentrations

Kofler, T.,

Originally presented on: 11/4/2013 17:20:00 - 17:45:00

Effect of the Solids Loading on the Uniflow Cyclone Performance at Low Solids Concentrations

T. Kofler[1], M. Kraxner[1], M. Pillei[1], R. Lanthaler[1], C. Illchmann[1], R. Goller[1], U. Muschelknautz[1]

[1]MCI – The Entrepreneurial School, Environmental, Process & Energy Engineering, Maximilianstraße 2, A-6020 Innsbruck, AUSTRIA

Uniflow cyclones have gas and particles passing through them in only one direction and are preferably used for applications with space limitations. Recently comprehensive information has been gained on the design criteria of uniflow cyclones by systematic experiments. However, limited knowledge about the impact of solids loading on the separation efficiency and the pressure drop requires a systematic investigation.

In standard reverse flow cyclone separators the solids loading of the entering gas flow is a key parameter strongly affecting their performance. The separation efficiency increases and the pressure drop decreases with growing solids loading. The effect can be calculated by a well proven mechanistic model which states that the gas can carry only a maximum amount of solids, called the critical loading. If the solids loading exceeds this critical loading, the solids are immediately separated from the gas in the entrance zone of the cyclone. The solids remaining in the gas are then separated in the inner vortex of the cyclone. The critical solids loading derived from this model typically comprises some grams solids per cubic meter gas. In most industrial applications of reverse flow cyclones the total separation efficiency is essentially affected by the separation due to exceeding the critical loading.

For our analysis of uniflow cyclones, we assumed that the gas separation is governed by the same physical principles as in reverse flow cyclones. Experimental studies on uniflow cyclones have been performed in order to identify the critical loading as a function of relevant parameters, such as the cyclone geometry and the particle size of the feed. It is shown that the values of the critical loading are within the same range compared to reverse flow cyclones. Furthermore, the experiments indicate that for particle loadings up to 60 g/m^3 the separation efficiency of uniflow cyclones increases considerably slower with rising solids loading than that of reverse flow cyclones.

(241a) Mathematical Modeling and Design of Layer Crystallization in a Concentric Annulus With and Without Recirculation

Zhou, L., Massachusetts Institute of Technology
Su, M., Massachusetts Institute of Technology
Benyahia, B., Massachusetts Institute of Technology
Singh, A., Massachusetts Institute of Technology
Barton, P. I., Massachusetts Institute of Technology
Trout, B. L., Massachusetts Institute of Technology
Myerson, A. S., Massachusetts Institute of Technology
Braatz, R. D., Massachusetts Institute of Technology
Originally presented on: 11/5/2013 8:30:00 - 8:52:00

In pharmaceutical manufacturing, the drug product is often separated from a liquid phase by crystallization, washing, filtration, and drying—each of which reduce the overall product yield. Poorly filtering crystals can result in bottlenecks in the downstream processing and can affect crystal product purity or reduce yield due to additional washes. These observations have motivated efforts to design continuous crystallization processes that are reliable for manufacturing a product that meets purity, yield, shape, and size requirements, and at the same time ensures efficient downstream processing.

One way to avoid slurry handling is to constrain crystal growth to occur on fixed surfaces so that the amount of solid particles in the bulk is minimized. Layer crystallization has been characterized as being a process in which coherent crystal layers grow on cooled surfaces of a specially designed tube or plate heat exchanger.[1,2] Layer crystallization from melts has been widely applied for the purification of organics and inorganics such as sucrose and milk fat, and in seawater desalination.[3-6] For layer crystallization, the liquid phase is always at a higher temperature than the solid phase, so that in melts, the temperature is required to be higher than the melting point, but layer crystallization from solutions can be operated at much lower temperatures with less energy consumption.[7]

This work proposes a layer crystallization process from a solution in a concentric annulus that avoids the need for washing, filtration, and drying.[8] A continuous flow of drug product could be generated by employing multiple cyclic annular crystallizers in parallel. Many past studies on mass[9-12] and heat transfer[12,13] in annular reactors have concentrated on obtaining correlations of dimensionless number, but these methods are limited by the configurations and operating conditions[13,14] and they are not able to predict the entire concentration or temperature field that is important for crystallization. To our knowledge, this work for the first time thoroughly examines heat and mass transfer of a crystallization process with solid-layer growth in an annulus with and without recirculation. The mass and/or heat transfer model with solid-layer growth on a cylindrical surface can be employed for many other applications, such as fouling in heat exchangers,[15] biofilms in petroleum pipelines,[16] and melt layer crystallization in a tank.[17,18]

A dynamic model is developed in the form of coupled partial differential equations to (1) predict the crystal layer thickness as a function of time and axial position, (2) update the nonuniform boundary temperature at the solid-liquid interface, which is affected by the local thickness of the crystal layer, and (3) calculate the temperature and concentration fields in the annulus. The model predictions are shown to closely track experimental data that were not used in the model's construction, and also compared to an analytical solution that can be used for quickly obtaining rough estimates when there is no recirculation loop. The effects of modifying the operating conditions are investigated, including variation in initial solute concentrations, cooling liquid jacket temperature, inlet solution velocities, core liquid temperature, and extent of mixing in the recirculation loop. The model can be used to optimize product yield and crystal layer thickness uniformity, with constraints on the supersaturation to avoid bulk nucleation by adjusting cooling temperatures in the core and jacket. Additionally, the model could be extended to predict molecular purity of the crystal layer, by considering more components.

References

1. Glade H, Ulrich J. *Melt Crystallization: Fundamentals, Equipment and Applications*. Aachen, Germany: Shaker; 2003.

2. Ulrich J, Bülau HC. 7 - Melt crystallization. In: Myerson AS, ed. *Handbook of Industrial Crystallization*. 2nd ed. Woburn: Butterworth-Heinemann; 2002:161-179.

3. Parisi M, Chianese A. The crystal layer growth from a well-mixed melt. *Chemical Engineering Science.* 2001;56(14):4245-4256.

4. Rich A, Mandri Y, Mangin D, Rivoire A, Abderafi S, Bebon C, Semlali N, Klein J-P, Bounahmidi T, Bouhaouss A, Veesler S. Sea water desalination by dynamic layer melt crystallization: Parametric study of the freezing and sweating steps. *Journal of Crystal Growth.* 2012;342(1):110-116.

5. Flesland O. Freeze concentration by layer crystallization. *Drying Technol.* 1995;13(8-9):1713-1739.

6. Peters-Erjawetz S, Ulrich J, Tiedtke M, Hartel R. Milk fat fractionation by solid-layer melt crystallization. *Journal of the American Oil Chemists' Society.* 1999;76(5):579-584.

7. Aumock NM. *A Process for Separation by Semi-Continuous Counter-Current Crystallization.* Cambridge, MA: Dept. of Chemical Engineering, Massachusetts Institute of Technology; 2011.

8. Zhou L, Su M, Benyahia B, Singh A, Barton PI, Trout BL, Myerson AS, Braatz RD. Mathematical modeling and design of layer crystallization in a concentric annulus with and without recirculation. *AIChE Journal.* 2013;59(4):1308-1321.

9. Ross TK, Wragg AA. Electrochemical mass transfer studies in annuli. *Electrochim. Acta.* 1965;10(11):1093-1106.

10. Mobarak AA, Farag HA, H.Sedahmed G. Mass transfer in smooth and rough annular ducts under developing flow conditions. *J. Appl. Electrochem.* 1997;27(2):201-207.

11. Ould-Rouis M, Salem A, Legrand J, Nouar C. Etude numÃ©rique et expÃ©rimentale des transferts de matière et de quantitÃ© de mouvement dans un Ã©coulement annulaire laminaire non Ã©tabli. *Int. J. Heat Mass Transfer.* 1995;38(6):953-967.

12. Ghosh UK, Upadhyay SN. Mass transfer to newtonian and non-newtonian fluids in short annuli. *AIChE Journal.* 1985;31(10):1721-1724.

13. Viskanta R. Heat transfer with laminar flow in a concentric annulus with prescribed wall temperatures. *Appl. Sci. Res.* 1964;12(6):463-476.

14. Mitrović J, Maletić B. Effect of thermal asymmetry on heat transfer in a laminar annular flow. *Chemical Engineering & Technology.* 2005;28(10):1144-1150.

15. Kirkpatrick JP, McIntire LV, Characklis WG. Mass and heat transfer in a circular tube with biofouling. *Water Res.* 1980;14(2):117-127.

16. Al-Darbi MM, Agha KQ, Islam MR. Estimation of the substrate concentration profiles in petroleum pipelines containing biofilms. *Pet. Sci. Technol.* 2006;24(11):1249-1265.

17. Chianese A, Santilli N. Modelling of the solid layer growth from melt crystallization—the integral formulation approach. *Chemical Engineering Science.* 1998;53(1):107-111.

18. Guardani R, Neiro SMS, Bülau H, Ulrich J. Experimental comparison and simulation of static and dynamic solid layer melt crystallization. *Chemical Engineering Science.* 2001;56(7):2371-2379.

(241b) Cocrystals: Latest Advances in 'in Silico' Methods for Predicting Cocrystal Coformers and Their Properties

Gallagher, D. A., CAChe Research
Loschen, C.,
Originally presented on: 11/5/2013 8:52:00 - 9:14:00

The formulation, design, and implementation of active pharmaceutical ingredients (API) is an area of great interest. Changing the structure and composition of an API by cocrystallizing it with a coformer can have a significant influence on the properties and bioavailability of the drug, and can even offer the potential for new patents and associated revenues. For example, changing the solubility can influence the formation of crystals, cocrystals or solvates, as well as the pharmacokinetics through dissolution during drug release.

The traditional experimental approach to screening for possible new coformers and measuring their properties is time-consuming and expensive. However, *in silico* methods are potentially orders of magnitude faster and are becoming more accurate, such that it is now possible to prescreen huge libraries of potential coformers in minutes or hours on a laptop computer, to determine the most likely candidates. This enables the chemical engineer to take into account a much wider range of possible candidates and then focus the experimental work on the most promising ones, so greatly improving the chances of success in the lab.

This presentation will review the capabilities and limitations of some of the available *in silico* tools for predicting cocrystal coformers and their properties, and compare them with experiment, including group contribution methods, quantum chemistry, and statistical thermodynamics. In addition, the latest algorithmic and method advances for improving the prediction of compounds that can form cocrystals and their properties in a range of solvents, will be discussed.

(241c) Determination of Critical Seed Loading From Nucleation Kinetics

Tseng, Y. T., National Taiwan University
Ward, J. D., National Taiwan University
Originally presented on: 11/5/2013 9:14:00 - 9:36:00

It has been observed experimentally that in at least some cases, secondary nucleation in batch crystallization can be suppressed by using an adequate seed loading. Kubota and coworkers[1–3] call this seed loading the critical seed loading and propose a correlation to predict the critical seed loading as a function of seed size. So far, critical seed loading has only been determined experimentally for a limited number of systems. In this contribution, a method is presented for preparing the critical seed loading diagram knowing only the nucleation kinetics for the system in question. This is advantageous because it is much less time-consuming than conducting experiments or developing and solving a complete batch process model including solubility expressions and growth and nucleation rate kinetics.

The results for 43 systems taken from the literature show that there is a great deal of variability in the critical seed loading, and a single correlation is inadequate to describe the relationship between critical seed loading and seed size. For a seed size of 10 microns, seed loading can effectively suppress nucleation in 92% of cases, while for seeds of size 50, 100 and 200 microns, seed loading can suppress nucleation in 74%, 61% and 39% of cases respectively.

Furthermore, the effects of changing process parameters (total batch time and the net crystal yield) can also be readily studied, although they are found to have a small effect compared to seed size. Finally the possibility of using a critical seed length or critical seed surface area for process development instead of critical seed mass is also investigated, however these alternatives are not found to be more attractive than the critical seed mass.

References

1. Doki N, Kubota N, Yokota M, Chianese A. Determination of critical seed loading ratio for the production of crystals of uni-modal size distribution in batch cooling crystallization of potassium alum. *J Chem Eng Jpn.* Jul 2002;35(7):670-676.
2. Doki N, Kubota N, Sato A, Yokota M, Hamada O, Masumi F. Scaleup experiments on seeded batch cooling crystallization of potassium alum. *AIChE J.* Dec 1999;45(12):2527-2533.
3. Jagadesh D, Kubota N, Yokota M, Sato A, Tavare NS. Large and mono-sized product crystals from natural cooling mode batch crystallizer. *J Chem Eng Jpn.* Oct 1996;29(5):865-873.

(241d) Design and Optimization of Continuous Crystallization Processes: Cascades of Mixed Suspension Mixed Product Removal Crystallizers

Vetter, T., ETH Zurich
Doherty, M. F., University of California, Santa Barbara
Originally presented on: 11/5/2013 9:36:00 - 9:58:00

Crystallization processes in the pharmaceutical industry have been carried out for decades as batch processes. It is recognized that this type of operation suffers from product variability from batch-to-batch and comparably high manufacturing costs [1]. While continuous processing is a proven technique in many large scale industries for overcoming both these deficiencies, the pharmaceutical industry was reluctant to adopt such an approach in the rapid growth era of the twentieth century. However, competition and regulatory demands are ever-increasing and there is now more focus on reducing manufacturing costs while maintaining high product quality.

While the design and optimization methodology for batch crystallization processes are relatively well understood for various combinations of cooling, anti-solvent or reactive crystallization [2] and for different optimization objectives [3], there are yet relatively few studies targeted on the design and optimization of continuous crystallization processes that keep the specific challenges posed to the pharmaceutical industry in mind [4,5].

In this work we present a comprehensive investigation of cascades of mixed suspension mixed product removal (MSMPR) crystallizers based on population balance equation models. The influence of the number of stages, temperature and residence time in each crystallizer on the particle size distribution of the product is ellucidated. Specifically, we show that for any number of MSMPRs and a constant production rate, there exists a clearly defined attainable region in a diagram of mean particle size vs. total residence time in the MSMPR cascade. This attainable region can be entirely traversed by altering the temperature and residence time in each MSMPR. We further report the influence of constraints (such as limitations in the maximally tolerated supersaturation in the MSMPRs and yield constraints) on the attainable region. By rephrasing well-established batch operating policies in the context of the attainable regoins, we highlight the underlying similarities between MSMPR cascades and batch processes and report design heuristics that can be used even in the absence of kinetic data. The concept of the attainable region and the associated heuristics could evolve into a convenient design tool for the conceptual design of continuous crystallization processes in a cost- and time-efficient manner.

Since the above mentioned investigations are dependent on the kinetics of crystal nucleation and growth, we report several case studies that employ different kinetic models and different ways to induce crystallization; among them the cooling crystallization of paracetamol from ethanol and the combined anti-solvent/cooling crystallization of aspirin from ethanol (solvent)/water (anti-solvent).

References:

[1] Randolph, A. D., Larson, M. A., Theory of Particulate Process: Analysis and Techniques of Continuous Crystallization, 2nd ed.; Academic Press: Toronto, Canada, 1988.

[2] Lindenberg, C., Krättli, M., Cornel, J., Mazzotti, M., Cryst. Growth. Des. 9, 1124-1136 (2009).

[3] Ward, J.D., Mellichamp, D.A., Doherty, M.F., AIChE J. 52, 2046-2054 (2006).

[4] Wong, S.E., Tatusko, A.P., Trout, B.L., Myerson, A.S., Cryst. Growth Des. 12, 5701-5707 (2012).

[5] Zhang, H., Quon, J., Alvarez, A.J., Evans, J., Myerson, A.S., Trout, B.L., Org. Process Res. Dev. 16, 915-924 (2012).

(241e) On the Fines Removal in a Continuous Plug Flow Crystallizer: Optimal Spatial Temperature Profile With Dissolution

Majumder, A., Loughborough University
Nagy, Z. K., Purdue University
Originally presented on: 11/5/2013 9:58:00 - 10:20:00

Fine crystals from a crystallizer can cause reduction of the efficiency of the downstream processes as well as final product quality. In this work, a systematic study is performed for obtaining the optimal temperature profile for the reduction of fines in a continuous plug flow crystallizer (PFC). The PFC consists of multiple modules or segments where the temperature of the each segment can be controlled separately. An optimization problem is formulated for a target crystal size distribution (CSD) (without fines) with the temperature of the PFC segments as decision variables. It is found that for the crystallization kinetics considered here, the optimal temperature profile also introduces dissolution steps so that the crystal fines due to secondary nucleation can be reduced significantly. A systematic study on the growth and dissolution kinetics is also performed. This study suggests that the key factor that determines whether the dissolution steps will be successful in reducing fines (without compromising the final size of the crystals from seed) is the size dependence of the growth and dissolution kinetics. If kinetics of the system is such that the larger crystals grow faster than the smaller ones and the smaller crystals dissolve faster than the larger ones, best results for fines removal is achieved. On the other hand, when both the growth and dissolution kinetics are independent of crystal size, fines removal found to be ineffective by temperature cycling.

(241f) Design and Operation of Crystallization-Based Enantioselective Processes: Enantioselective Crystallization of Chiral Compounds On An Enzymatic Reactive Environment

Encarnacion-Gomez, L. G., Georgia Institute of Technology
Bommarius, A. S., Georgia Institute of Technology
Rousseau, R. W., Georgia Institute of Technology
Originally presented on: 11/5/2013 10:20:00 - 10:42:00

Crystallization is a widely used separation and purification technique for a vast number of specialty chemicals and pharmaceutical compounds. However, its application on enantioselective separations can become troublesome. Enantiomeric compounds have the same physiochemical properties (other than optical rotation); as a result, an excess of the desired enantiomer is necessary to obtain enantiopure crystals.

To achieve the necessary enantiomeric excess, we are performing enantioselective enzymatic reactions on supersaturated solutions. To increase the productivity of the process, reaction and separation are performed in parallel in a single vessel. Preliminary work has shown the possibility of performing such a process utilizing chemoenzymatic reactions and kinetic resolutions. Present work is focused on using reaction and crystallization kinetics to design processes in which the operational time is minimized. As a case study, we are investigating the enantioselective deacylation of N-acetyl-DL-methionine by L-aminoacylase. The solubility of the DL-methionine has been measured as a function of substrate concentration, enantiomeric excess, and temperature. The information is being used to set the boundaries of the operation in which an enantiopure crystal can be recovered. The concentration in the reactive crystallizer is analyzed online by combining optical rotation, refractive index, and focused beam refractive measurements (FBRM). These signals can be used to monitor solution and crystals properties. Hence, we can control the chemical and enantiomeric purity of the crystals and their respective size distribution.

(241g) From Batch to Continuous Crystallisation: Nucleation and Growth Under Various Flow Conditions

Schacht, U., University of Strathclyde
Briggs, N., University of Strathclyde
Florence, A. J., University of Strathclyde
Sefcik, J., EPSRC Centre for Innovative Manufacturing in Continuous Manufacturing and Crystallisation University of Strathclyde
Originally presented on: 11/5/2013 10:42:00 - 11:04:00

Crystallisation processes are widely used for separation and purification in chemical and pharmaceutical industries. In order to develop continuous crystallisation of desired polymorphic forms, it is crucial to better understand the influence of mixing and flow conditions on nucleation of various polymorphs from supersaturated solution. For example, both supersaturation and post-mixing flow conditions can be used to control competitive nucleation of polymorphic forms in reactive crystallisation of L-Glutamic acid at constant temperature. We investigated various flow units for treatment of supersaturated L-Glutamic acid solutions prepared under identical initial mixing conditions in either reactive or antisolvent crystallisation.

We obtained respective crystal forms of L-Glutamic acid (either metastable α or stable β) by exposing supersaturated solutions to various flow units, such as a stirred tank, a magnetically stirred crystalliser, a peristaltic pump recycle loop or an oscillatory baffled crystalliser, operating at constant temperature. Using suitable flow units as nucleators to produce seeding suspensions of desired polymorphs, we designed a continuous crystallisation process by connecting a mixing unit (such as a static or in-line mixer) with a nucleator unit (a flow unit) and a continuous crystal growth unit (based on cooling crystallisation in a continuous oscillatory baffled crystalliser).

This set up was successfully used for continuous crystallisation of the stable β polymorph of L-Glutamic acid, achieving a good control over the final crystal size with a narrow particle size distribution.

(264a) Investigation of a Pendular Roller Mill

Wheeldon, M.,

Originally presented on: 11/5/2013 8:30:00 - 8:48:00

Due to new applications, such as high performance concrete, the demand for very fine particles at high throughputs is increasing rapidly. At the same time a growing concern for our planet's resources has raised the demand for energy efficient processes.

By being able to produce particle size distributions with a $d_{97} < 10$ μm while evidencing comparatively low power consumption values and high solids throughputs Roller mills in general and specifically pendular roller mills are up for the challenge.

The design of the pendular roller mill is thus far relying heavily on the experience gathered from industrially utilized mills. Due to the lack of experience in the new and coming fields of application a new, physically sound scale-up method must be developed to guarantee a successful design process. In order to determine scale-up rules that are valid for all product particle size distributions the operating behavior of pendular roller mills must first be fully understood.

The process occurring in roller mills can be subdivided into the following steps: comminution between the rollers and the grinding ring, solids transport between milling zone and classifier and classification of the crushed solids. In order to establish a reliable scale-up method all process steps must be completely characterized.

While both the classification and conveying process were investigated by measuring the pressure loss and the respective power consumption special attention was given to the comminution process. For that reason a new measuring method was developed with which the milling gap between the rollers and the grinding ring and the rotation speed of the rollers can be determined simultaneously.

The measurements were carried out on a pilot scale pendular roller mill during coarse and fine grinding processes to investigate the influence of process settings on the comminution, conveying and classification conditions.

By measuring the milling gap the correlations between parameters such as energy input, solids through put, solids hold-up or pressure loss on the one hand and the milling gap on the other hand can be acquired for the first time. With these correlations a physical model of the comminution process can be derived. Measurements of the rotation speeds of the rollers make it possible to detect slippage between the rollers and the grinding ring, allowing for a better evaluation of the efficiency of the comminution process.

With the gathered information a new scale-up rule is developed which accounts for the mills different operating behavior at different product particle size distributions.

After giving an insight into the working principle of pendular roller mills at different operating conditions the presentation will introduce a new scale-up procedure for pendular roller mills.

(264b) Computational Study of Needle-Shaped Particle Breakage Under the Agitation By Rotating Blades

Guo, Y., University of Florida
Wassgren, C. R., Purdue University
Ketterhagen, W. R., Pfizer Inc.
Hancock, B. C., Pfizer Worldwide Research and Development
Curtis, J. S., University of Florida
Originally presented on: 11/5/2013 8:48:00 - 9:06:00

The prediction of particle breakage is critical in some solids handling processes, such as filtration, drying, blending, and transport, in order to achieve the desired Particle Size Distribution (PSD) for granular materials. Recently, the Freeman FT4 Powder Rheometer has been applied to test the friability of Active Pharmaceutical Ingredients (APIs). In the tests using the FT4 powder rheometer, the powder bed under the compression of an applied weight on the top is agitated by rotating blades. After a certain time of agitation, the reduction in the mean particle size is used to classify the API powders as easy, medium, or hard to break. In this work, the Discrete Element Method (DEM) is employed to numerically model needle-shaped particle breakage in FT4 powder rheometers. The needle-shaped particle in DEM is formed by connecting a number of spheres in a straight line using elastic bonds, and the particle breaks at the bonds where the tensile or shear stresses exceed the corresponding strengths. In the numerical modeling, the evolutions of average particle size, PSD, and number of particles with each aspect ratio can be monitored. Therefore, the breakage kernel and daughter distribution function can be determined for the population balance model in order to describe the effect of particle breakage on the PSD.

The DEM simulation results show that the particle breakage rate depends on the friction between the particles and the base and top flat walls, the applied weight on the top of powder bed, and the blade rotational speed. With small particle-wall friction, the particle bed rotates like a solid body along with the blades and particle breakage rarely occurs. While with large particle-wall friction, retarded flow zones are obtained close to the base and top walls, and the particles are broken due to strong interactions with the blades. As a larger weight is applied to compress the powder bed, the interactions between the particles and walls/blades increase, leading to an increase in breakage rate. The breakage rate increases as the rotational speed of blades increases, due to the increased frequency of particle-blade interaction. However, the breakage per blade revolution is independent of rotational speed. It is also found that the increased number of particles and the average particle size follow power law relationships with the input work done by the blades, which is linearly proportional to the blade rotational speed, elapsed time, and the applied weight.

(264c) Modeling the Breakage of Crystalline Materials in Air Jet Mills

Starkey, D., Eli Lilly
Mecholsky, J., University of Florida
Iacocca, R.,
Powers, K., University of Florida
Svoronos, S., University of Florida
Originally presented on: 11/5/2013 9:06:00 - 9:24:00

Air jet mills are important tools in the size reduction processing of pharmaceutical powders. The benefits of the air jet mill to the pharmaceutical industry are its sanitary design (no moving parts or media) and ability to produce narrow particle size distributions (PSDs). Due to the high-value nature of active pharmaceutical ingredients (APIs), a trial-and-error approach to obtain optimal milling conditions would not be cost-effective.

A novel PBE (population balance equation)-aided approach was proposed for modeling the continuous milling of a spiral jet mill. It involves: (a) use of inexpensive, readily available excipient powders to predict mill-related model parameters; and (b) use of micro-identation or micro-milling to measure the material-dependent properties of the high value APIs (with only small quantities being consumed). The model requires PSD of the feed and mill operating conditions (grinding pressure, feed pressure, and feed rate) to predict the product PSD of brittle crystalline materials. The advantage of this approach is that it can be utilized to describe many self-classifying mills. In order to obtain optimal milling conditions for a new mill, mill-dependent model parameters can be obtained from inexpensive excipient powders. Quick material characterization experiments can be made with limited powder consumption on expensive powders to determine powder-dependent mill parameters.

(264d) Fundamental Studies of Comminution

Harriott, G., Air Products and Chemicals
Boland, T. E., Air Products and Chemicals
Zheng, S.,
Malik, R., North Carolina State University
Trembley, J.,
Originally presented on: 11/5/2013 9:24:00 - 9:42:00

Comminution is an important step in the production of many materials made from polymers for selective laser sintering, powder coatings and rotational molding compounds. A longstanding problem in grinding technology involves predicting the operating conditions such as temperature and minimum impact velocity for effective comminution. Although many materials break readily at low temperatures, some remain resistant to grinding even when immersed in liquid nitrogen. Here we outline a novel approach to characterize grinding based on 1-dimensional mechanics of particle-mill collisions. Collision of a slab particle with a rigid mill surface generates compression and tension waves that sweep across the particle at sonic velocity. If at any time, deformation exceeds the coherence limit, the particle ruptures. Large amplitude stress-strain measurements, conducted from ambient to cryogenic temperatures, provide the basis for calculation of minimum impact velocity for fracture. Further, a minimum particle size may be determined at which collisions, tempered by ambient resistance, are too slow to cause fracture. Cryogenic grinding tests on polyethylene and ethylene copolymer resins reveal differences in failure mode (ductile/brittle) consistent with model predictions.

(264e) Fragmentation and Bounce of Nanosized Agglomerates Due to the Impaction

Ihalainen, M., University of Eastern Finland
Lind, T., Paul Scherrer Institut
Torvela, T., University of Eastern Finland
Ruusunen, J.,
Lahde, A., University of Eastern Finland
Tiitta, P., University of Eastern Finland
Jokiniemi, J., VTT Fine Particles
Originally presented on: 11/5/2013 9:42:00 - 10:00:00

The impaction behaviour of aerosol particles has applications in fields such as powder processing, nuclear safety, and health research. Two major phenomena may occur during the impaction of agglomerated particle onto a surface. The first one is the plastic deformation of the agglomerate which may lead to fragmentation of the agglomerate if there is enough impaction energy available. The second process is the bounce of the particles whenever the adherence between the particle and surface can be overcome by the energies stored to elastic deformations.

To achieve better understanding on how these processes work and affect each other, the fragmentation and the bounce should be observed simultaneously. In this study, these processes were investigated for two agglomerate materials with varying agglomerate properties.

A method to study the simultaneous fragmentation and bounce of the particles was developed and it is described in detail by Ihalainen et al. (2012). The basic principle of the method was as follows. The aerosols in study were first size classified by their electrical mobility with a Differential Mobility Analyzer (DMA) to achieve relatively monodispersed aerosol population. This population was then introduced to a single stage Micro Orifice Uniform Deposit Impactor (MOUDI) in which the particles impact onto the impaction plate due to the inertia of the particles. The impaction velocity depended on the aerodynamic diameter of the particle and the air flow conditions at the impactor. To be able to estimate this impaction velocity, computational fluid dynamics simulations (CFD) coupled with Lagrangian method to analyze the particle trajectories were made. The sample flow from the MOUDI was collected to a low pressure sampling chamber which enables the analysis of the bounced with both online and offline methods. The deposited particles were collected both from the impaction plate and bounced fraction and analyzed using a Transmission Electron Microscopy (TEM). In addition to the TEM, the bounced particles were characterized with a scanning mobility particle sizer (SMPS). The mass fraction of the particles which bounce was estimated from the SMPS data using size dependent effective density. The effective density of the particles was evaluated using a combination of aerosol particle mass analyzer (APM) and SMPS.

The agglomerated particles of TiO_2 and Fe were generated using atmospheric pressure chemical vapor synthesis (APCVS; LÃ¤hde et al, 2011). Titanium tetra isopropoxide was used to generate TiO_2 agglomerates and iron penta carbonyl to produce iron agglomerates. The properties of the agglomerates were manipulated during the generation to find out how these changes affected the fragmentation and bounce. The properties varied for TiO_2 particles were the primary particle size and the degree of sintering, i.e. the bridging between the particles. The average primary particle sizes used here were 27 and 16 nm. The oxidation state of the iron particles was varied by changing the amount of oxygen available during the particle formation phase. The oxygen concentrations used during the generation were 21 and 0.05 %.

The size distributions of the bounced particles for each particle type after impaction at about 150 m/s are presented in Figure 1. It shows that the original monodispersed size distribution has transformed to a wider distribution of smaller particles due to the fragmentation. Fig. 1 also shows that not all of the particles did break-up and some fractions of the particle populations remained intact during the impaction. This was pronounced at lowest impaction velocities for the TiO_2 particles with higher degree of sintering and for the iron oxide particles. It can also be observed from Fig. 1 that the changing the agglomerate properties affected the fragment size. Increasing of TiO_2 agglomerates sintering increased the size of the fragments meaning that fewer bonds per agglomerate broke up. This was due to the sintering which made the bonds between the primary particles stronger. The reduction of the primary particle size also resulted in larger fragments however the change was not as significant as in the case of sintering. The fragments of the iron oxide particles were larger than in the case of TiO_2 at the same velocities and the oxidation level of the iron oxide particles was not found to have a significant effect on the size of the fragments.

The mass fractions of the fragmented and bounced particles increased as the impaction velocity increased in almost all of the cases. With TiO_2, the smaller primary particle size reduced the bounce fraction compared to the case with larger primary particle size. There was no significant difference in bounced mass fractions between the iron oxide cases.

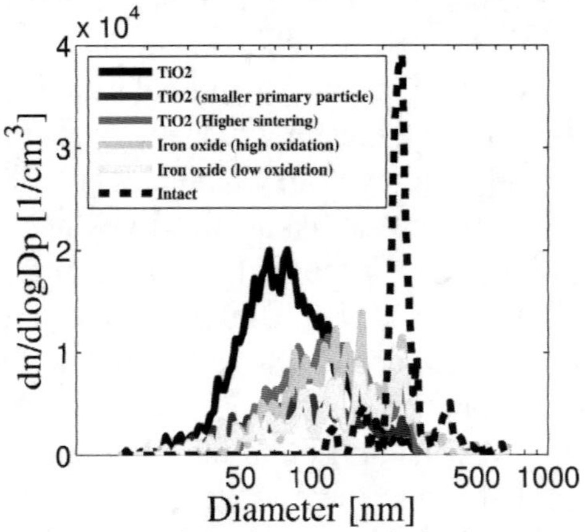

Figure 1. The number size distributions of the intact and the bounced particles at similar impaction conditions.

This work was carried out in support of the international ARTIST II program which is acknowledged for funding this work. The authors express their gratitude to those members of the Laboratory for Thermal-Hydraulics at PSI and of the Fine Particle and Aerosol Technology Laboratory at the University of Eastern Finland that participated in the experimental work and construction of the facility.

Ihalainen, M., Lind, T., Torvela, T., Lehtinen, K.E.J. and Jokiniemi, J. (2012) *Aerosol Sci. Technol.* **46:9,** 990-1001.

Lahde, A., Kokkonen, N., Karttunen, A., Jaaskelainen, S., Tapper, U., Pakkanen, T.A., Jokiniemi, J (2011) *J. Nanopart. Res.* **13,** 3591-3598.

(264f) Modeling the Amorphous Transformation Kinetics of APIs During Dry Ball Milling

Capece, M., New Jersey Institute of Technology
Dave, R. N., New Jersey Institute of Technology
Bilgili, E., New Jersey Institute of Technology
Originally presented on: 11/5/2013 10:00:00 - 10:18:00

Amorphous active pharmaceutical ingredients (APIs) can attain higher solubility than their crystalline counterparts resulting in higher dissolution rates and bioavailability. While amorphous solids offer a method to engineer efficacious dosage forms, particularly for BCS Class II drugs (low solubility, higher permeability), the proper processing knowledge is lacking. Among the various methods to prepare amorphous solids, dry milling is among the most common techniques used to induce the transformation from crystalline to amorphous in processes known as amorphization. While the phenomenon of particle breakage is well understood and characterized in ball milling, amorphization lacks similar modeling and predictive tools. Particle breakage is traditionally modeled as a first order rate process [1] and the same methodology is extended in this study to describe amorphization. This study investigates the dynamics in the amorphization process of the model API acetaminophen using a vibrational ball mill. The transformation dynamics are modeled using a modified form of the "event-based" milling model commonly used in the population balance modeling of particle breakage. It is found that the transformation dynamics can be modeled using this approach and may be used as a predictive tool in amorphization processes. Mechanical alloying (i.e. milling) acetaminophen with stabilizing polymers is also investigated to improve the stability of the amorphous API and is also incorporated into the model. It is found that stabilizing polymers may be necessary to decrease the re-crystallization rate of the API in order to achieve complete amorphous transformation and allow amorphization processes via milling to be successful.

[1] Prasher, C.L., 1987. Crushing and Grinding Process Handbook. Wiley, Chichester.

(264g) Robustness & Scaling of Slurry Milling Operations Via PAT

Tvetan, J.,
Huggins, S., Amgen
Originally presented on: 11/5/2013 10:18:00 - 10:36:00

Slurry milling can be an important element of pharmaceutical production to achieve particle size requirements for drug product performance. The work to be outlined and discussed will focus on a robust, scaleable method using real time feedback via FBRM for acquiring a consistent final PSD with the use of a slurry mill for particle size reduction. The FBRM probe was placed inline with the milling loop to simplify the impact of hydrodynamics on chord length distribution measurement. Milling samples were pulled at discrete time intervals and analyzed off-line to correlate PSD to a FBRM chord length distribution which can be used as a target during milling. Robustness of impact to variation in PSD at milling start point and transfer of CLD target across scales will be discussed.

(264h) Discussion of Effects of Grinding Modes On Particle Shape During Comminution
Xi, Z., Tel-Tek
Originally presented on: 11/5/2013 10:36:00 - 10:54:00

This investigation is a study of particle shape of carbon anthracite grinded by a pin mill. Different grinding modes were discussed in the study to understanding the effects on the particle shape during pin milling using 2 types of pins. The particle shape is quantified in terms of sphericity, aspect ratio, symmetry and convexity, based on the results measured and analyzed by particle image processing. The preliminary results show that attrition is the most profound effect determing the particle shape in pin milling.

(273a) The Interplay of Component Size, Morphology and Gelatinization On the Flow Performance of An Encapsulated Pharmaceutical Blend

Nakagawa, H., Merck & Co., Inc
Goldfarb, D., Research done at Exxon Corporate Research
Park, E.,
Conway, S. L., Merck & Co., Inc.
Originally presented on: 11/5/2013 8:30:00 - 8:48:00

Understanding the flow properties of a pharmaceutical powder blend intended for formulation in a gelatin capsule is crucial for proper process design and robustness, and avoidance of weight uniformity issues that can ultimately impact safety and efficacy. We conduct physical characterization of severl component lots and blend compositions to determine that the flowability for a series of prototype formulations is highly variable. In attempting to understand the underlying causes of the variability, we discover unexpected transtions in flow properties as a result of changing drug load. The interactions between the active pharmaceutical ingredient and the excipients in the blend are shown to cause these transtions, and optimal concentrations are identified. The potential roles individual components as glidants is shown to depend on factors such as gelatization extent, particle morphology and particle size distribution. General blend design strategies for improved flow performance are explored.

(273b) An Experimental Study and Phenomenological Model of Powder Flow Through a Moving Orifice

Hecht, J., Procter & Gamble
Murthy, V., Procter and Gamble
Stoltz, C. G., Procter & Gamble Company
Originally presented on: 11/5/2013 8:48:00 - 9:06:00

This presentation describes an experiment to study powder flow through a moving orifice. The results and underlying science are rich and counterintuitive. Experimental observations, discrete-element modeling, and granular-temperature theory will be used to describe and explain the phenomena. The device also presents an opportunity to solve the common practical problems of funnel flow, ratholing, and cohesive arching during discharge from silos.

(273c) Effect of Particle Size and Density On Flow of Powder Through Feed Frames On Rotary Tablet Presses

Baserinia, R., University of Leicester
Woldu, A. B., Merck&Co., Inc.
Kheiripour, M., Merck&Co., Inc.
Rajniak, P., Merck Sharp and Dohme
Sinka, I. C., University of Leicester
Originally presented on: 11/5/2013 9:06:00 - 9:24:00

Pharmaceutical tablets are manufactured using rotary tablet presses. In a typical press the powder formulation is placed in a hopper above the press which is connected through a vertical pipe to a feed frame which usually involves rotating paddles that transfer the powder horizontally over the die opening. The flow behaviour of the powder through the feed frame is critical to achieve high throughput of the press and maintain weight uniformity.

In this paper the flow of powders through a lab scale rotary feeder is examined. The system was presented elsewhere [1]. Five grades of microcrystalline cellulose are used to study the effect of particle size and particle density for the same base powder material. The flow rate of powders through the feeder is measured for different rotational speeds (25-250 rpm) and exit diameters of the feeder (5-20 mm). The flowability of the powders is discussed in relation to conventional powder flow measures (such as angle of repose and critical orifice diameter) as well as flow measures determined using a model shoe-die system for the same powders [2].

References

[1] Singh G., Givand J., Rajniak P., Chern R., Reynolds S.D. and Sinka I.C. , 2012. Effect of paddle shape on flow of powder through feed frames on rotary tablet presses. Paper presented at: AIChE Annual Meeting, October 28 – November 2, Pittsburgh, PA.

[2] Mills L.A. and Sinka I.C., 2013. Effect of particle size and density on the die fill of powders. European Journal of Pharmaceutics and Biopharmaceutics. Accepted for publication.

(273d) Experimental and Multi-Scale Modelling to Understand Electrostatic Behavior of Pharmaceutical Powders

Naik, S. S., University of Connecticut
Sarkar, S., University of Connecticut
Hancock, B. C., Pfizer Worldwide Research and Development
Yu, W., Pfizer
Abramov, Y., Pfizer
Gupta, V., University of Connecticut
Chaudhuri, B., University of Connecticut
Originally presented on: 11/5/2013 9:24:00 - 9:42:00

In pharmaceutical formulation processes, particle charging is often a nuisance and can cause problems in the manufacture of products, such as affecting powder flow, and reducing fill and dose uniformity. For a fundamental understanding of the powder triboelectrification, it is essential to study charge transfer under well-defined conditions. Hence all experiments were conducted in a humidity controlled glove box at 20%rh. To understand tribocharging materials, different surface viz. PVC, aluminum, Teflon, PMMA were used along with two pharmaceutical excipients and two API's. The experiments were performed in V-blender at a prescribed speed of 13 rpm. A triboelectric series of pharmaceutical materials was generated. For the pharmaceutical materials the work function was determined using MOPAC. MOPAC is a general-purpose semi-empirical molecular orbital package for the study of solid state and molecular structures and reactions providing band gap and ionization potential of molecules. In addition a discrete element method (DEM), including charge transfer and screened coulombic forces was developed for the V-blender system. There are different competing mechanisms involved in tribocharging. The two most important ones are electron transfer and ion transfer. Electron transfer has been well established for metal-metal and to some extent for insulator-metal contacts. There is still an ongoing debate for insulator-insulator charging whether the charge carriers are electrons or ions. As per work function hypothesis or established tribo-electric series powders should charge negatively against Alumimium, PMMA surfaces and positively against Teflon surfaces and PVC surfaces. According to the experiments in V-blender both the excipients (lactose and MCC) and API's charged negatively against PMMA and Aluminum and positively against Teflon and PVC surfaces

(273e) Comparisons Between Solid Plug Model Predictions for a Powder Roll Compactor Screw Feeder and Experimental Measurements

Patterson, T. J., Purdue University
Wassgren, C. R., Purdue University
Originally presented on: 11/5/2013 9:42:00 - 10:00:00

Comparisons between solid plug model predictions for a powder roll compactor screw feeder and experimental measurements

T. Patterson and C. Wassgren, Purdue University

Abstract

Solid plug models previously developed to predict the pressure distribution in extruded plastic pellets fed by a screw feeder are evaluated in the present work for feeding pharmaceutical powders, specifically to the inlet of the nip region in a roll compactor.Â Experimental measurements of the screw torque required to feed the powder at a given feed rate, the pressure at the outlet of the feeder, and the powder bulk density distribution along the axis of the screw are compared to corresponding predictions from the solid plug models.Â The screw rotational speed and outlet pressure are varied in the experiments. The goal of the work is to use the measured screw feeder torque, a parameter often available in commercial roll compactors, and the solid plug model to predict the stress at the exit of the feed screw, and use this stress as an input to Johanson's one-dimensional model for predicting the ribbon density exiting the roll compactor.Â

(273f) Methodology for Identification of Interaction Parameters for DEM Studies

Santhanam, P. R., New Jersey Institute of Technology
Dreizin, E. L., New Jersey Institute of Technology
Originally presented on: 11/5/2013 10:00:00 - 10:18:00

Discrete Element Method (DEM) is rapidly expanding as a valuable approach to describe various unit operations in the pharmaceutical, food and mining industries such as mixing, milling, fluidization, transport, etc. However, accurate identification of interaction parameters (friction and restitution coefficients) remains a challenge in all these applications. The issue is even more challenging considering that new materials are being rapidly developed with broad particle size distributions, varying particle shapes, as well as broad range of bulk material properties. Direct experimentation using specific unit operation of interest with each individual powder yielding trustworthy results is possible, but is expensive both in terms of facilities and labor required. At the same time, results of experiments with a restricted range of experimental configurations and selected materials cannot be justifiably extended to predict behavior of powders that have not been tested directly.

The objective of this work is to develop a simple and easy to use methodology to quantify the particle-particle and particle-geometry interaction parameters for DEM studies. Direct experimental validation is the key aspect of this approach. To enable the direct comparisons of experiments and calculations we work with a carefully designed experimental configuration with a relatively small number of particles (~20,000-30,000), for which systematic changes of powder processing parameters can be readily implemented experimentally as well as described in DEM. The experimental set-up consists of a small vibrating hopper in which the powder is loaded. The number and size of the particles exiting the hopper can be tracked illuminating the powder stream with a laser sheet and using a photomultiplier tube to register the pulses of scattered light. For these initial validation experiments, we use spherical powders with different surface properties, densities, and particle size distributions, including a monodispersed powder.

A parametric study investigating behavior of different powders both experimentally and computationally will be discussed. The particle size distribution is directly represented in DEM. In both model and experiment, the oscillation frequency and amplitude are varied and respective feed rates and sizes of the particles fed in real time are monitored. The sensitivity of the DEM predictions to specific values of friction and restitution coefficients is investigated. The effectiveness of different particle interaction models in accurately predicting the experimental data is tested.

(273g) Estimating Effects of Agglomerate Properties On Dust Generation During Handling

Halt, J., Michigan Tech University
Kawatra, S. K., Michigan Technological University
Originally presented on: 11/5/2013 10:18:00 - 10:36:00

Dust is a significant concern in raw material processing industries as evidenced by increasing levels of regulations. Potential sources for dust generation and emission includes shipping and handling high-tonnage products, such as coal, fertilizer, iron ore, and limestone. Agglomerates made from these materials can fragment and abrade during impact and abrasion and generate particulates across a wide range of sizes. Certain sizes, such as particulate matter with aerodynamic diameters less than 10 and 2.5 micrometers (PM_{10} and $PM_{2.5}$), are able to stay suspended in air for long periods of time and are undesired for health and environmental reasons. The purpose of this research was to investigate the dominant factors controlling the dustiness of iron ore agglomerates. The theory and development of the chosen method designed to measure material dustiness is presented, as well as results from testing both industrial and laboratory-made agglomerates. Factors investigated ranged from raw material inputs to agglomerating conditions.

(273h) Design and Operation of Continuous Rotary Calciners

Emady, H. N., Purdue University
Anderson, K. V., Rutgers University
Glasser, B. J., Rutgers University
Muzzio, F. J., Rutgers, The State University of New Jersey
Borghard, W. G., ExxonMobil
Cuitino, A., Rutgers University
Originally presented on: 11/5/2013 10:36:00 - 10:54:00

In catalyst manufacturing, continuous rotary calcination is a thermal treatment process in which particles flow through a rotating drum with heated walls. Although this process is widely used, there is a lack of predictive models for material flow and heat transfer, particularly upon scale-up. This work aims at a fundamental understanding of the effects of operation, design, and material properties on residence time and temperature distributions in rotary calcination processes to provide a methodology for scale-up.

For successful calciner performance, the residence time of the particles in the calciner must be longer than the time required for heating and calcination of the particles. This problem can be separated into axial and radial processes. In the axial direction, residence time and dispersion will be important, requiring experiments in full calciners. In the radial direction, heat transfer and dispersion will be important, requiring experiments and discrete element method (DEM) simulations in cross-sections of the calciner. This presentation will discuss results of studies on the competing axial and radial processes and their implications for scale-up of continuous rotary calciners.

(274a) Howard Littman and the Belgrade School of Fluidization
Vunjak-Novakovic, G., Columbia University
Jovanovic, S.,
Originally presented on: 11/5/2013 8:40:00 - 9:00:00

Howard Littman and the Belgrade School of Fluidization

Gordana Vunjak-Novakovic, Columbia University and Stevan Jovanovic, The Linde Group

In 1970s, Howard Littman came to the Belgrade University's Department of Chemical Engineering as a Fulbright Professor, to leave one of the most remarkable legacies in the field of multiphase transport and fluidized systems. His collaboration with Professor Vukovic, who was a mentor to many of us involved in fluidization research, resulted in a long-term NSF-funded research that yielded numerous publications describing new theory and reporting new data. This work transformed some of the existing paradigms and became known as the Belgrade School of Fluidization. We reflect here on the intellectual leadership of Professor Littman to this new direction of work, and his generous support of this important effort. We will wrap up our presentation by highlighting the personal legacy Professor Littman has created by his ultimate moral integrity and his enduring compassion for others.

(274b) Tunable Porosity Gradients in Photocatalytic Titania Nanostructures for Air and Water Purification

Plawsky, J. L., Rensselaer Polytechnic Institute

Originally presented on: 11/5/2013 9:00:00 - 9:20:00

Titanium dioxide nanoparticles have been used extensively a photocatalyst to breakdown organic compounds when exposed to ultraviolet light. However, it is difficult to control the morphology and structure of a collection of nanoparticles. Oblique angle deposition allows for the growth of catalytic nanostructures on solid surfaces and the ability to tailor the shape, structure, and density of the material as a function of deposition conditions. The efficiency of porous TiO_2 films developed for this purpose is limited by light intensity losses due to absorption as well as the ability of the reactant to diffuse into the pores. The interplay of light absorption and reactant diffusion in films grown by oblique angle deposition was investigated to find the optimal reaction conditions. High surface area nanorods of titanium dioxide were grown on a transparent substrate to investigate their effectiveness as photocatalytic agents for the destruction of organic contaminants in air and water. Optical transmission measurements were made that allowed for an estimation of the porosity of the film (75%-78%). The photocatalytic degradation of an indigo carmine dye solution and a gas mixture containing isobutylene on the porous films was shown to depend on film thickness and annealing conditions. The effectiveness of the film was assessed by observing the change in absorbance of the dye at 610nm over time and quantifying the film performance using a pseudo-first-order reaction rate model. Reaction rates increased as the film thickness increased from 600nm to 1000nm, but leveled out or decreased at thicknesses beyond 1500nm. A transport/reaction model was used to show that there exists an optimal geometry that maximizes the overall reaction rate and that such a geometry can be simply produced using glancing angle deposition. The nanorod films were benchmarked against nanoparticle films and were shown to perform as well as 0.73g/L of 25nm diameter anatase nanoparticles having a surface area of $50m^2/g$.

(274c) Control of a Pressure-Regulated Circulating Riser System for Running a Conversion Process

Paccione, J., SUNY at Albany, School of Public Health
Originally presented on: 11/5/2013 9:20:00 - 9:40:00

The pressure-regulated circulating riser (PCR) system provides several significant control and operational capabilities over other generically similar systems. The PCR has the ability to support the operation of up to three different physical and/or chemical reactions in the riser, particle separator, and the counter current moving packed bed sections. The unique capability of independently controlling the pressure drop across and the fluid flow rate through the riser provides a means of optimizing the solids fraction and particle velocity in the riser when running a process there. The design and operation of a PCR is based on a two-fluid model that was previously used to demonstrate how the operation of a riser requires only two independent variables to be specified in order to determine the conditions there. The PCR has a total of three degrees of freedom that define its overall operation, which reduces the complexity of controlling a process. The unique control system provides the basis for ready sizing of PCR systems over a wide range of scales from pilot to that needed for industrial processes.

In this paper, a liquid-solid process is modeled in which a stripping operation takes place in the moving packed bed and a pseudo first order reaction which is dependent on the solids fraction and residence time is run in the riser. Modeling equations are presented that show how the control system is configured to accommodate / optimize a first order reaction in the riser which is dependent on both solids fraction and residence time. The modeling also provides a means of determining whether a system is adsorption or reaction limited and demonstrates how resources can be allocated to the limiting step in the process. This model may be adapted to a variety of chemical separation, conversion and treatment operations.

(274d) Tumble Mixing of Flat Particles

Hook, B. D., The Dow Chemical Company
Hook, D. T., Austin College
Originally presented on: 11/5/2013 9:40:00 - 10:00:00

Particle Shape plays a consideration in tumble mixing. For flat or disc shaped particles, understanding when the mechanism of particles sliding dominates over rolling has an impact on flow regime and the mixing time of the segregated mixture. Data and models will be presented.

(274e) Calling Unruly Fluidized Beds to Order

Coppens, M. O., Rensselaer Polytechnic Institute

Originally presented on: 11/5/2013 10:00:00 - 10:20:00

The multiphase hydrodynamics of gas-solid fluidized beds are very complex. Nonlinear interactions between gas and clustering solid particles lead to heterogeneous structures at multiple length and time scales, with broad, non-uniform cluster and void distributions that could span the entire bed. This may have unexpected and undesired repercussions on product quality, and even on safety due to hot spots, whether it is in the drying or coating of solid particles, or in exothermic reactions catalyzed by the solids.

Useful insights in fluidized bed hydrodynamics are obtained by analyzing high-frequency pressure fluctuations, which are easy to measure also in industrial fluidized beds. The analysis of pressure fluctuations can also be used to monitor changes in bed dynamics, e.g., due to undesired agglomeration, so that an operator could interfere before the problems become irreversible.

There are also exciting opportunities to change the hydrodynamics of fluidized beds in a controlled way – to make them more uniform and to facilitate scale-up – by drawing lessons from nature. Our *nature-inspired engineering* approach consists in identifying fundamental mechanisms (rather than imitation out of context) underlying desirable properties in natural systems, such as scalability, robustness or efficiency, and using these principles to guide the rational design of materials and processes with similar properties.

In the context of fluidization, one way is to use a distributed feed of gas, via a so-called fractal injector, the structure of which is inspired by branching lungs and trees. The fractal injector circumvents some of the complex, chaotic heterogeneity of typical fluidized beds, and is easily scalable. Bubbles are smaller and uniformly distributed, while gas is directly injected within the emulsion phase. This helps to control product distribution.

Another way is to pulse part of the gas flow above minimum fluidization. Within a range of frequencies of a few Hertz, pulsing may lead to remarkably regular, rising bubble patterns up to heights of the same order as the bed width. Pressure fluctuation analysis shows increased bubble sizes, and improved homogeneity. This is particularly useful to fluidize cohesive particles and more rapidly and uniformly dry wet particles.

(274f) Reflections on a Stellar Teacher and Researcher: Dr. Howard Littman, Professor Emeritus
Morgan, M.,
Originally presented on: 11/5/2013 10:20:00 - 10:40:00

Reflections on a Stellar Teacher and Researcher: Dr. Howard Littman, Professor Emeritus

I am sure that many of Professor Littman's former students (disciples) can speak of being indoctrinated into the Littman research philosophy of "A & E"; no not "Arts and Entertainment" but rigorous "analysis" and arduous "experimentation". Professor Littman was a tough taskmaster who demanded much of his graduate students but who was always kind, considerate and supportive in the management of us. For me personally, I contribute much of my professional success and growth to his research philsophy, sage advice, guidance and a heavy dose of Jewish history
that Howard bestowed on many occasions.

In the time allotted for my presentation I will highlight a few of Howard's seminal papers on the stability of spouted bed systems that speaks to his research approach ("A & E"), his focus on developing a clear fundamental understanding of the underlying physics at play in such systems and the significance of his research to fluid-particle systems in general.

(298a) Applying Crystallization Modeling to Improve the Understanding of a Batch Cooling Process of An Agrochemical Active Ingredient

Brent, R.,
Parmar, M., Syngenta
George, N., Syngenta
Sillers, P.,
Bryce, M.,
Clelland, G.,
Mumtaz, H., Process Systems Enterprise
Mitchell, N., Process Systems Enterprise
Bermingham, S. K., Process Systems Enterprise
Originally presented on: 11/5/2013 12:30:00 - 12:52:00

The current production process of a Syngenta active ingredient isolates the batch by a cooling crystallization process. A major bottleneck of this process is the isolation time of the product after crystallization, which can vary significantly between batches. From observation, the key variable that dictates the isolation time of the product is the width of the particle size distribution of the crystals obtained; the wider the particle size distribution, the slower the filtration time observed. Typically for this product, two distinct environments of fine particles and large single crystals are observed within a single batch. In this work an experimental scale-down study was carried out to replicate the plant process and to assess whether the filtration properties could be improved by modifying the conditions in the crystallizer. Seeding the batch with different quantities of seed with a narrow particle size and modification of the cooling profile where carried out, with limited success in narrowing the particle size distribution significantly. Crystallization modeling using gCRYSTAL was employed to rationalize the experimental observations and to identify the key crystallization phenomena that are dominant in the process. It was observed that the controlling phenomena for this crystallization where the favorable kinetics for crystal growth and also attrition of the crystals, once attaining crystals that had grown to a critical size. Having identified this from the modeling, it was then possible to devise additional experiments to reduce the width of the particle size distribution of this material with the potential to improve plant operations.

(298b) Generic Modelling Framework for Design and Analysis of Crystallization Operations

Meisler, K. T., Technical University of Denmark (DTU)
Samad, N. A. F. A., Technical University of Denmark (DTU)
Gernaey, K. V., Technical University of Denmark (DTU)
Gani, R., Technical University of Denmark
Originally presented on: 11/5/2013 12:52:00 - 13:14:00

A systematic model-based approach to design and analysis of crystallization operations as a first step could result in reduction of time and experimental resources. To this end a generic and systematic model-based framework for the design of a process monitoring and control system to achieve the desired crystal size distribution (CSD) and crystal shape for a wide range of crystallization processes has been developed. This framework combines a generic multi-dimensional modelling framework, tools for design of supersaturation set-point, for design of PAT systems as well as options to perform the uncertainty and sensitivity analysis of the PAT system design. Through this framework, it is possible for a wide range of crystallization processes, to generate the necessary problem-system specific models, the necessary supersaturation set-point using an extended analytical CSD estimator and/or a response surface method (RSM) and a PAT system design including implementation of monitoring tools and control strategies in order to produce a desired product with its corresponding target properties. The problem-system specific models involve generic elements and system specific elements. The system specific elements, such as kinetic constitutive models (models for growth, nucleation, agglomeration and breakage) are stored in libraries, from which they can be extracted, analyzed, used and re-used. This allows models to be analyzed against other models for specific systems and the possibility of comparing models for different systems in a fast and flexible manner. An advantage of this is the possibility to study the effect of different constitutive model parameters on the kinetic phenomena associated with different chemical systems together with the crystallization operational parameters and employ this knowledge to plan and design crystallization operations. Uncertainty and sensitivity analysis is also possible within the framework, and can be carried out to determine, for example, Â theÂ kinetic model parameters in the growth rate model that are the most sensitive This presentation highlights the advantages of utilizing the generic modeling framework to support crystallization operation modeling and the use thereof in design and analysis of crystallization operations. This is illustrated through simulation studies of a potassium dihydrogen phosphate (KDP) and a paracetamol crystallization operation. The operation of KDP batch cooling crystallization is simulated with varying seeding policies and temperature profiles in order to establish the effects of different operational policies on the final product. This includes the effects of uncertainties in the input, for example in the kinetic model parameters. The advantage of using a framework for which models can be re-used is highlighted as the user is guided in specifying the needed data under the assumptions stated in the modeling problem through the formulation and solution of the model equations representing the operation scenario under investigation. The advantage of the generic framework is further highlighted when the chemical system is switched to paracetamol for a similar operation scenario. Here, switching the kinetic models within the operation scenario model allows the study of paracetamol crystallization operation, in one as well as two-dimensions.

(298c) Crystallization Process Optimization Using Derivative-Free Optimization Algorithms

Zhao, H.,
Vouzis, P., Carnegie Mellon University
Sahinidis, N., Carnegie Mellon University
Originally presented on: 11/5/2013 13:14:00 - 13:36:00

Polymorphism is of key importance in the pharmaceutical and fine-chemicals industry. Modeling of polymorphism phenomena is based on Population Balance Equations (PBEs), which are sets of partial differential equations whose analytical solution is not always available [1]. Moreover, when optimization is to be done over a simulator, it is often difficult to calculate derivatives of variables of PBEs either analytically or numerically. This renders most derivative-based optimization methods of little or no use. Derivative-free optimization (DFO) algorithms are specifically designed for this kind of problem [2]. In this work, we use g-CRYSTAL as a black box simulator to perform simulation-based optimization of crystallization processes. DFO is used to optimize one particular polymorph and minimize the time of producing a desired particle size distribution. This goal is achieved by optimizing process variables, including temperature and cooling profile of reactor, super saturation of seed molecule, flow rates, and concentrations of species. We present extensive computational results, where we compare the results of 28 DFO algorithms on several case studies.

References

[1] Rudiyanto Gunawan, Irene Fusman, and Richard D. Braatz. High resolution algorithms for multidimensional population balance equations. AIChE Journal, 50(11):2738{2749, 2004.

[2] Luis Miguels Rios and Nikolaos V.Sahinidis. Derivative-free optimization: A review of algorithms and comparison of software implementations. Journal of Global Optimization, DOI 10.1007/s10898-012-9951-y, 2012.

(298d) Stable Configurations in the Horizontal Ribbon Growth Process

Oliveros, G. A., Carnegie Mellon University
Ydstie, B. E., Carnegie Mellon University
Sridhar, S., Carnegie Mellon University
Originally presented on: 11/5/2013 13:36:00 - 13:58:00

In the Horizontal Ribbon Growth (HRG) process, a thin crystal ribbon is produced and extracted continuously from a molten substrate. The advantage of this technique is its capacity to achieve large rates of production compared to conventional crystal growth processes, such as the Czochralski and Bridgman. However, despite its potential and efforts to develop the technology for more than 50 years, many technical and experimental challenges have been reported regarding the difficulty in trying to maintain a continuous and stable ribbon production.

One of the principal causes of disruption is the freezing of the ribbon to the end of the crucible. The reason for this is that there is no melt present between the ribbon and the crucible. The solid surface of the crucible offers plenty of nucleation points for the ribbon to crystallize onto as it is being pulled. The opposite problem can also arise: an excess of melt flow that causes the melt to spill over the crucible. In an ideal operation, a space between the ribbon and the lip of the crucible must be maintained to avoid undesired nucleation points, and a meniscus must be pinned between the crucible lip and the ribbon to avoid melt spill-over.

In this work, we present a theoretical analysis to assess these technical difficulties. We analyze the conditions under which a meniscus exists between the ribbon and the crucible and stays pinned to its corner, as well as the allowable range of melt levels that give rise to stable menisci. Assuming that the existence of the meniscus is defined by hydrostatics, we use a theoretical model based on the governing Young-Laplace equation to solve analytically for the shape of the meniscus. Then, we show the range of operating conditions that are physically viable in a silicon system, and construct the possible stable configurations. We show that these stable configurations depend mainly on the pulling angle, the melt level, and the type and shape of the crucible.

(298e) Using ATR-FTIR to Track the Crystallization of Multiple Solutes

Griffin, D., Georgia Institute of Technology
Grover, M., Georgia Institute of Technology
Kawajiri, Y., Georgia Institute of Technology
Rousseau, R. W., Georgia Institute of Technology
Originally presented on: 11/5/2013 13:58:00 - 14:20:00

In-situ measurements of concentration play a large role in developing and implementing control schemes for crystallizers [1, 2]. In recent years Attenuated Total Reflectance Fourier Transform Infrared (ATR-FTIR) technology has been widely used to track the concentration profiles of a single solute during crystallization processes [3-5]. However, if the parent solution is complex and contains multiple potentially crystallizing solutes (e.g. nuclear waste) multiple solutes must be tracked simultaneously. This can be a challenging task if the characteristic infrared absorbance peaks for the individual solutes are overlapping. In this work a supervised learning approach was taken to construct a correlation between the infrared absorbance spectrum and the concentrations of multiple solutes in the solution. Furthermore, to avoid over-fitting and establish robust correlations, the training set was carefully selected and projection-based regression techniques [3, 6] were applied—of these, Partial Least Squares Regression (PLSR) performed the best.

This methodology was used to monitor crystallizations of complex electrolytic solutions representative of nuclear waste stored at Hanford, WA [7-9]. In particular, four solutes (carbonate, sulfate, nitrate, and nitrite) were tracked during batch cooling crystallizations. By monitoring a number of crystallizations, operating zones for crystallizing particular salts were identified. Furthermore, in future work, multi-solute concentration monitoring will be used to facilitate a direct design approach [10] to control the crystallization of clean (non-radioactive) sodium salts from nuclear waste mixtures.

1. Nagy, Z.K. and R.D. Braatz, *Advances and New Directions in Crystallization Control.* Annual Review of Chemical and Biomolecular Engineering, Vol 3, 2012.

2. Zhou, G.X., et al., *Direct design of pharmaceutical antisolvent crystallization through concentration control.* Crystal Growth & Design, 2006. **6**(4): p. 892-898.

3. Togkalidou, T., et al., *Solute concentration prediction using chemometrics and ATR-FTIR spectroscopy.* Journal of Crystal Growth, 2001. **231**(4): p. 534-543.

4. Lewiner, F., et al., *On-line ATR FTIR measurement of supersaturation during solution crystallization processes. Calibration and applications on three solute/solvent systems.* Chemical Engineering Science, 2001. **56**(6): p. 2069-2084.

5. Dunuwila, D.D. and K.A. Berglund, *ATR FTIR spectroscopy for in situ measurement of supersaturation.* Journal of Crystal Growth, 1997. **179**(1-2).

6. Xiaobo, Z., et al., *Variables selection methods in near-infrared spectroscopy.* Analytica Chimica Acta, 2010. **667**(1–2): p. 14-32.

7. Herting, D.L., *Clean Salt Process - Final Report*, E. Management, Editor. 1996.

8. Nassif, L., et al., *Pretreatment of Hanford medium-curie wastes by fractional crystallization.* Environmental Science & Technology, 2008. **42**(13): p. 4940-4945.

9. Rajbanshi, A., B.A. Moyer, and R. Custelcean, *Sulfate Separation from Aqueous Alkaline Solutions by Selective Crystallization of Alkali Metal Coordination Capsules.* Crystal Growth & Design, 2011. **11**(7): p. 2702-2706.

10. Fujiwara, M., et al., *First-principles and direct design approaches for the control of pharmaceutical crystallization.* Journal of Process Control, 2005. **15**(5): p. 493-504.

(298f) Operation and Optimization of Seeded Batch Crystallization When Crystal Growth Rate Is Size-Dependent

Ting, W. L., National Taiwan University
Wang, H. Y.,
Ward, J. D., National Taiwan University
Originally presented on: 11/5/2013 14:20:00 - 14:42:00

In this work, the operation and optimization of seeded batch crystallization processes is investigated when crystal growth rate is size-dependent using a dimensionless model[1] and model reported in the literature for the crystallization of potassium alum.[2] Many of the results are qualitatively similar to the case when growth rate is size independent: The seed mass and average seed size affect the results in a similar way. An approximate analytical trajectory that works well for size-independent growth also works well when the growth rate is size dependent. The optimal supersaturation trajectory is qualitatively similar to the case when growth rate is size independent. Finally, we show that an approximate, size-independent model can be used to determine the optimal supersaturation trajectory with little error. This is advantageous because optimizing the size-independent model requires much less computational effort.

References

1. Ward JD, Yu CC, Doherty MF. A New Framework and a Simpler Method for the Development of Batch Crystallization Recipes. *Aiche J.* 2011;57(3):606-617.
2. Tai CY, Lu EL, Yu KH. Determination of Secondary Nucleation Kinetics Using a Batch Crystallizer. *J. Chin. I. Ch. E.* 1991;22(4):231-239.

(298g) Effects of Several Heat Shield On the Crystallization Front of Sapphire Crystal Grown By Kyropoulos Method

Song, D. W., LG Siltron
Kim, H., University of Seoul
Ageyeva, S., CDT Equipment LLC
Originally presented on: 11/5/2013 14:42:00 - 15:04:00

In the growth of sapphire crystals, the critical density of dislocations determined by variations of the ratio of G/V affects on the quality of crystal structure. Where G is the axial temperature gradient and V is the velocity of crystallization front (CF). In general, it is not the initial temperature gradient is so important for structure control but rather the degree of its constancy maintaining during growth. However in Kyropoulos (KY) method the CF, has formed the inverse shape of cone, is gradually down to crucible bottom. Therefore it is more difficult to control its consistency during growth in KY method than other methods.

In our experiments, we have studied the influences of heat shield (HS) on the formation of low angle block or grain boundaries (LAGB) and micro-bubbles. We have estimated the angle and velocity of convex of CF in several HS configurations. The shapes of CF are determined by the patterns of micro-bubbles generated by the power down-and-up test (PDU) in quick time. And the defects such as LAGB and micro-bubbles are inspected in the vertical slug which had been cut in center line of grown crystals by polarized microscope. The velocities are measured in PDU-tests by the evaluation data of vertical slugs and calculated by SK Forecast (developed by Alyuda.com) and our simulators.

The defects free shape of CF is depended on the HS configurations with respect to time. It means that there exist an optimal tendency of CF in each HS. The LAGB is related to the acceleration of growth rate and the micro-bubble is related to the oscillation of growth rate that will be discussed in detail. The difference between our simulators and SK Forecast are come from the consideration of energy conservation and initial guessing value of angle and velocity of CF.

(300a) Breakage Functions for Single Impact of Spheres

Hill, P. J., Mississippi State University

Originally presented on: 11/5/2013 12:30:00 - 12:49:00

In modeling particle breakage, it is often assumed that all particles have the same shape factor. While this is often a good approximation for the original parent particles, the shape of the child particles frequently are quite different. Other models that do account for changes in shape focus on how a cuboid crystal changes shape due to the corners being chipped, but excludes the shapes of the particles formed by the chipped off material [1]. Later models are more general and use shapes other than cubes, but still use the attrition approach [2]. However, not all breakage is by attrition. Single impact of spheres causes fragmentation that results in different size and shape distributions as shown experimentally [3]. These experimental results cannot be accounted for by the attrition models.

This work presents recent developments in modeling fragmentation of spheres based on the experimental results. Since the number of particles formed depends on the force of impact, the models allow the user to adjust the number of particles formed. The models are general in that they can be applied to any unit operation where particles are impacted such as crystallizers.

1. Briesen, H., “Two-dimensional population balance modeling for shape dependent crystal attrition,” *Chem. Eng. Sci.*, 64, 661-672 (2009).
2. Reinhold, A., and H. Briesen, “Convex geometry for the Morphological Modeling and Characterization of Crystal Shapes,” *Part. Part. Syst. Charact.*, 28, 37-56 (2011).
3. Salman, A.D., G.K. Reynolds, J.S. Fu, Y.S. Cheong, C.A. Biggs, M.J. Adams, D.A. Gorham, J. Lukenics, and M.J. Hounslow, “Descriptive classification of the impact failure modes of spherical particles,” Powder Tech., 143– 144, 19–30 (2004).

(300b) Ligands Dominate Agglomerate Morphologies of Unpolar Nanoparticles

Born, P.,
Kraus, T., ETH Zurich / IBM Research GmbH
Originally presented on: 11/5/2013 12:49:00 - 13:08:00

Agglomeration is a ubiquitous feature of nanoparticles. Surprisingly little is known on the factors that dominate the morphology of agglomerates formed by technologically relevant, unpolar nanoparticles. In this contribution, we discuss experiments that suggest a dominating effect on agglomerate morphologies. Temperature-dependant agglomeration studies on alkylthiol-stabilized gold nanoparticles show that ligand solubility dominates the kinetics of agglomeration and the structure at the micrometer scale. Precipitation induced by adding poor solvents at different temperatures indicate that the molecular structure of the ligand shell affects kinetics and structure at the nanometer scale.

We used dynamic light scattering and small-angle x-ray scattering (SAXS) to follow the agglomeration of alkylthiol-coated gold nanoparticles with 6 nm core diameter *in situ* and analyzed the data using modified Smoluchowski models. Systematic variations of the ligand chain lengths between 12 to 18 carbon atoms indicated a temperature-dependant solvation of the ligands. We identified critical temperatures at which the agglomeration changed from reaction- to diffusion limited mechanisms [1]. The resulting agglomerates had markedly different geometries as visible in SAXS and transmission electron microscopy (TEM).

The same particles were precipitated by the addition of a more polar solvent, 1-propanol, that is less solvating towards the alkyl chains of the ligands. We found a sharp transition in the nanoscale geometry of the formed agglomerates: at temperatures below the melting point of the dry ligand monolayer, randomly dense packed agglomerates formed. At temperatures above the melting point, the agglomerates were dense and crystalline, with sharp facets of the superlattice [2]. Brownian dynamics simulations were performed to interpret this behavior. We believe that the short-range interactions between the particles are strongly affected by the molecular structure of the ligand layer. Molten ligands facilitate the relative motion or particles inside agglomerates.

In summary, this contribution presents experimental results indicating a crucial role of the ligands in the structure formation of unpolar nanoparticle agglomerates. The core material has less influence on the structure of the superstructures. We therefore believe that rational design of agglomerate morphologies for the use in hybrid materials is possible.

[1] P. Born and T. Kraus: Ligand-dominated temperature dependence of agglomeration kinetics and morphology in alkyl thiol-coated gold nanoparticles. *Physical Review E*, under review (2013)

[2] T. Geyer, P. Born and T. Kraus: Switching between crystallization and amorphous agglomeration of alkyl thiol-coated gold nanoparticles. *Physical Review Letters* 109 (2012) 128302

(300c) Restructuring of Aggregates and Their Primary Particle Size Distribution During Sintering

Eggersdorfer, M., Swiss Federal Institute of Technology (ETH Zurich)

Pratsinis, S. E., Swiss Federal Institute of Technology (ETH Zurich)

Originally presented on: 11/5/2013 13:08:00 - 13:27:00

Particle suspensions and films often exhibit unique properties that depend on particle size distribution, like the opacity of titania pigments, the color of quantum dot or plasmonic nanoparticle suspensions and the superparamagnetism of iron oxide nanoparticles to name a few. Typically a narrow size distribution facilitates harvesting these effects from paints to solar cells, bio sensors and light emitting devices.

Aerosol processes allow rapid and scalable production of nanoparticles over a wide range of sizes. In such processes, however, particles grow typically by Brownian coagulation that places a lower limit to the width of the size distribution, the so-called self-preserving size distribution with a geometric standard deviation of about 1.45. Such a constraint would limit aerosol-made particles in applications requiring narrow size distributions. Furthermore, when primary particles (PP) with different high temperature residence time histories (e.g. from different reactor streamlines or reaction rates) are mixed, fractal-like particles with quite polydisperse PP size distributions (PPSD) are obtained. This polydispersity, however, can be reduced by PP sintering that proceeds inversely proportional to particle size, similar to condensation that is routinely used in generation of monodisperse aerosols.

Sintering can lead to a narrower PPSD than their parent aggregates. The narrowest PPSD was obtained when TiO_2 aggregates were produced by $TiCl_4$ or titanium isopropoxide oxidation followed by coagulation and sintering (Heine and Pratsinis, 2007). This was observed also with flame-made ZnO nanocrystalline aggregates exhibiting a blueshift of their absorption spectrum with decreasing ZnO crystallite size from about 8 to 1.5 nm (Mädler et al., 2002). A narrow crystal size distribution is required for this quantum-size effect.

During sintering of aggregates of polydisperse PP, restructuring takes place, the average PP size increases and the PPSD narrows affecting particle performance in a number of applications (Fig. 1, Eggersdorfer and Pratsinis, 2013). Here, aggregate sintering by viscous flow (Eggersdorfer et al., 2011), lattice and grain boundary diffusion (Eggersdorfer et al., 2012) is simulated by multiparticle discrete element methods focusing on PP growth dynamics and elucidating the restructuring of aggregates during their coalescence. The effect of initial PPSD and sintering mechanisms on the evolution of PP polydispersity and surface area mean diameter are presented. Each sintering mechanism results in a distinct evolution of PPSD but quite similar growth in average PP diameter. Grain boundary diffusion has the strongest impact among all sintering mechanisms and rapidly results in the narrowest PPSD as it has the strongest dependence on PP size. During sintering of aggregates with initially monodisperse PPs, the PPSD goes through a maximum width before narrowing again as PPs coalesce. A power law holds between projected aggregate surface area and number of PPs regardless of sintering mechanism and initial PP polydispersity. This law can be readily used in aerosol reactor design and for characterization of aggregates independent of material composition, initial PP polydispersity and sintering mechanism.

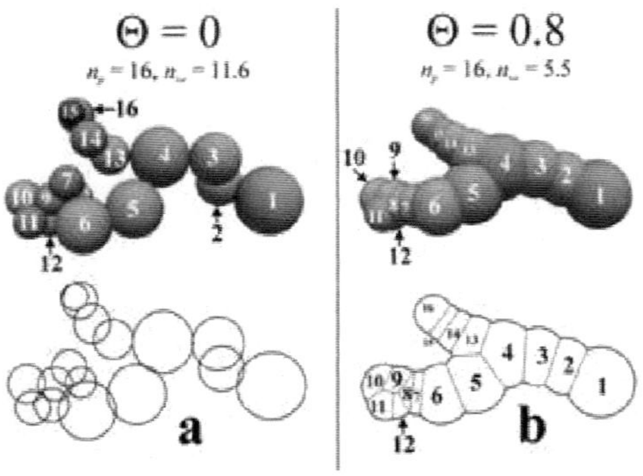

Figure 1. The geometric standard deviation Rg of PP size distribution for a) agglomerates and b) aggregates is calculated with the diameters d of their constituent PPs at dimensionless time . The sinter neck or grain boundaries between neighboring particles are sketched with thin lines. For aggregates, the d represents the maximum extension of the PP or crystal size (truncated sphere). The number of PPs, np, remains constant while the average PP number, nva, decreases during sintering due to the decreasing surface area of the aggregate.

Financial support by ETH Research Grant (ETHIIRA) ETH-11 09-1 and the European Research Council under the European Union's Seventh Framework Programme (FP7/2007-2013)/ERC grant agreement n° 247283 is gratefully acknowledged.

Eggersdorfer, M.L., Kadau, D., Herrmann, H.J. and Pratsinis, S.E. (2011) Langmuir 27, 6358-6367.

Eggersdorfer, M.L., Kadau, D., Herrmann, H.J. and Pratsinis, S.E. (2012) J. Aerosol Sci. 46, 7-19.

Eggersdorfer, M.L., Pratsinis, S.E. (2013) AIChE J. in press.

Heine, M.C. and Pratsinis, S.E. (2007) J. Aerosol Sci. 38, 17-38.

Mädler, L., Stark, W.J. and Pratsinis S.E. (2002) J. App. Phys. 92, 6537-6540.

(300d) Experimental and Theoretical Study of the Emergence of Single Chirality in Attrition-Enhanced Deracemization

Blackmond, D. G., The Scripps Research Institute
Gherase, D.,
Originally presented on: 11/5/2013 13:27:00 - 13:46:00

The single-handedness of biological molecules has fascinated scientists since Pasteur's first painstaking physical separation of the mirror-image crystals of a tartrate salt. The emergence of single chirality requires sustaining and propagating a small imbalance induced by the initial symmetry breaking in a presumably racemic prebiotic environment. Viedma's striking demonstration that a racemic mixture of enantiomorphic crystals of the achiral salt $NaClO_3$ under dynamic phase equilibrium evolves inexorably to crystals of a single solid enantiomorph when mechanically stirred in the presence of glass beads has sparked considerable research into what is now known as "Viedma Ripening" or attrition-enhanced deracemization. Experimental studies help to deconvolute the driving forces for crystal growth during attrition-enhanced deracemization, demonstrating an interplay between crystal size and crystal number in the emergence of homochirality. A modified population balance model is presented based on considerations of the solubility driving force, as outlined by the Gibbs-Thomson rule, and the total interfacial surface area between solid crystals and the solution phase.

(300e) Separation of Enantiomers By Temperature Swing Viedma Ripening

Iggland, M.,
Mazzotti, M., ETH Zurich
Originally presented on: 11/5/2013 13:46:00 - 14:05:00

Chiral resolutions are an important process step in the pharmaceutical industry. The term 'chiral resolutions' is a collective name for a number of methods with the aim of obtaining optically pure substances from mixtures of two enantiomers. Notable examples include chiral chromatography, preferential crystallization and resolution via formation of diastereomers. Most such methods require either auxiliary chiral substances, which are usually expensive, or detailed knowledge of thermodynamics and kinetics as well as exact control of process conditions in order to be operated successfully.

Viedma ripening is a process in which a suspension of a racemic or scalemic conglomerate in contact with a solution containing a racemizing agent is subjected to grinding by, for example, glass beads. Upon stirring for some time, this process results in the conversion of the whole solid phase to one of the enantiomers. In other words, one obtains an enantiomerically pure solid phase at a high yield through a relatively simple experimental protocol, without using auxiliary chiral substances. The first demonstrations of this process were published for sodium chlorate (NaClO), which is an achiral substance in solution but forms optically active crystals [1] and for a small organic molecule which is induced to racemize by the organic base DBU [2].

The mechanisms behind this process are believed to be racemization, growth and dissolution due to the size-dependence of solubility, agglomeration and breakage. The racemization reaction is often induced by a catalyst, such as an acid or a base. In the case of sodium chlorate, there is no racemization reaction since the substance is achiral in solution. The growth and dissolution is caused by the same thermodynamic phenomenon which causes Ostwald ripening: the higher solubility of small crystals compared to larger ones [3]. Agglomeration and breakage are processes which affect the crystals and crystal size distributions, but not the solution phase directly.

We have previously presented a mathematical model for Viedma ripening based on population balance equations incorporating all of the mechanisms mentioned above [4,5]. This model includes two particle size distributions and two population balances describing the change of these due to action of the various mechanisms. The solution phase is modelled using a solute mass balance. The mechanisms of breakage and agglomeration are governed by the choice of the constitutive equations, however, the general behaviour of the simulations is not affected by the exact form of these equations, or by the exact values of the rate parameters used. This lets us conclude that the simulations are generally valid and thus that a size-dependence of solubility, agglomeration and racemization are indeed responsible for the observed behaviour. Breakage serves to speed up the process. Simulations carried out using this model are able to reproduce experimentally observed behaviour. For example, it is possible to direct the evolution of the system towards one of the enantiomers by starting with an enantiomeric excess by mass in the solid phase, or by starting differently sized crystals for the two enantiomers [1,2,6,7]. Our model also successfully reproduces the observation that an increase in the breakage rate leads to faster deracemisation [1,6].

The use of glass beads for grinding is a barrier to the implementation of Viedma ripening industrially. The beads can introduce impurities, which is a problem especially in the pharmaceutical industry. Furthermore, the glass beads need to be separated from the solid product after the process is completed. Recently, it has been shown that temperature swings are another alternative to grinding: cycles of heating and cooling lead to the same exponential increase in the enantiomeric excess as observed with grinding. A change in temperature affects the growth and dissolution of crystals by changing the solubility and critical size, in contrast to grinding which affects the crystal size directly.

The use of temperature swings would overcome the problems caused by the glass beads, but poses other challenges. There are, in principle, infinitely many possible temperature profiles possible, varying in the temperature range which is exploited and in the rate of heating and cooling. Choosing an optimal temperature profile requires the understanding of how temperature changes affect the particle size distributions of the two enantiomers and the solution phase concentration. Other factors that need to be considered are the thermal stability of the substance of interest, which may limit the maximum temperature which can be applied. Furthermore, the optimal rate of heating or cooling is affected by growth and dissolution kinetics, as well as by nucleation kinetics – since nucleation should be avoided, this effectively limits the rate of cooling applicable. The number of heating and cooling cycles, the absolute temperature difference and the rate of heating and cooling all affect the energy input required. Besides minimizing the time required to reach an optically pure solid phase, this energy input should of course be kept as low as possible.

We have extended our model to be able to simulate the effects of temperature changes. We use a combination of experiments and simulations in order to improve our understanding of how Viedma ripening by temperature swings works and how a process can be optimally designed in order to minimize the time required to reach an optically pure solid phase.

[1] Viedma, Phys. Rev. Lett. 94 (2005) 065504
[2] Noorduin et al., J. Am. Chem. Soc. 130 (2008) 1158
[3] Iggland and Mazzotti, Cryst. Growth Des. 12 (2012) 1489
[4] Iggland and Mazzotti, Cryst. Growth Des. 11 (2011) 4611
[5] Iggland and Mazzotti, CrystEngComm 15 (2013) 2319
[6] Cheung et al., Chem. Commun. (2008) 987
[7] Noorduin et al., Org. Process Res. Dev. 14 (2010) 908

(300f) "Collision Theory Re-Interpretation of Kinetic Data for the Growth of Organic Crystal Surfaces; Implications for the Precipitation of Pharmaceuticals in Trans-Critical Environments?"

Rosner, D. E., Yale University
Originally presented on: 11/5/2013 14:05:00 - 14:24:00

We expect that the challenge of interrelating kinetic data on the growth of organic crystal faces across a wide variety of environmental conditions (from liquids, to gas-expanded liquids, to moderate- and low- density gases) will be simplified by uniformly adopting a molecular "collision-theory" viewpoint-----*ie.*, interpreting/reporting single crystal face $\{klm\}$ growth rate data in terms of dimensionless effective overall "*incorporation probabilities*", e, bounded by 1/ Z, where Z (often 1, 2 or 4) is the number of growth molecules per unit cell of the relevant crystal structure. In effect, this is equivalent to the generally accepted notion that, at any given departure from interfacial equilibrium and surface temperature, "only a fraction of the total (surface) sites are available for attachment" and incorporation. Remarkably, however, this valuable perspective is rarely mentioned for CG from *melts*[1], or from *liquid solutions*[2], and is now even uncommon for the physical *vapor* growth of low molecular weight organics[3] ((*eg.*, "surrogates" for active (pharmaceutical) ingredients (APIs)). Motivated, in part, by the sparsity of reliable published growth rate information on actual APIs, especially under "compressed gas-expanded liquid" processing conditions of current pharmaceutical interest[4,5], we therefore illustrate here a tractable method to re-interpret (in terms of abovementioned dimensionless overall "incorporation probabilities") available growth rate data for a selection of organic compounds [with the stoichiometry: CaHbOcNd ranging from *succinic acid* (with a=4) to *rubrene* (a=42), including *acetaminophen* (a=8), and the 28-atom stereo isomer: *ethylene diamine d-tartrate*(a=6)]. For instructive quantitative illustrations we consider only interface-controlled, face-specific experimental rate data obtained on single crystals immersed in dilute otherwise single component carrier fluids at supersaturations much smaller than the threshold for homogeneous nucleation, and for crystals large enough to be outside the domain of interface curvature-shifted equilibrium solute composition. We close with a brief discussion of implications of such overall e(S,T; $\{klm\}$)-estimates for: a) readily identifying the particle size range at which fluid-phase diffusional limitations will inevitably set in, and, more ambitiously: b) estimating the growth rate of structurally similar API-surrogates in, say, expanded liquid (gas-anti-solvent) or dense-carrier vapor precipitation conditions------ of current interest for the production of "micronized" drugs.

Prepared for presentation at *AIChE 2013 Annual Mtg*., San Francisco,CA; Nov.4-8, 2013
May 10, 2013

* *LW Jones Jr.* Professor of Chemical Engineering, Director: *Sol Reaction Engineering Group*

daniel.rosner@yale.edu

References:

1. Rosner, DE, "Collision Theory Re-Interpretation of Kinetic Data for the Growth of Organic Crystal Surfaces; Part I. Melt Growth; *Implications for the Precipitation of Pharmaceuticals in Trans-critical Environments?"; Crystal Growth & Design*(ACS)(to be submitted); March 2013

2. Rosner, DE, "Collision Theory Re-Interpretation of Kinetic Data for the Growth of Organic Crystal Surfaces; Part III. Solution–Growth". *Crystal Growth & Design* (ACS) (to be submitted); May 2013

3.. Rosner, DE, "Collision Theory Re-Interpretation of Kinetic Data for the Growth of Organic Crystal Surfaces; Part II. .Physical Vapor–Growth". *Crystal Growth & Design*(ACS) (to be submitted), June 2013

4. Martin, A. and Cocero, MJ, "Micronization Processes With Supercritical Fluids: Fundamentals and Mechanisms", *Advanced Drug Delivery Reviews* (Elsevier),vol **60**, pp 339-350(2008)

5 Rosner, DE and Arias-Zugasti, M, "Theory of Pharmaceutical Powder 'Micronization' Using Compressed Gas Anti-solvent (Re-) Precipitation"; Paper # WG10S1O1, *European Aerosol Conference*; Granada, Spain , September 6, 2012; *J Supercritical Fluids*(to be submitted) August, 2013

(300g) Evaporation of Emulsion Droplets for the Templated Assembly of Spherical Particles: A Population Balance Model

Chang, E. P., Massachusetts Institute of Technology
Braatz, R. D., Massachusetts Institute of Technology
Hatton, T. A., Massachusetts Institute of Technology
Originally presented on: 11/5/2013 14:24:00 - 14:43:00

The emulsion droplet solvent evaporation method is used in the preparation of spherical particles, which form due to processes such as the clustering of nanocrystals or precipitation of polymers as the volume of solvent in the droplets decreases. A population balance model is presented to describe this transport of solvent from nanocrystal- or polymer-laden droplets in an emulsion that flows through a pervaporation unit or down an inclined surface. The solvent transport and lateral migration of droplets was simulated using a high-resolution finite-volume algorithm, which provided a smooth solution with second-order accuracy. In a pervaporation fiber, concentration gradients in the continuous phase become prominent when the resistance to solvent transport in the continuous phase dominates that in the membrane; in contrast, with the membrane resistance controlling the overall transport rate, a lumped capacitance assumption can be made and a simpler plug flow model would be sufficient. The simulations also indicate that the particle size distributions are generally bimodal, and are broader for low dispersed-phase volume fractions and very low solvent solubilities. Furthermore, for both the pervaporation and inclined surface systems, diffusion of the particles can occur to a significant degree, and the diffusional behavior is different depending on the temperature and flow configuration. Such simulations offer insight into how solvent is removed from emulsion droplets and should be useful in the design of solvent evaporation systems for that purpose.

(300h) Determining The Metastable Limit Of Nucleation From Gauge-CELL Monte Carlo Simulations

Tomassone, M. S., Rutgers University
Clark, M., Rutgers University
Originally presented on: 11/5/2013 14:43:00 - 15:02:00

The phenomenon of primary homogeneous nucleation continues to be mysterious because it is a dynamic evolution of many non-equilibrium steps. Of greatest interest is the metastable limit, the lowest supersaturation at which nucleation is spontaneous, because predicting the metastable limit allows the design and optimization of nucleation experiments. Toward that goal, we elucidate the non-equilibrium thermodynamics of nucleation with a recent simulation technique, Gauge-Cell Monte Carlo, which collects the non-equilibrium chemical potentials of all nucleation steps into an N-μ isotherm. From this isotherm we calculate accurately the solubility, the surface tension, and critically the metastable limit of the model mixture. We construct several isotherms for a large number solutes, with solubilities ranging from 1% to 0.0001%, at temperatures from 0°C to 100°C. We find a correlation for all these isotherms which show convergence on a simple relationship: 1/(metastable limit) ~ L*·log(solubility); where L* is a characteristic length related to the number of nuclei that are generated, representing the average inter-nuclear distance at nucleation. This heuristic equation is sufficiently general, and accurately describes all our simulation results, thus shedding light on the most relevant variables to the nucleation phenomenon and suggesting how to design and optimizating new nucleation studies. We additionally examine the interplay between stable and unstable clustering; describe how the "critical nucleus" depends on environmental conditions, and assess the free energy barriers to nucleation for several isotherms.

(319a) Core-Shell Structured Composite Microparticles With Ability to Store a Chemical Payload and Release a Defined Quantity "On Demand"

Kovacik, P., Prague Institute of Chemical Technology
Stepanek, F., Institute of Chemical Technology, Prague
Singh, M.,
Originally presented on: 11/5/2013 12:30:00 - 12:55:00

Core-shell structured composite microparticles with ability to store a chemical payload and release a defined quantity "on demand"

P. Kovacik, M. Singh, F.Stepanek

ICT Prague, Laboratory of chemical robotics, Technicka 5, 166 28 Prague,
Tel.: 2 2044 3046, e-mail: kovacikp@vscht.cz

The aim of this work was the preparation of biocompatible microparticles with a hollow interrior. The microparticles must be able to encapsulate and store a chemical payload for a certain time, followed by the release on demand of this payload. In our previous work [1] we have prepared silica microparticles with a hollow core and found that the diffusion across the mesoporous silica shell was strongly dependent on temperature. In this work, we used this dependence and attached iron oxide nanoparticles on the surface of the silica shell to create composite iron oxide/silica microparticles. The iron oxide nanoparticles were able to heat up in the presence of an alternating magnetic field. This property allowed us to use magnetic field as a tool for remote control of diffusion across the microparticle shell. To avoid spontaneous leakage of encapsulated payload in time, we have modified the surface of the composite microparticles with a layer of palm oil. Palm oil is a phase change material which is solid under 37 Â°C. We showed that the resultant composite microparticles are able to store a payload for several months and release a defined quantity on demand by the application of a magnetic field. The particles were characterised in shape, size, heating ability and their mass transport properties.

[1] Kovacik P., Kremlackova Z., Stepanek F., ?Investigation of radiofrequency induced release kinetics from magnetic hollow silica microspheres?, Micropor. Mesopor. Mater. 159, 119-125 (2012)

(319b) ZnO Nanorods Hermetically Encapsulated By a Nanothin Amorphous SiO2 Coating: Toxicological Profile and Optical Properties

Sotiriou, G. A., Harvard University
Murdaugh, K.,
Cohen, J., Harvard University, School of Public Health
Watson, C.,
Elder, A.,
Demokritou, P., Harvard School of Public Health
Originally presented on: 11/5/2013 12:55:00 - 13:20:00

Zinc oxide (ZnO) is a wide-bandgap metal oxide semiconductor that can be excited at room temperature by UV irradiation, and finds applications in a variety of fields and products including paints, pigments, batteries, photocatalysis, foods, to name a few. When in nanometer size range, it may be used in polymer nanocomposites and sunscreens as an efficient UV-filter with high transparency in the visible wavelength range. However, the photocatalytic activity of ZnO causes the degradation of the surrounding polymer rendering it unsuitable for long-term employment. Furthermore, ZnO nanoparticles are highly toxic and may pose risks to the public and the environment. Here, ZnO nanorods are made by scalable flame synthesis [1,2] and are in-situ encapsulated by an amorphous nanothin SiO_2 layer [3,4]. The as-prepared nanoparticles are characterized by electron microscopy (EM), X-ray diffraction (XRD), X-ray photoelectron spectroscopy (XPS) and N_2 adsorption. The hermetic nature of the SiO_2 coating is evaluated by a detailed dissolution study and the photocatalytic activity is monitored by the decomposition of methylene blue dye. The presence of SiO_2 facilitates the dispersion of these nanorods in relevant solutions. The core ZnO nanorods exhibit the characteristic optical properties as determined by UV/vis and luminescent spectrometry, while the SiO_2 coating eliminates the photocatalytic activity and minimizes the toxicity and genotoxicity to human cells, as illustrated in cellular studies using multiple cell lines. Therefore, these flame-made ZnO nanorods may be safely used as fillers in polymer UV-filter nanocomposites and in sunscreens exhibiting superior performance while mitigating their impact on environmental health.

References

[1] Sotiriou, G. A., Sannomiya, T., Teleki, A., Krumeich, F., Vörös, J. and Pratsinis, S. E. Non-toxic dry-coated nanosilver for plasmonic biosensors. *Adv. Funct. Mater.* **20**, 4250-4257, (2010).

[2] Sotiriou, G. A., Hirt, A. M., Lozach, P. Y., Teleki, A., Krumeich, F. and Pratsinis, S. E. Hybrid, silica-coated, Janus-like plasmonic-magnetic nanoparticles. *Chem. Mater.* **23**, 1985-1992, (2011).

[3] Sotiriou, G. A., Franco, D., Poulikakos, D. and Ferrari, A. Optically Stable Biocompatible Flame-Made SiO_2-Coated Y_2O_3:Tb^{3+} Nanophosphors for Cell Imaging. *ACS Nano* **6**, 3888-3897, (2012).

[4] Gass, S., Cohen, J. M., Pyrgiotakis, G., Sotiriou, G. A., Pratsinis, S. E. and Demokritou, P. Safer Formulation Concept for Flame-Generated Engineered Nanomaterials. *ACS Sustainable Chem. Eng.* **in press**, DOI: 10.1021/sc300152f, (2013).

(319c) Single-Pot Synthesis of Uniform Glucan Multilayers On Oxide Particles

Jankolovits, J.,
Gazit, O., University of California Berkeley
Nigra, M., University of California, Berkeley
Katz, A., University of California, Berkeley
Originally presented on: 11/5/2013 13:20:00 - 13:45:00

Title: Single-Pot Synthesis of Uniform Glucan Multilayers on Oxide Particles

Authors: Joseph Jankolovits, Oz M. Gazit, Michael M. Nigra, Alexander Katz

Department of Chemical and Biomolecular Engineering

University of California, Berkeley

Polymer coatings are used widely to tune the surface chemistry of solids in biomaterials, paints, and other materials. Covalently grafted polymer coatings provide advantageous stability, though their prevailing syntheses involve multiple step procedures, low solid yields and surface coverage, and/or non-uniformity on the nanoscale. A novel method for uniformly grafting polymer multilayers on inorganic oxide particles in a single-pot process will be demonstrated using a wide variety of characterization techniques. Implications of the grafted polymer multilayer coatings on inorganic oxide particles for the stabilization of paint pigment dispersions will be discussed. Efforts to prepare grafted polymer multilayers on nanoparticles will be described. The promise of these grafted polymer multilayer coatings on inorganic oxide particles in biomaterials and nanomedicine will be highlighted.

(319d) Continuous Polymer Coating of Nanoparticles: A Novel Method

Sirkar, K. K., New Jersey Institute of Technology
Chen, D., New Jersey Institute of Technology
Singh, D., New Jersey Institute of Technology
Pfeffer, R., Arizona State University
Originally presented on: 11/5/2013 13:45:00 - 14:10:00

Nanoparticle-based drug delivery systems are of significant interest in controlled release of drugs, delivery of anticancer drugs and imaging agents to tumors, tuberculosis treatment and as non-viral gene delivery vehicles. To develop controlled release of the drug as well as to protect the drug from demanding environments, appropriate coatings are employed on the drug nanoparticles. Conventional methods for coating or encapsulating micron–sized and nanoparticles utilize dry or wet approaches. Dry methods include physical vapor deposition, plasma treatment, chemical vapor deposition and pyrolysis of polymeric organic materials. Wet methods cover sol-gel processes, emulsification and solvent evaporation techniques. Supercritical fluid (SCF) processes such as Rapid Expansion of Supercritical Solutions (RESS), Supercritical Anti-Solvent (SAS), and Gas Anti-Solvent (GAS) processes employing supercritical CO_2 are alternative methods for nanoparticle coating or encapsulation of ultrafine particles. These SCF-based processes have many shortcomings. Some processes require demanding operating conditions, suffer from low polymer solubility etc. Most of these techniques are also batch processes. Fluidized bed-based processes, which can be continuous, face difficulties with nanoparticles caused by van der Waals and other interparticle forces. In all such processes, scale-up is also quite demanding. We will illustrate a technique for continuous polymer coating of nanoparticles that is easily scalable. The nanoparticles used were of silica. Polymers used for coating were Eudragit and PLGA.

(319e) Coating FINE Particles With ULTRA Thin FILMS Using Atomic Layer Deposition

Meesters, G., DSM
Valdesueiro, D., Delft University of Technology
van Ommen, J. R., Delft University of Technology
Originally presented on: 11/5/2013 14:10:00 - 14:35:00

Atomic Layer Deposition (ALD) is a gas-phase coating technique that can provide fine particles with ultrathin films, typically 1-50 nm. Due to the large value of the specific surface area, fine particles –diameter ranging from 1 nm to 50 µm– are used in a wide range of industrial fields such as electronics, pharmaceutics, energy storage and catalysis. The aim of the coating is to improve the performance of these materials, either by the activation of the surface of the particles, or protecting them against the environment.

ALD can deposit uniform inorganic layers [1] by a set of two self-limiting reactions (Fig. 1). The precursors, which are fed into the reactor in subsequent steps, react on the surface of the particles to form a very thin layer of coating material. The final thickness of the coating is controlled by repeating the cycle a desired number of times. This layer-by-layer growth mechanism can also be used to deposit polymeric compounds (Molecular Layer Deposition) [2] or inorganic-organic hybrid compounds [3]. The coating process is carried out in a fluidized bed reactor, which offers high mass and energy transfer, and the possibility to process large amounts of solids.

The production of core-shell particles via ALD has been already applied in fields such as the Li-ion batteries [1] (Fig. 2) or photo-catalysis [4]. Our current work aims at the production of non-permanent coatings to be applied in the controlled release of compounds for food and pharmaceutical applications.

Fig. 1. Schematic representation of the two ALD half-reactions.

Fig. 2. Li-ion cathode material (LiMn2O4) coated with 5 cycles of Al2O3 using ALD [1]

1. Van Ommen, J.R., et al., *Scalable gas-phase processes to create nanostructured particles.* Particuology, 2010. **8**(6): p. 572-577.

2. Adamczyk, N.M., A.A. Dameron, and S.M. George, *Molecular layer deposition of poly(p-phenylene terephthalamide) films using terephthaloyl chloride and p-phenylenediamine.* Langmuir, 2008. **24**(5): p. 2081-2089.

3. Yoon, B., et al., *Molecular layer deposition of hybrid organic - Inorganic alucone polymer films using a three-step abc reaction sequence.* Chemistry of Materials, 2009. **21**(22): p. 5365-5374.

4. Zhou, Y., et al., *Optimal preparation of Pt/TiO2 photocatalysts using atomic layer deposition.* Applied Catalysis B: Environmental, 2010. **101**(1-2): p. 54-60.

(319f) Surface Engineered Quantum Dots for Light Selective Polymer Films

Mumin, M. A., University of Western Ontario
Allan, J. M., Western University
Charpentier, P. A., Western University

Originally presented on: 11/5/2013 14:35:00 - 15:00:00

We have investigated the optimization of the optical properties of polymer films integrating engineered quantum dot (QD) nanocrystals to manipulate and control light transmission for applications in light selective polymer films. Highly luminescent CdS QDs and CdS/ZnS core/shell QDs with 5 nm size were synthesized using a facile approach based on the pyrolysis of single molecule precursors. After capping the CdS QDs with a thin layer of ZnS to reduce toxicity, the photoluminescence and photostability of the core-shell QDs was significantly enhanced. The ZnS shell was found to passivate the surface defects of the CdS dots while confining the electron-hole pair to the CdS core region.

To make both the bare and core/shell structure QDs more resistant against photochemical reactions, a mesoporous silica layer was grown on the QDs through a reverse microemulsion technique based on hydrophobic interaction and surface silanization. Mesoporous silica can protect completely both type of QDs from photobleaching and also improves the quantum yield by providing much stronger resistance to oxidation and Oswald ripening of QDs. The encapsulated QDs were then functionalized using silane coupling agent and integrated to poly ethylene vinyl acetate (pEVA) network both by melt mixing (using twin screw extruder) and grafting from polymerization in supercritical CO_2. Morphology, UV absorption, visible light transmission, and IR retention were observed and compared to that obtained from commercial pEVA greenhouse plastics. Mesoporous silica encapsulated QDs into EVA polymer films were found to provide simultaneous UV protection, infrared retention, and light selectivity of solar-polymer films.

(331a) Derek Geldat and the A/B Transition

Sundaresan, S., Princeton University

Originally presented on: 11/5/2013 12:30:00 - 12:49:00

Geldart's 1972 paper [1] on classification of particles in terms of their fluidization characteristics is easily at or near the top of the most influential publications on the subject of fluidization. According to his classification, group A particles manifest a small window of stable bed expansion upon fluidization before bubbling sets in. In contrast, Group B particles bubble immediately upon fluidization. The boundary separating these two groups of particles, commonly referred to as the A/B transition, has been the subject of a great deal of research and controversy. Several authors have attempted to explain the A/B transition as a manifestation of a stability boundary, while others have been critical of such studies; (for example see the review article by this author [2]). Experiments with particles of different sizes [3] suggest that the window of stable bed expansion becomes smaller and smaller as the particle size increases, which provides a simpler interpretation.

In this talk, the history of this debate will be reviewed. Questions remaining about this A/B transition will be described.

References

[1] D. Geldart, Powder Technol., 7 (1973) 285.

[2] S. Sundaresan, Annu. Rev. Fluid Mech., 35 (2003) 63.

[3] P. N. Loezos, P. Costamagna and S. Sundaresan, *Chem. Eng. Sci.* , 57 (2002) 5123.

(331c) Particles, Bubbles and Electrostatics in Gas-Solids Fluidized Beds

Bi, X. T., University of British Columbia

Originally presented on: 11/5/2013 12:49:00 - 13:08:00

This presentation will discuss recent advances in several important areas of fluidization which Derek Geldart pioneered or contributed. First, the transition velocity Uc for group A powders will be discussed based on a simple mechanics pressure fluctuation model by assuming that the bubble reaches a maximum stable size as predicted by the Geldart equation, the comparison between the predicted and measured Uc values will be presented. Secondly, the fluidization behaviour of sawdust particles as observed in a vibrating fluidized bed will be discussed based on the measured angle of repose and Geldart powder classifications. Thirdly, the effect of particle charging and charge distribution on bubble and particle movements, a phenomenon first explored by Geldart in early 1970s, will be discussed based on our recent simulation and measurement using both collision and induction electrostatic probes. The reconstructed charge distribution around gas bubbles will be compared with early observations of Boland and Geldart (1971/1972).

(331d) Coal Direct Chemical Looping Process: Geldart Group D Particle Applications

Fan, L. S., The Ohio State University

Originally presented on: 11/5/2013 13:08:00 - 13:27:00

Chemical looping processes have developed into a promising technology for the efficient conversion of carbonaceous fuels to electricity and/or high value energy carriers with minimal carbon emissions. These processes utilize an oxygen carrier solid material to indirectly supply oxygen to the fuel source eliminating the need for costly and energy intensive gas-gas separation techniques. As one of the most advanced chemical looping units developed to date, the Ohio State University (OSU) 25 kW_{th} sub-pilot coal direct chemical looping (CDCL) process for solid fuel conversion has been successfully demonstrated for over 570 hours of operation achieving >99% fuel conversion and nearly 100% CO_2 capture efficiency. The OSU CDCL process consists of a packed moving-bed reducer reactor with a counter-current gas-solid contacting pattern and a bubbling/turbulent fluidized-bed combustor reactor. Geldart Group D metal oxide particles are used in this demonstration unit as the oxygen carrier material. The hydrodynamic characteristics of an interconnected moving-bed and fluidized-bed reactor process significantly deviate from conventional circulating fluidized bed (CFB) systems.

In this study, the hydrodynamic gas-solid flow characteristics and reactant conversion of the 25 kW_{th} CDCL process are examined. Particularly, the counter-current gas-solid flow and reactions in the moving bed reducer are investigated in terms of its advantages for full fuel conversion and high purity of CO_2 concentration. Further, the interrelationship between the stoichiometric oxygen demand, reaction residence time, and superficial gas velocity requirements that ensure oxygen carrier regeneration in the combustor are analyzed. The application of non-mechanical valves for flow control of oxygen carriers is also studied.

(331e) Powders With Class: Beyond Fluidization

Davies, C. E., Massey University

Originally presented on: 11/5/2013 13:27:00 - 13:46:00

Fluidization in the context of industrial practice involves the deliberate and controlled aeration of a body of particulate solids, and the Geldart classification provides general insights into the behaviour of these fluidized solids, on the basis of their density and size. There are many other situations, both in industry and in nature, where bulk particle-fluid contact and interaction take place, or are caused to take place, and not surprisingly, the Geldart Classification offers a potentially useful frame of reference for the behaviour of such systems.

Included in the presentation will be a look at pneumatic conveying where the Geldart classification first found application outside fluidization and still endures as a useful indicator of suitability for dense phase or other conveying modes, and consideration of the behaviour of particulate materials in rotating drums, where powders from different groups display distinctly different avalanching behaviour. Also, I will give a brief account of an industrial case study where the work of Derek Geldart facilitated the quantification and explanation of a problem involving the handling and transport of a material with a wide size distribution, and finally the talk will visit the natural world.

(331f) Periodically Pulsed Fluidized Beds â€“ Taming a Complex System

Coppens, M. O., Rensselaer Polytechnic Institute

Originally presented on: 11/5/2013 13:46:00 - 14:05:00

By periodically pulsing the gas flow introduced to bubbling fluidized beds, remarkably regular bubble patterns may form within a certain range of frequencies (e.g., a few Hz for Geldart B particles) and amplitudes. This is observed in both quasi-2D and 3D fluidized beds. Viewed from the top, some of the 3D patterns are reminiscent of those formed in vibrated granular systems, and the regular ripples on dunes and beaches. However, they are conveyed to greater heights, which is especially clear in quasi-2D beds.

Fundamentally, such patterns could be used as a way to characterize a gas-solid system – and we suggest them as a method to test CFD codes. From a practical standpoint, pulsing has several advantages, such as the regular structuring of the bed (within a range of frequencies and amplitudes), which facilitates scale-up, as well as (more generally) avoiding channelling and making the hydrodynamics more uniform. This is particularly relevant when drying cohesive particles and for other processes for which local overheating would be a safety concern or detrimental to product quality.

References

M.-O. Coppens, M.A. Regelink and C.M. van den Bleek, *World Conference in Particle Technology* (2002).

M.-O. Coppens and J.R. van Ommen, *Chem. Eng. J.* **96**, 117-124 (2003).

A. Akhavan, J.R. van Ommen, J. Nijenhuis, X.S. Wang, M.-O. Coppens and M. Rhodes, *Ind. Eng. Chem. Res.* **48**, 302-309 (2009).

B. Hadi, J.R. van Ommen and M.-O. Coppens, *Ind. Eng. Chem. Res.* **51**, 1713-1721 (2012).

(331g) Fluidization and Subsequent Processing of Geldart Group C Powders

Dave, R. N., New Jersey Institute of Technology

Originally presented on: 11/5/2013 14:05:00 - 14:24:00

As shown in Prof. Derek Geldart's seminal work, fine powders are cohesive and they cannot be easily fluidized and fall under what was empirically classified as Group C powders. It has been also shown that the particle size, solid and fluidizing medium densities, and particle adhesion forces govern the transition between Group C and A. From fundamental considerations, if the powder material and fluidizing medium are fixed and cannot be altered, other than changing the particle size, there are only two other ways to shift the transition between Group C and A. These two factors are the gravity force or the body weight and the cohesion forces. Interestingly, these two factors define what is known as the granular Bond number, which is a ratio of particle cohesion force and body force, which is usually its weight. In this work, we consider both these factors and show how either by changing the fluidization mode from conventional gravity driven to a centrifugal mode, or by changing particle cohesion through manipulating the surface roughness or the size of asperities and/or changing surface energy. For the first case, we show how use of rotating fluidized bed allows fluidization and subsequent processing of fine powders. For the second case, considered in more detail due to its practicality, dry particle coating is used to deposit a very small amount of nano-sized particles (as low as 0.01 % by weight) with a high degree of precision onto the surface of cohesive, Geldart group C powders to make them fluidize like group A powders. A model taking into account the effect of the size of the guest and host particles as well as Surface Area Coverage (SAC) of the coated nano-sized particles is developed to predict the effect of coating on the adhesion reduction of cohesive powders. A series of experiments are performed to investigate the improvement in the fluidizability of dry particle coated group C powders (e.g., cornstarch and aluminum), and the effect of various parameters such as, surface area coverage (SAC), guest particle size and host particle size are systematically investigated. The results clearly show the effect of each of these parameters on the fluidization behavior of cohesive powders, and also validate the model. The study also indicates that a critical SAC is required to make the coated cornstarch fluidize, which is about 5 %; the smaller the guest size, the better its effect on improved fluidizability, although the improvement is reduced if the guest size is smaller than about 10 nm; and if the conditions regarding the SAC and guest size are satisfied, dry particle coating will significantly improve the fluidization of cohesive particles even as small as 5-10 microns.

(331h) CFD Simulation of CO2 Sorption Using Solid Sorbents and Fluidized Bed Process

Arastoopour, H., Illinois Institute of Technology
Abbasi, E., Illinois Institute of Technology
Gadirian, E.,
Originally presented on: 11/5/2013 14:24:00 - 14:43:00

Climate Change, which has been associated with the increasing concentration of greenhouse gases, mainly carbon dioxide (CO_2), is regarded as one of the key environmental issues in the 21st century.

Cost effective capturing and storage of CO_2 before it is emitted to the atmosphere will result in use of coal as one of the major fuel source for electricity generation without affecting economic growth while still reducing carbon emissions to the atmosphere. One of the most promising approaches is use of solid sorbents and fluidized bed process, which currently is not ready for implementation in coal-based power plants. One of the major challenges is the fact that it is still in the lab or bench scale stages. To successfully scale-up such processes, a powerful tool, i.e. Computational fluid dynamics (CFD), is needed to shorten the gap between lab/bench scale and large scales needed for demonstration. Analysis using CFD complements and reduces physical testing, and it can result in a significant time and cost savings. Therefore, multiphase Computational Fluid Dynamics (CFD) provides an attractive option to study and design of the gas-solid processes required for continuous CO_2 removal using solid sorbents in a systematic and economical way. In this presentation Simulation of bubbling and circulation fluidized beds for CO_2 removal will be presented Using kinetic theory based constitutive equations and modified shrinking core reaction model.

(331b) Geldart Classification and Limitations of the Kinetic Theory of Granular Flow in Solid-Gas Modelling

Makkawi, Y., Aston University
Ocone, R., Heriot-Watt University
Originally presented on: 11/5/2013 14:43:00 - 15:02:00

Numerical simulation of the hydrodynamic features of a bubbling bed over a broad range of operating conditions is presented and validated against experimental measurements. The two-fluid model (e.g. Ocone et al. AIChE J., **39**, 1261-1271, 1993) is solved using MFIX and employing the constitutive relations derived by the kinetic theory of granular flow (KTGF). To validate the theory experiments were conducted in a cold conventional gas-solid bubbling column of 13.8 cm internal diameter. Different particle groups, in the size range of 125 – 800 μm (belonging to Geldart group A/B, B and D), were fluidised over a broad range of gas velocity. Electrical Capacitance Tomography (ECT) was used to record the cross-sectional particle-gas distribution over 70 sec operations. The measurements were then used to estimate the bubble characteristics.

It is shown that the model predictions significantly improve as the particle size increases. The accuracy of the quantified bubble characteristics (e.g. velocity, frequency) is very sensitive to the particle size; improvement is noticed at increased gas velocity. A systematic analysis is then undertaken by studying the model response to changing the gas drag law, the particle restitution coefficient and the numerical grid size. Additionally, the particulate phase stress tensor is investigated by modifying the inter-particle forces by including cohesion and liquid-bridge forces in slightly wet particulate flow. The conclusions from the work are in support of the recent growing concern about the validity of the classic kinetic constitutive relations in capturing the mechanism of particle-particle interactions in bubbling fluidised beds.

(159a) Modeling the Effect of Operational Parameters On High Shear Wet Granulation

Mota-Aguilar, D. A., University of Puerto Rico - Mayaguez Campus
Velazquez, C., University of Puerto Rico - Mayaguez Campus
Originally presented on: 11/5/2013 15:15:00 - 15:33:00

High shear wet granulation is mainly aimed to improve the flowability of materials in solid dosage manufacturing processes. It also decreases segregation, reduces dust and modifies the appearance and properties of final product. However, it is a sensitive method that has not been fully understood since the particle growth dynamics are the result of different mechanisms that may take place simultaneously inside a granulator (e.g. wetting, coalescence, attrition). Therefore, it is important to implement the adequate modeling approach that could be used to describe the phenomena and propose process control algorithms.

Recent experimental findings have shown the effect of operational and design parameters on the high shear granulation performance. This work is focused on the study of the impeller velocity and design, which plays a fundamental role as it determines the way in which the energy is transmitted to the material and the collisions between particles. The objective was to propose and implement a modeling method that takes into account this design parameter.

Validation experiments were performed in a laboratory-scale high shear granulator using a lactose-based formulation as model system and a diluted Povidone solution as binder. A factorial design was implemented using three-bladed impellers at different blade angles (42°, 90° and 127°) and agitation velocities (450 rpm, 620 rpm and 807 rpm). Early-stage growth kinetics was determined performing granulations at 6 different massing times from 15 s to 150 s after wetting stage.

Regime maps were proposed and a mathematical model was developed based on a population balance approach to describe and quantify the influence of the blade angle and its interaction with the impeller velocity.

(159b) Comparison of Liquid Bridge Volume Models in DEM Simulations

Easo, L. A.,
Wassgren, C. R., Purdue University
Originally presented on: 11/5/2013 15:33:00 - 15:51:00

Liquid bridge forces play an important role in many processes that handle particulates, such as wet granulation and spray coating. Although there is little disagreement on models for predicting the normal force and the maximum extensional length of liquid bridges, there has been considerably less work investigating how much liquid participates in a liquid bridge. In many DEM simulations that include liquid bridges, it is assumed that liquid covers the surface of the particle uniformly and that all of the liquid on the particle instantly participates in a newly formed liquid bridge. If subsequent liquid bridges form, the liquid on the particle is frequently assumed to instantly divide so that equal volumes of liquid participate in all of the liquid bridges. Another recent model also assumes that liquid is uniformly distributed on particles, but that the volume participating in a liquid bridge is a function only of the geometry of the contact, with the liquid bridge volume from the partnering particle consisting of the liquid volume contained within the angle subtended by the particle center and the tangent surface of the partnering particle. Still other models have proposed that liquid does not uniformly coat a particle's surface, but instead is distributed in drops. This talk will describe the implementation of these various liquid bridge volume models in separate DEM simulations and then compare the distributions of liquid volume as functions of time for the various models.

(159c) Transport Rates of Adhering Liquid Films in Sheared Particle Beds

Mohan, B., Graz University of Technology
Khinast, J. G., Graz University of Technology
Radl, S., Graz University of Technology
Originally presented on: 11/5/2013 15:51:00 - 16:09:00

Wet granular materials show complex flow features and are key for numerous industrial processes such as mixing, drying, granulation, or hopper flows. The development of simple models to describe these flows as a continuum is still a challenging task.

Here we focus on a detailed simulation of the formation and rupture of liquid bridges in a sheared granular flow. Our ultimate goal is the establishment of a mesoscopic, continuum-based model for the effective liquid transfer rate. We use the Discrete Element Method (DEM), supplemented with a variety of microscopic liquid transfer models that model liquid transfer in a single particle-particle collision.

We present results for the transport of liquid through the particle bed by analyzing in detail the mode of liquid transfer (i.e., the conductive and convective flux of liquid). Finally we present correlations of the effective liquid flux with key flow parameters (e.g., the conductive stress and the collision frequency) to guide our development of a mesoscopic liquid transport model

(159d) Experimental Investigation of Binder Distribution During Reactive Granulation

Schongut, M., Institute of Chemical Technology, Prague
Stepanek, F., Institute of Chemical Technology, Prague
Originally presented on: 11/5/2013 16:09:00 - 16:27:00

Binders used in wet granulation can be grouped according to the mechanism by which they solidify as either solution binders (e.g. aqueous solutions of polymers such as PVP or HPC, or water itself), melt binders, and reactive binders. While the kinetics of binder setting in the former two cases can to some extent be controlled by the process parameters (drying rate or cooling rate), in the reactive case the rate of phase change is more or less spontaneous and driven by the kinetics of local reaction-diffusion processes near the binder-particle interface. Investigation of granulation-related phenomena such as binder spreading, particle collisions, nucleation and granule growth are thus further complicated by the simultaneously occurring chemical reaction and the change of physico-chemical properties of the binder that are associated with it. A two-way coupling exists between granulation and reaction kinetics: the evolving structure of the granules being formed influences the solid-liquid contact area to which the overall reaction rate is proportional, and the reaction influences the viscosity and other properties of the binder, which control its spreading on primary particles and ability to dissipate the kinetic energy of inter-particle impacts.

The system of dodecyl-benzenesulfonic acid as a binder and sodium carbonate powder as primary particles is considered in this work [1,2], motivated by the so-called dry neturalisation process used in the manufacture of powder detergents. The measurement of dry neutralization kinetics in a laboratory-scale mechanically agitated granulator is described. The volume of evolved carbon dioxide, which is a by–product of the neutralisation reaction, has been used for following the reaction kinetics, and the effect several parameters on the progress of the reaction has been investigated. The parameters include process temperature, agitation rate, size distribution (thus specific surface) of the primary particles, and the mode of binder addition (step-wise addition). The overall conversion of the reaction and the properties of the formed granules (size distribution, internal structure and porosity) have been evaluated in each case. A mathematical model of the process has been formulated and used for testing hypotheses about the rate-limiting steps in each stage of the process, which can be rate of binder spreading, the availability of fresh solid surfaces, the diffusion rate of reaction components near the interface, or the intrinsic reaction kinetics. Knowlege of the rate-limiting steps then serves for making rational process control decisions.

(159e) Physical Gel Versus Pozzolanic Reactions for Agglomerating of Fine Particulates

Mcdonald, J.,
Kawatra, S. K., Michigan Technological University
Originally presented on: 11/5/2013 16:27:00 - 16:45:00

Inorganic agglomerates of concentrated iron ores are typically made using a sodium bentonite as a binding agent. This paper examines the effects of adding corn starch and fly ash to bentonite for making concentrated iron ore agglomerates. The use of a mixture of bentonite and fly ash revealed a distinctly different binding mechanism than the use of a mixture of corn starch and bentonite. Fly ash bentonite mixtures showed pozzolanic activity as the principle binding mechanism as the concentration of fly ash in the mixture approached 100%. At lower concentrations of fly ash in the mixture, the competing mechanisms of bentonite and fly ash diminished the strength of the agglomerates. Corn starch – bentonite mixtures showed physical gelling and expulsion of moisture as the principle binding mechanism as the concentration of corn starch in the mixture approached 100%. Examination of these binding mechanisms and their roles in agglomeration will be discussed.

(159f) Spherical Agglomeration of API: Screening Protocol, Process Design and Scale-Up

McKeown, R. R.,
Ward, D., GlaxoSmithKline

Originally presented on: 11/5/2013 16:45:00 - 17:03:00

Spherical agglomeration is a technique to grow agglomerates from already formed particles using a L-L-S system where one liquid phase preferentially wets the surface of the solid phase.
Active pharmaceutial ingredient (API) is suspended in the continuous phase, an immiscible (or partially miscible) bridging liquid is added that preferentially wets the surface of the API particles. The process is mechanistically similar to wet granulation, including the processes of particle wetting, agglomerate growth, consolidation, and agglomerate attrition. The process can utilize existing API particles that are resuspended or API particles can consectively be produced in-situ and agglomerated (typically with a ternary solvent system).

Spherical agglomeration can be used to produce granular API with improved bulk density and flow properties as well as improved filterability. A generic screening protocol is presented to allow for rapid screening of spherical agglomeration conditions for water insoluble compounds. For Compound A, process design at 100 mL scale will be discussed. The scale up of spherical agglomeration for Compound A to 1 L and 50 L scale will then be presented and discussed.

(159g) Characterizing the Porosity-Performance Relationship of Matrix Tablets Containing Intra- and/Or Extra-Granular HPMC Materials With Different Morphological Attributes

Hewlett, K., The Dow Chemical Company
Balwinski, K., The Dow Chemical Company
Hoelzer, B., The Dow Chemical Company
Theuerkauf, J., The Dow Chemical Co.
Rogers, T., The Dow Chemical Company
Originally presented on: 11/5/2013 17:03:00 - 17:21:00

The purpose of this work is to use mercury porosimetry as a tool to characterize pore sizes and distribution within tablets. We examine how tablet porosity corresponds with other tablet physical and performance properties and changes with HMPC morphology, HPMC location (intra- and/or extra-granular), and compression force.

Two model APIs were chosen: gliclazide, a low-dose, high-potency very slightly soluble drug, and metformin HCl, a high-dose, low-potency highly soluble drug. Formulations were roller compacted using 2 tons of force, milled, and subsequently compressed at forces ranging from 3,000 to 8,500 lbf depending on the API. Either a granular or fibrous morphology of HPMC was included in the formulation intra- or extra-granularly, and a combination of intra- and extra-granular HPMC was also investigated. Roller compacted ribbons were evaluated visually. The milled granules were analyzed for density, flow, and particle size distribution. Tablets were characterized primarily for porosity, physical properties, and modified-release performance. Tablet pore size and distribution was measured using mercury intrusion porosimetry. Intrusion pressures ranged from 0.5 to 60,000 psia, allowing for detection of pore sizes from 0.003-400 μm in diameter.

Results showed that friability was acceptable in all cases, and was always below 0.60% regardless of compression force, HPMC morphology, or HPMC location. Porosity ranged from 9 to 15% and average pore diameter ranged from 0.27 to 0.75 μm. A higher tablet compression force resulted in reduced porosity. Lower tablet porosities and smaller average pore sizes were obtained with fibrous HPMC. Lower porosities also corresponded with higher tablet hardness. Drug release ranged from 29 to 60% at 12 h and 48 to 85% at 24 h, depending somewhat upon HPMC morphology and compression force, but more significantly upon the location of HPMC. More details will be presented on the relationships observed between HPMC morphology and location, formulation process variables, and tablet physical and performance properties.

These findings assist the formulator in understanding the relationships between tableting processing variables, raw material characteristics, and the resulting tablet physical and performance attributes.

(159h) Granulation of Lime Pretreated Switchgrass Powders

Mani, S., University of Georgia
Yandapalli, V., University of Georgia
Originally presented on: 11/5/2013 17:21:00 - 17:39:00

Densification of switchgrass will reduce handling, transport and storage issues in the biorefinery due to low bulk density and poor flowability by improving the physical and flow characteristics. Conventionally, switchgrass is densified into pellets and briquettes, but the densification process requires high pressure, energy and cost for compaction. Hence, a novel low-pressure wet granulation method of densification was proposed to evaluate the quality of densified products. Wet granulation is a process of agglomerating fine particles into granules by addition of liquid binders and application of shear/vibrating forces. In this study, switchgrass was ground into fine powders and pretreated with calcium oxide at three loading rates (0.05, 0.1, 0.2 g/g of biomass) at 121°C for 30 min and at room temperature (23-25°C) for 72 hours. Pretreated powders were granulated using cornstarch and sodium alginate combination binder in a pan granulator with standard operating procedures. Structural changes occurred during pretreatment of switchgrass were analyzed by scanning electron microscopy and autofluorescence microscopy. The physical, flow and chemical properties of pretreated switchgrass granules were determined and compared with untreated switchgrass granules. Switchgrass granules made from pretreatment of calcium oxide (0.2 g/g of switchgrass powder) at both autoclave and room temperature conditions had higher single granule density and lower binder to powder percentage compared to untreated switchgrass granules. Granulation of biomass can be a viable alternative to improve biomass quality compared to energy intensive pelleting technology.

(362a) Motion of a Large Solid in Bubbling Fluidized Bed

Tsuji, T., Osaka University
Okuyama, Y., Osaka University
Higashida, K., Osaka University
Tanaka, T., Osaka University
Originally presented on: 11/5/2013 15:15:00 - 15:37:00

In the practical applications of fluidized bed such as gasification, combustion, separation and granulation, the size of solid materials is not uniform and large solids are coexisting with small emulsion solids. The large size-difference induces characteristic motion of solids in the beds and results in segregation phenomena. It is difficult to reproduce the behavior of these systems by the existing numerical models and a new numerical model is proposed in the present study. The model is based on the DEM-CFD mesoscopic model and extended by introducing an idea appeared in Volume penalization method (Kolomensky & Schneider 2009).In the present study, the behavior of single large solid in bubbling fluidized beds is investigated by using the proposed model. Forces working on the large solid during fluidization can be decomposed into pressure gradient, viscous and contact forces. In the study, the relations between these forces and the motion of the large solid are discussed in detail.

(362b) A New Mathematical Model for Bubbling Fluidized Bed Reactors

Jain, M. P., Bhabha Atomic Research Centre (BARC)
Sathiyamoorthy, D., Bhabha Atomic Research Center (BARC)
Rao, V. G., Indian Institute of Technology Bombay
Originally presented on: 11/5/2013 15:37:00 - 15:59:00

Various mathematical models have been proposed in the past for estimating the conversions of reactant gases in fluidized bed reactors. A new mathematical model is being proposed in this paper that gives relatively better results compared to the prevailing models for bubbling fluidized bed reactors. The new model is named as JSR (Jain, Sathiyamoorthy, Rao) model and it is a modified version of bubble assemblage model of Kato and Wen (1969). This paper discusses the development of JSR model and its verification by using experimental data from hydrochlorination of silicon in a fluidized bed reactor. The new model has also been verified by taking data from chemical engineering literature for the case of ammoxidation of propylene, hydrogenation of ethylene, decomposition of nitrous oxide and oxidation of ammonia. The JSR model was compared with other conventionally popular models like Kunii and Levenspiel (K-L) model and Kato and Wen (K-W) model.

(362c) Liquid Spreading Rates On Gas Fluidized Particles

Wasylkiewicz, M. J., University of Alberta

Originally presented on: 11/5/2013 15:59:00 - 16:21:00

Fluidization is used extensively in industry for heat and mass transfer applications, including fluidized bed reactors, dryers, fluid catalytic converters and fluid cokers. Fluid coking involves the injection of a heavy hydrocarbon liquid into the fluidized bed through feed nozzles. The liquid hydrocarbon chains are cracked at the particle surface, forming lighter hydrocarbon gases and rejected carbon solid (coke). This makes the efficiency of the coker dependent on the spreading of liquid. Spreading of liquid far from the nozzles occurs through collisions of wet and dry particles and their motion due to fluidization.

Simulations of fluidized beds are challenging due to the multiple scales of importance, ranging from the overall unit dimensions down to the gas dynamics and particle-liquid interactions. To get a full understanding of the system, a multi-scale approach is necessary. We present particle-resolved simulations of ~1000 particles. A sample flow field for a small simulation is shown in the figure. While these simulations allow for the direct modeling of gas-solid interactions, they are computationally expensive, making the addition of a liquid phase impractical. Instead of resolving the liquid films, we track the volume of liquid on each particle.

In this work we examine the spreading of liquid through a fluidized bed through particle-particle collisions, without (at this stage) considering the inter-particle forces due to liquid bridges and their agglomerating effects. We study multiple solids volume fractions, and consider the effect of liquid surface tension and viscosity. From these simulations, we determine effective liquid diffusion rates both horizontally and vertically. The goal of this work is to enhance our knowledge of liquid spreading through a fluidized bed, and to assess the dependence of the spreading rates on liquid properties and solids volume fraction.

Figure 1: Cross-section of simulation domain with solids volume fraction 0.3. Coloring indicates normalized non-dimensional velocity.

(362d) Developing a Cohesive Model to Describe Channeling in Fluidized Behavior

Johanson, K. D., Material Flow Solutions Inc.

Originally presented on: 11/5/2013 16:21:00 - 16:43:00

Fluidization is an important unit operation critical to drying processes, catalyst cracking processes, and coating processes. In many instances fluidization of cohesive materials is difficult to obtain and control. Models have been developed in an attempt to introduce cohesive effects into fluidization models, but these have been largely semi-empirical in nature. This paper describes a more fundamental approach; it uses a limiting state analysis similar to the limiting state rathole analysis used by Jenike to describe the stability of channels where gas pressure gradients and cohesive flow properties control the stability of channels. The methodology sets allowable size ranges on the channels based on measureable cohesive effects. The cohesive effects are measured at very low solids consolidation stress values, comparable to those expected in fluid bed operation. These cohesive measurements are conducted utilizing a new tester that measures the bulk strength of powder material at consolidation stresses as low as 30 Pa. An energy minimization principle is applied to gas flow through channeling bulk samples to determine the size, number, and spacing of channels in the material as a function of total gas flow. Fluidization data using typical FCC catalyst material, coupled with strength data measured with this new tester, is presented to validate the cohesive model describing channeling behavior. It is expected that this new model will find applicability in fluidization unit operations as well as prediction of de-aeration times for cohesive material in processes where gas is injected to induce flow.

(362e) A Favre-Averaged Two Fluid Turbulence Model for Gas-Particle Flows, Including the Kinetic Theory of Granular Flow

O'Brien, T. J., National Energy technology Laboratory

Originally presented on: 11/5/2013 16:43:00 - 17:05:00

Multiphase CFD calculations are very costly since they are invariably transient. No turbulence model for the gas-particle two-fluid equations, which would allow a direct calculation of the time-steady solution, has been widely accepted. A long transient calculation is usually performed from which the time-steady behavior is extracted by averaging. A general RANS approach has not been extensively used because of the many non-linear terms in the two-fluid theory equation set, which lead to need to specify many constitutive laws (Elghobashi and Abou-Arab, 1983; Dasgupta *et al.*,1994; Hrenya and Sinclair, 1997; Zheng *et al.*, 2001; Gao, 2012). In order to reduce the required number of closure relationships, Besnard and Harlow (1988) used a generalization of Favre averaging, which Favre introduced to address this same issue for compressible single phase flows (Favre, 1965; 1992). This approach has been followed for the description of bubbly flows, mainly motivated by nuclear reactor safety issues (Lahey, 2005; Krepper *et al.*, 2005, 2009). However, an equivalent development has not occurred to describe gas-solid flows.

The two-fluid equations are themselves derived by an averaging procedure which introduces several constitutive relations, the most important being the phase-interaction (drag) term and the granular phase stress tensor. This presents an additional complication for a multiphase RANS theory since many different forms of these relations have been presented in the literature. In the initial development of a RANS model for gas-solid flow (O'Brien, 2013), the full equation set of MFIX (Syamlal, *et al.*, 1993) was truncated to describe isothermal, non-reactive flow and a fixed viscosity was used (Miller and Gidaspow, 1992) which avoided the use of the granular pseudo-temperature equation. In this presentation, this assumption is lifted and a Favre-averaged granular temperature is introduced in addition to the Favre-average of the phasic velocities are introduced previously (O'Brien, 2013), the full granular temperature equation is considered.

As in single phase turbulence theory, transport equations can be derived for the turbulent stresses and turbulent kinetic energies, which include dissipative terms. The energy of the flow for each phase can be partitioned into two quadratic terms, one involving the square of the Favre-averaged velocity and the other being the turbulent kinetic energy. Thus, the energy of the flow can be partitioned between a term involving the mean flow, including the effects of the fluctuations associated with volume fractions, and the turbulent kinetic energy which accounts for the energy of the remaining fluctuations. It is clear from these equations that the turbulent energy cascade is much more complicated than single phase flow since fluctuation energy can be exchanged between phases. Most likely, the energy introduced in the fluid (carrier) phase is transferred to the granular phase and dissipated there. Since the granular temperature equation is now included, this dissipation can be either viscous or due to inelastic collisions.

References:

Besnard, D.C., and F .H. Harlow, "Turbulence in multiphase flow," *Int. J. Multiphase Flow* **14**, 679-699, 1988.

Dasgupta, S., R. Jackson and S. Sundaresan, "Turbulent gas-particle flow in vertical risers," *AIChE J.* **40**, 215-228, 1994.

Elghobashi, S.E., and T.W. Abou-Arab, "A two equation turbulence model for two-phase flows," *Phys. Fluids* **26**, 931-938, 1983.

Favre, A., "Équations des gaz turbulents compressibles I.- Formes gÃ©nÃ©rales," Journal de MÃ©canique **4**, 361-390, 1965.

Favre, A., "Formulation of the statistical equations of turbulent flows with variable density," in *Studies in Turbulence*, T.B. Gatski, S. Sarkar and C.O. Speziale, eds., Springer-Verlag, 324-341, 1992.

Gao, X., L.-J. Wang, C. Wu, Y.-W. Cheng and X. Li, "Steady-state simulation of core-annulus flow in a circulating fluidized bed (CFB) riser," *Chem. Eng. Sci.* **78**, 98–110, 2012.

Hrenya, C., and J. Sinclair, "Effects of particle-phase turbulent in gas-solid flows," *AIChE J.* **43**853-869, 1997.

Krepper, E., D. Lucas and H.-M. Prasser, "On the modelling of bubbly flow in vertical pipes," *Nucl. Eng. Des.* **235**, 597–611, 2005.

Krepper, E., M. Beyer, T. Frank, D. Lucas and H.-M. Prasser, "CFD modelling of polydispersed bubbly two-phase flow around an obstacle," *Nucl. Eng. Des.* **239**, 2372–2381, 2009.

Lahey Jr., R.T., "The simulation of multidimensional multiphase flows," *Nucl. Eng. Des.* **235**, 1043-1060, 2005

Miller, A., and D. Gidaspow, "Dense, vertical gas-solid flow in a pipe," *AIChE J.* **38**, 1801-1815, 1992.

O'Brien, T.J., "A multiphase turbulence theory for gas-solid flows: I. Continuity and momentum equations with Favre-averaging", *Powder Tech*, to be published 2013.

Syamlal M., W. Rogers and T.J O'Brien, "MFIX documentation: theory guide," Technical Note, DOE/METC-94/1004, NTIS/DE94000087, Springfield, VA: National Technical Information Service, 1993.

Zheng, Y., X. Wan, Z. Qian, F. Wei and Y. Jin, "Numerical simulation of the gas–particle turbulent flow in riser reactor based on $k–\varepsilon–k_p–\varepsilon_p–\Theta$ two-fluid model," *Chem. Eng. Sci.* **56**, 6813–6822, 2001.

(362f) Measurement the Solid Flux in a Circulating Fluidised Bed By ECT and Microwave Dopplor Method

Wang, H.,
Qiu, G.,
Ye, J.,
Wu, Z.,
Ge, R.,
Jiang, Y.,
Originally presented on: 11/5/2013 17:05:00 - 17:27:00

Circulating fluidised bed (CFB) combustion technology combined with a high efficiency gas-solid separation circulation is one of the key factors to reducing air emissions and the effect on global warming. To scale up a CFB from small scale to large scale, it is important to investigate the solids distribution in the CFB system. In this paper, electrical capacitance tomography (ECT) and Microwave Doppler method are used to measure the solids flux in the standpipe in a circulating fluidized bed with four cyclone separators. An ECT sensor with eight electrodes microwave antenna was mounted outside of the standpipe to measure the solids concentration and solids average velocity respectively. The total solids flux re-circulated to the bed was calculated based on the solids concentration and velocity. The operation parameters including fluidization flux and initial materials load were investigated. The measured result is compared with conventional solids flux measurement methods.

(362g) Electrostatic Charging and Entrainment Behaviors of Binary and Quaternary Particulate Systems in Multiphase Flows

Zhou, X. H., Columbia University
Park, A. H. A., Columbia University
Originally presented on: 11/5/2013 17:27:00 - 17:49:00

A number of commercial fluidization processes involve particulate systems with different particle sizes and densities. During the operation of multiphase systems such as fluidized beds, electrostatic charges are generated majoritarily via triboelectric or frictional charging due to the dielectric nature of the materials. The accumulation of electrostatic charge within the system can impact the fluidization behavior of the system and in some cases can be operationally hazardous. In this study, the electrostatic charge generation and accumulation are investigated for binary and quaternary particulate systems using a faraday cup system and an on-line electrostatic probe system. Specifically, the effect of the addition of two different fine iron ores (i.e., hematite and magnetite) in a fluidized bed is studied in terms of particle-particle interactions and entrainment rates. The behaviors of different particulate systems are found to highly depend on the chemical and physical properties of particles such as size, density, hydrophobicity and even magnetism. The results suggest that the entrainment behaviors of complex particulate systems (i.e., mineral processing units) cannot simply be predicted based on particle sizes and densities.

(362h) Fluidization of Particles and Agglomerates in Wet Gas-Fluidized Beds

Girardi, M., Princeton University
Radl, S., Graz University of Technology
Sundaresan, S., Princeton University
Originally presented on: 11/5/2013 17:49:00 - 18:11:00

Gas-fluidized beds of dry particles exhibit complex behavior with instabilities giving rise to inhomogeneities that span a wide range of time and length scales. Adding liquid to the system such that particles support a liquid film coating introduces further complexity, allowing for the fluidization of both particles and agglomerates.Â Constitutive models for the effective fluid-particle drag force in such systems must account for inhomogeneous structures[1], such as agglomerates.Â Thus, our primary goal is to develop an effective drag force model for use in large-scale simulations of wet gas-fluidized beds. We perform Euler-Lagrange simulations of a fluidized bed using a 3D periodic domain 2 cm^3 in size with as many as 250,000 particles.Â The domain-average slip velocity serves as a quantitative measure of the inhomogeneous structure within the system.Â We find that the system behavior may be characterized by a modified Bond number: Bo Φ^{α} , where α is a constant and Φ is the liquid loading level.Â $Bo = 3\gamma/(2r^2\rho g)$, where γ is the surface tension, ρ is the liquid density, and r is the particle radius.Â As shown in Figure 1, the collapse of the domain-average slip velocity with respect to modified Bond number is achieved regardless of domain size or whether liquid spreading is permitted.Â This observation suggests that the approximation of constant liquid bridges is valid[2,3]; moreover, the robust collapse suggests that the drag in a wide variety of systems may be predicted with a simple model.

[1] Igci, Y., Andrews, A. T., Sundaresan, S., & O'Brien, T. (2008). Filtered Two-Fluid Models for Fluidized Gas-Particle Suspensions. *AIChE Journal*, *54*(6), 1431?1448.

[2] Mikami, T., Kamiya, H., & Horio, M. (1998). Numerical simulation of cohesive powder behavior in a fluidized bed. *Chemical Engineering Science*, *53*(10), 1927?1940.

[3] Shi, D., & McCarthy, J. J. (2008). Numerical simulation of liquid transfer between particles. *Powder Technology*, *184*(1), 64?75.

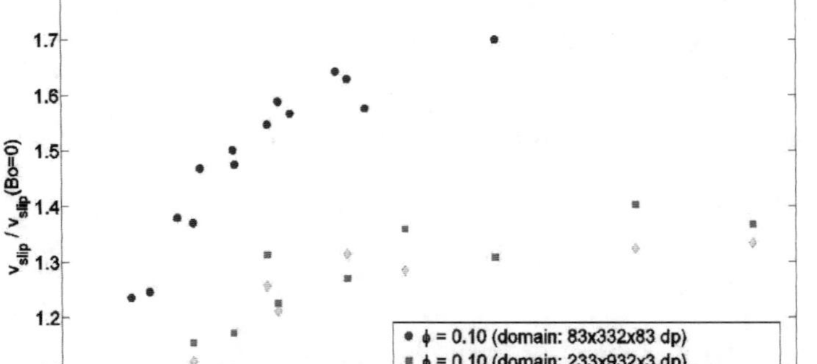

Figure 1: Domain-average slip velocity vs. modified Bond number.
Domain-average slip velocity is scaled against that of a comparable dry system. Bo is Bond number, φ is solid volume fraction, and Φ is liquid loading level.

(374a) Simulation Tools for Nanoparticle-Based Composite Processing and Property Prediction

Bolintineanu, D. S., University of Minnesota
Lechman, J. B., Sandia National Laboratories
Schunk, P. R., Sandia National Laboratories

Originally presented on: 11/5/2013 15:15:00 - 15:40:00

We present a set of tools for the modeling of various aspects of solution-based nanomanufacturing processes. First, we discuss several discrete element modeling (DEM) techniques for the simulation of nanoparticle/colloidal suspensions, including efficient algorithms for the inclusion of hydrodynamic interactions, capillary forces, long-range colloidal interparticle forces, granular contact forces, solvent evaporation and highly aspherical particles. We apply these techniques to the simulation of a battery electrode manufacturing process, and show that the predicted microstructures provide relevant insights for the processing parameters. In the second half of the talk, we discuss methods for the prediction of the bulk properties of disordered, particle-based composites. We present several approaches that rely on a finite element model (FEM) representation of a sample of the material, which can be extracted from a DEM simulation or taken from a tomographic image of the real composite. Overall, this work presents an extensive, multiscale simulation toolset for connecting particle-based nanomanufacturing processing conditions to resulting microstructures and ultimately to material properties.

(374b) Tin Oxide Nanowires and Their Hybrid Architectures for Kinetically Fast Redox Couples in Dye-Sensitized Solar Cells

Vendra, V. K., University of Louisville
Nguyen, T., University of Houston
Druffel, T., Conn Center for Renewable Energy Research
Sunkara, M., University of Louisville
Amos, D. A., University of Louisville
Originally presented on: 11/5/2013 15:40:00 - 16:05:00

The redox electrolyte is an important component of the dye-sensitized solar cells. Iodide/tri-iodide electrolyte used in high efficiency cell is not suitable for practical application due to its corrosive nature, toxicity, competition between dye and tri-iodide for light aborption. One of the major drawback of using alternate redox couples is their fast recombination kinetics and mass transport limitations.

In this study, we show that tin oxide nanowires show a substantial improvement in the photovoltaic performance when compared to titania nanoparticles, when electrolytes having fast recombination kinetics are employed in dye-sensitized solar cells (DSCs). State of the art methods in literature employ titania nanoparticles use passivating layers of Al_2O_3 or poly(methyl siloxane) to lower the electron recombination rate. There have been reports investigating the performance of nanowire in conjunction with alternate redox couples. The single crystalline nature allows for fast electron transport in tin oxide nanowires. Further, prior work from our group has also shown that tin oxide nanowires exhibit two orders of magnitude higher electron lifetimes when compared to titania nanoparticles when iodide/tri-iodide is used as an electrolyte. In this study we employed tin oxide nanowires as photoanodes for DSCs and investigated their performance when kinetically fast electrolytes such as ferrocene/ferrocenium,tempo/tempo,ferricyanide /ferrocyanide are used as redox electrolytes.

Surface modifications such as $TiCl_4$ treatment and atomic layer deposition of titania and alumina were performed to further lower electron recombination in tin oxide nanowires. Fundamental properties such as electron transport time scales and electron recombination time scales were determined for tin oxide nanowires and hybrid architectures where the surface of tin oxide nanowires is coated with titania and alumina.

(374c) Multiscale Sustainability Assessment in Nanocoating Material Design and Manufacturing

Song, H., Wayne State University
Uttarwar, R., Wayne State University
Huang, Y., Wayne State University
Originally presented on: 11/5/2013 16:05:00 - 16:30:00

Nanopaint has drawn great attention in the coatings industry, as it may offer surface coatings with significantly improved performance or new functionalities and properties, such as surface texture alteration control, self-healing, toxic chemical/acid/corrosive agent repelling, glass opaqueness control, building exterior's infrared reflecting/absorbing properties control, and even pathogen growth control. Nevertheless, it is under scrutiny for the environmental and health risks which may pose in nanopaint manufacturing and nanocoating realization. Therefore, it is crucial to analyze the sustainability performance of nanocoating materials and prepare a roadmap for their successful development and application in the future.

In this paper, we introduce a systematic sustainability assessment methodology by resorting to multiscale science and sustainability science. Different from known methodologies focusing on the assessment of manufacturing process and manufactured product, this methodology initiates the assessment starting at the nanomaterial formulation design stage. A variety of material design parameters, such as those about polymer system, nanomaterial type, shape, size, and orientation, solvent and additives, are analyzed using a number of sustainability indicators, which will be incorporated into the analysis of coating manufacturing stages. Multiscale models are developed to predict nanoparticle emission severity in nanopaint spray and curing, nanocoating properties, product realization process performance on quality assurance, waste reduction, and operational safety. The efficacy of the multiscale sustainability assessment methodology will be demonstrated by the case studies on nanocoatings with super-scratch resistance performance and barrier property. The introduced methodology is generic and can be applied to study the sustainability performance of a wide range of nanocoating materials in their early design stage.

(374d) Continuous Nanoparticle Sizing and Characterization Via Microfluidic Interfacial Fluorescent Complexation

Meng, F., Texas A&M University
Ugaz, V. M., Texas A&M University
Originally presented on: 11/5/2013 16:30:00 - 16:55:00

A variety of characterization tools are available to sensitively analyze collected nanoparticles (e.g., condensation particle counters, differential mobility analyzers, scanning mobility particle sizers (SMPS), ion chromatography, mass spectrometry, and electron microscopy). But these methods often require a dedicated laboratory infrastructure to operate optimally, and are therefore not straightforward to adapt for online use. We have developed a new microfluidic-based approach that overcomes many of these limitations by harnessing the inherently steep chemical gradient established at the interface between co-flowing streams containing a nanoparticle-laden suspension and a fluorescent dye. This sharp mismatch acts to localize adsorptive dye-nanoparticle complexation interactions within a narrow interfacial zone, instantaneously producing an intense and easily detectable fluorescence signature. These interactions are inherently dominated by phenomena occurring at the nanoparticle surface, introducing the exciting possibility to extract information about particle size and morphology from the fluorescence profile. We demonstrate this capability experimentally using ZnO and TiO_2 nanoparticles, and introduce a reaction model that enables interfacial fluorescence to be predicted so that kinetic parameters associated with the underlying surface complexation can be quantified. Our approach offers potential for continuous operation that is particularly attractive in manufacturing settings where robust on-line characterization tools are needed.

(374e) Safer Formulation Concept for Flame-Generated Engineered Nanomaterials

Sotiriou, G. A., ETH Zurich
Gass, S.,
Cohen, J., Harvard University, School of Public Health
Pyrgiotakis, G., University of Florida
Pratsinis, S. E., ETH Zurich
Demokritou, P., Harvard School of Public Health
Originally presented on: 11/5/2013 16:55:00 - 17:20:00

The likely success or failure of the nanotechnology industry depends on the environmental health and safety of engineered nanomaterials (ENMs). While efforts toward engineering safer ENMs are sparse, such efforts are considered crucial to the sustainability of the nanotech industry. A promising approach in this regard is to coat potentially toxic nanomaterials with a biologically inert layer of amorphous SiO_2. Core-shell particles exhibit the surface properties of their amorphous SiO_2 shell while maintaining specific functional properties of their core material. A major challenge in the development of functional core-shell particles is the design of scalable high-yield processes that can meet large-scale industrial demand. Here, we present a safer formulation concept for flame-generated ENMs based on a one-step, in flight SiO_2 encapsulation process, which was recently introduced by the authors as a means for a scalable manufacturing of SiO_2 coated ENMs. Firstly, the versatility of the SiO_2-coating process is demonstrated by applying it to four ENMs (CeO_2, ZnO, Fe_2O_3, Ag) marked by their prevalence in consumer products as well as their range in toxicity. The ENM-dependent coating fundamentals are assessed and process parameters are optimized for each ENM investigated. The effects of the SiO_2-coating on core material structure, composition and morphology, as well as the coating efficiency on each nanostructured material, are evaluated using state-of-the-art analytical methods (XRD, N_2 adsorption, TEM, XPS, isopropanol chemisorption). Finally, the biological interactions of SiO_2-coated vs. uncoated ENMs are evaluated using cellular and animal bioassays, providing valuable evidence for reduced toxicity for the SiO_2-coated ENMs. Results indicate that the proposed 'safer by design' concept bears great promise for scaled-up application in industry in order to reduce the toxicological profile of ENMs for certain applications.

References

[1] Sotiriou G.A., Sannomiya T., Teleki A., Krumeich F., Vörös J. and Pratsinis S.E. Non-toxic dry-coated nanosilver for plasmonic biosensors. *Adv. Funct. Mater.* **20**, 4250-4257, (2010).

[2] Sotiriou G.A., Hirt A.M., Lozach P.Y., Teleki A., Krumeich,F. and Pratsinis S.E. Hybrid, silica-coated,

Janus-like plasmonic-magnetic nanoparticles. *Chem. Mater.* **23**, 1985-1992, (2011).

[3] Gass S., Cohen J.M., Pyrgiotakis G., Sotiriou G.A., Pratsinis S.E. and Demokritou P. Safer Formulation Concept for Flame-Generated Engineered Nanomaterials. *ACS Sustainable Chem. Eng.* **in press**, DOI: 10.1021/sc300152f, (2013).

[4] Demokritou P., Gass S., Pyrgiotakis G., Cohen J.M., Goldsmith W., McKinney W., Frazer D, Ma J, Schwegler-Berry D, Brain J, Castranova V., An in vivo and in vitro toxicological characterisation of realistic nanoscale CeO_2 inhalation exposures *Nanotoxicology* **in press**, DOI: 10.3109/17435390.2012.739665 (2013).

(374f) Green Pathways for Development of Nanostructured Aerogel Photocatalysts Effective At Both UV and Visible Range

Li, H., University of South Florida
Sunol, S. G., University of South Florida
Sunol, A. K., University of South Florida
Originally presented on: 11/5/2013 17:20:00 - 17:45:00

Nano-structured highly porous $TiO_2/WO_3/Fe^{3+}$ aerogel photocatalyst are prepared, characterized, and tested for model photo-catalytic reactions. The catalyst structure is tailored to capture environmental pollutants and destruct them under both u-v and visible light through oxidation.

A novel and green method is utilized to prepare the unique surfactant templated aerogel photo-catalyst that has highly porous nanostructure with high surface area and tailored pore size distribution. Sol-gel process is combined with supercritical extraction, deposition and drying. Supercritical drying with heat treatment results in titanium dioxide anatase crystal form. Templates used further enable retention and tuning of the nano-pore structure and surface properties.

The catalyst was characterized using AFM, SEM, FIB, XRD, XPS, BET and porosimetry prior to using in reactions. The band gap of the catalyst particles were determined using diffuse reflectance.

Resulting aerogel $TiO_2/WO_3/Fe^{3+}$ has similar photo-catalytic capability compared to highly optimized commercial Degussa P25 under UV exposure and offer much superior photo-catalytic capability under visible light exposure. The model reaction utilized employed Methylene Blue photo oxidation under Visible and UV light.

(376a) "Dry Particle Coating for Tailor-Made Product Properties: Coating and Decoating Processes"

Bluemel, C., Friedrich-Alexander University of Erlangen-Nuremberg
Wirth, K. E., Friedrich-Alexander University Erlangen-Nuremberg
Originally presented on: 11/5/2013 15:15:00 - 15:35:00

Today tailor-made particle properties of highly dispersed powders are of increasing interest in industrial applications. Theses powders consist of particles with small diameters (≤ 10 µm). With decreasing particle size, the influence of van-der-Waals forces in comparison to their weight-force is increasing resulting in agglomeration. The outcome is bad flowability of the powders leading to problems in further processing.

Dry particle coating is a cost-efficient and easy-handling mixing process for assembling tailor-made composite particles. Therefore nanoscale guest particles are adhered to microscale host particles. This process mainly bases on van-der-Waals forces. Within the adhesion of nanoparticles the distance between the host particles is increased. This results in decreased adhesion forces and increased flowability. Additionally properties of the particles can be modified or changed, for example wettability, solubility, adsorption properties or higher cycle stability of electrode materials. These tailor-made particle properties influence the resulting product properties and enable the production of complex devices.

The processing of such composite particles can for example be conducted in a tumbling mixer or a rotary drum mixer for understanding the major influences to create scale-up criteria. The number of host particle contacts as well as the applied energy have major impact on the dry particle coating results. The guest particle agglomerates should be deagglomerated ideally whereas the host particles should not be comminuted. There is an optimum distribution of guest particles on the surface of the host particles. To increase the applied energy, mixing agents can be added to the host particles and guest particles in order to improve the mixing intensity and thereby deagglomeration of the cohesive host and guest particles.

Longer process times respectively great applied energy lead to decoating processes of the host particles. Guest particles are removed from the host particles surface and agglomerate again to agglomerates with low porosity. A model has been developed for the coating process as well as for the decoating processes. The variation in time can be described as a first order reaction. The quality of the coating has been characterized by several methods, for example: SEM and TEM.

Flowability measurements of the host particles and composite particles were conducted in Schulze ring shear tester; Zimmermann tensile strength tester and AFM. These three methods enable to measure complementary data ranging from the single particle level to bulk as well as from small to higher loads.

The surface properties of the host particles and guest particles are important for the dry particle coating process. Hydrophobic guest particles only adhere to hydrophobic host particles as well as hydrophilic guest particles only adhere to hydrophilic host particles.

The dry particle coating process in a tumbling mixer is quite complicated, because rotation, translation and inversion are taken place simultaneously. Since mixing is conducted in a closed vessel, in-situ-observations are of great interest. Positron Emission Particle Tracking (PEPT) measurements enable such observations and therefore were conducted to learn more about the dry particle coating process. Via these measurements, a single mixing agent (diameter: 1 mm) was tracered with a positron (β^+) emitter. Thereby the mixing intensity can be determined for example by velocity distributions, movements and collisions of the mixing agents. The applied energy is proportional to the velocity of the glass beads and can be calculated in this way. The quality of the coating depends on the velocities of the mixing agents.

The composite particles assembled in this process open new possibilities in different applications, for example: selective laser beam melting (SLM), light weight construction as well as in lithium ion batteries. The fundamentals of the dry particle coating process, the description of the occurring mechanisms of coating and decoating as well as examples for applications will be given in the presentation.

(376b) Surface Functionalization of Polymeric Particles Using Low Temperature Atmospheric Pressure Plasma

Gilliam, M., University of Missouri - Columbia
Farhat, S., Kettering University
Stubbs, B., Kettering University
Magyar, M., Kettering University
Zand, A., Kettering University
Garner, G.,
Originally presented on: 11/5/2013 15:35:00 - 15:55:00

Atmospheric plasma was used to modify the surfaces of polymeric particles and alter the surface properties of the particles. The low temperature process was run under continuous operation with a fast treatment time. With the addition of chemical precursors to the plasma, the process has a capability to create a variety of surface functional groups, thus enabling tailoring of the surface properties for a given particle type and application. In particular, the use of polymeric particles in medical technology and life sciences is growing very rapidly. For many applications, a hydrophilic surface or a surface with affinity for specific biomolecules are often desirable.

The apparatus and methods for surface treatment of the particles are presented, as well as results of the process applied to polymeric particles, including polymethylmethacrylate (PMMA) and polytetrafluoroethylene (PTFE). Chemical precursors, including water vapor, glycerol, and hydroxyethylmethacrylate (HEMA), were added to the plasma process to introduce new surface functional groups. The effects of the plasma processing parameters on the properties of the particles were investigated. The treated particles were evaluated using X-ray photoelectron spectroscopy (XPS), contact angle of the pressed particles, and UV-visible spectroscopy, which was used in a manner to evaluate stability of dispersions of treated particles in water and settling over time. Plasma processing conditions and the choice of chemical precursor significantly influenced the type of functional group and hydrophilicity of the particles.

(376c) Surface Modification of Carbon Hard Sphere Via Surface Initiated Atom Transfer Radical Polymerization

Kamboj, L., Tulane University
Ejaz, M.,
Pesika, N., Tulane University
Originally presented on: 11/5/2013 15:55:00 - 16:15:00

Lubricants play an integral role in the operation of several technologies and in biology also, ranging from moving parts in machinery to the biolubrication of artificial joints. We have found that a colloidal dispersion consisting of easily synthesized highly spherical and uniform carbon particles coupled with surface grafting of polymers that results in a very effective water soluble lubricant with low friction coefficient and excellent surface wear protection. We propose that these particles would increase the overall friction coeffient than without the grafted polymer. These particles use a rolling mechanism similar to nano –or microscale ball bearing under confinement. The effect of several parameters of lubrication including the thickness of the grafted polymer layer, type of polymer grafted, along with the size and porosity of the carbon particles will be presented.

(376d) Characterization of Gas-Expanded Liquid-Deposited Nanofilms With Size-Controlled Gold Nanoparticles

Zhong, J., Auburn University
Roberts, C. B., Auburn University
Ashurst, W. R., Auburn University
Originally presented on: 11/5/2013 16:15:00 - 16:35:00

Alkanethiol-stabilized gold nanoparticles (AuNPs) were deposited via a gas-expanded liquid (GXL) technique utilizing CO_2-expanded hexane onto substrates of different surface energy. The different surface energies were achieved by coating silicon (100) substrates with various organic self-assembled monolayers (SAMs). The films were characterized and the rms surface roughness values were obtained to determine the nanoparticle film deposition and growth characteristics. Based on the critical surface tension of the SAMs, the film growth indicates a shift between layer-by-layer deposition and island formation, which also results in a change in the surface roughness. Moreover, the surface roughness is influenced by the concentration of the nanoparticle solution, the average size, the size distribution of AuNPs and the nature of the AuNP surface chemistry. The manner in which the surface roughness was affected was investigated by using AuNPs with different sizes. Based on this information, the potential exists to engineer nanoparticle films with desired morphologies and characteristics.

(376e) Engineering the Scattering Properties of Thin, Particle-Doped Coatings

Rogers, W. B., University of Pennsylvania
Manoharan, V. N., Harvard University

Originally presented on: 11/5/2013 16:35:00 - 16:55:00

Coatings and thin films of small particles suspended in a matrix possess optical properties that are important in several industries from cosmetics and paints to polymer composites. Many of the most interesting applications require coatings that produce several bulk effects simultaneously, but it is often difficult to rationally formulate materials with these desired optical properties. Here, we focus on the specific challenge of designing a thin nanoparticle film that maximizes both diffuse and total hemispherical transmission. We demonstrate that these bulk optical properties follow a simple scaling with two microscopic length scales: the scattering and transport mean free paths. Using these length scales and Mie scattering calculations, we generate basic design rules that relate scattering at the single particle level to the film's bulk optical properties. These ideas will be useful in the rational design of future optically active coatings.

(376f) Self-Assembled Nanoparticle Anti-Glare Coatings

Askar, K.,
Jiang, P., University of Florida
Originally presented on: 11/5/2013 16:55:00 - 17:15:00

Light reflection from an optical surface, such as automotive dashboards and flat panel displays, could pose safety hazards, impair the legibility of displays, and degrade performance of optical systems with multiple optical components. Here we present a simple and scalable bottom-up technology for assembling close-packed nanoparticle monolayers on both sides of a glass substrate as high-quality anti-glare coatings. Optical measurements show that monolayer coatings consisting of 110 nm silica nanoparticles can significantly reduce optical reflectance and enhance specular transmittance of the glass substrate for a broad range of visible wavelengths. Both experiments and numerical simulations based on a finite-difference time-domain (FDTD) model reveal that the anti-glare properties of the self-assembled colloidal monolayers are significantly affected by the size of the colloidal nanoparticles. We further demonstrate that the self-assembled nanoparticle arrays can be used as structural templates for fabricating nanoporous polymer membranes with similar anti-glare performance. Tunable anti-glare coatings have also been achieved by using templated nanoporous membranes consisting of solvent-sensitive shape memory polymers

(376g) Environmental Factors Affecting Spin-Processed Antireflection Coatings

Nolte, A. J., Rose-Hulman Institute of Technology
Chongson, R. B.,
Syed, M.,
Originally presented on: 11/5/2013 17:15:00 - 17:35:00

We have shown previously that spin-coated colloidal anti-reflection (AR) coatings offer several advantages among AR technologies, including low cost, speed, and ease of processing. In this talk, we will review recent efforts to create silicon dioxide nanoparticle coatings with tunable refractive index and thickness in order to enhance their compatibility with the widest possible range of optical systems and materials. Because such coatings are essentially porous silica frameworks, as-processed coatings remain hydrophilic and therefore have optical properties that are sensitive to the ambient humidity and temperature. We demonstrate the capability of such systems for environmental sensing, and discuss strategies for mitigating environmental sensitivity where optical property stability is desired.

(377a) Co Particle ALD Facilitated On Metallic W ALD

Van Norman, S. A., University of Colorado at Boulder
Falconer, J. L., University of Colorado
Weimer, A. W., University of Colorado at Boulder
Originally presented on: 11/5/2013 15:15:00 - 15:40:00

Efficient use of precious metal catalysts is essential to enhancing catalytic productivity. Traditional catalyst impregnation proceeds through an oxide phase of the metal followed by reduction. However, with traditional methods the metal is only partially reduced limiting the fraction of active material or the metal sinters under reduction conditions limiting active surface area. The novel deposition of metallic phase Co particles and their properties is discussed.

In this study, metallic Co deposition is performed via a bi-molecular atomic layer deposition (ALD) reaction of cobaltocene and ammonia. Surface adsorbed radicals are formed on a metallic W surface produced by ALD facilitate the reaction. The radical interacts with the metal precursor resulting in chemically bonded Co to the surface. Gas phase reaction is tracked via mass spectrophotometry. Elemental analysis confirms deposition of Co. Metal loading as a function of ALD cycles performed is also explored. The reaction is also shown for oxide surfaces which do not show activity for metallic Co ALD under these conditions.

(377b) Effect of Microporous Layer On Sintering and Dispersion of Pt

Izar, A., Instituto Tecnologico y de Estudios Superiores de Monterrey
Hollenbaugh, R., University of Colorado, Boulder
Weimer, A. W., University of Colorado at Boulder
Originally presented on: 11/5/2013 15:40:00 - 16:05:00

The prevention of sintering is one of the most desired goals in the catalyst industries since its effects are hard to be reversed, leading to huge economic losses. In this paper we analyze the effect of a microporous layer on the sintering of platinum dispersed in silica. This layer was made via molecular layer deposition (MLD) with trimethylaluminum, ethanolamine and maleic anhydride, which after calcination creates a porous layer with an average pore size of 0.8 nm. We studied two cases. In the first case the platinum was dispersed on the silica first, followed by coating with the MLD process. In the second case the platinum was dispersed after the porous layer was created. In both cases the particles were annealed at different temperatures to determine the effect of the coating in the sintering of the platinum. CO oxidation was used to monitor the active life of the catalyst as well.

(377c) Surface Modification of Transition Metal Carbide Nanoparticles Via Atomic Layer Deposition For Energy Applications

Li, D., University of Wyoming
Saha, S., University of Wyoming
Originally presented on: 11/5/2013 16:05:00 - 16:30:00

This work centers around surface modification of transition metal carbide (TMC) via Atomic Layer Deposition (ALD) for energy applications such as serving as catalysts for electrolyzers or proton exchange fuel cells. Transition Metal Carbides (TMCs) (W, Mo, V, Ti, Ta) are known to have similar electronic and geometric structure as Pt-group metals [1]. Among the TMCs, Mo_2C has shown good stability in acidic environment and sufficient hydrogen evolution reaction (HER) activity [2]. Therefore, Mo_2C could be a potential alternative to the conventional Pt. However, Mo_2C has not shown the same electrocatalytic activity as Platinum (Pt). Data in our group has further demonstrated that Mo_2C has order of magnitude higher Tafel slopes than Pt for HER and oxygen reduction reaction (ORR), which corresponds to a slower kinetics. For this reason the past decade or has seen burgeoning interest of modifying Mo_2C surface using Pt. For example, Pt supported on carbon nanotube has shown higher electrocatalytic activity than conventional Pt supported on Powder carbon [3]. Pt supported on Mo_2C is expected to show higher catalytic activity, thanks to its similar electronic structure to Pt and synergistic chemistry between Mo_2C and Pt. In this study ALD technique has been adopted to deposit Pt nanoparticles onto Mo_2C nanotubes, due to its simplicity, reproducibility, and molecular or even atomic level control on the deposited particle size and distribution.

Despite of the progress made in recent years on minimizing Pt loading, the challenge of low surface area of TMC support still persists. To address the challenges, the objectives of the present study are to modify the surface of TMCs with Pt and further investigate the synergetic effect between the nanotube TMC support and nanoparticle Pt deposited onto TMC nanotube via ALD. The hypothesis is that the nanosize of the support and deposited metal will allow maximized active areas of both Pt atoms and synergetic interaction areas between Pt nanoparticles and the TMC nanotube supports.

To screen the Pt-TMC co-catalyst activity, HER activity of both unmodified and modified TMCs were determined by cyclic voltammetry in N_2 saturated 0.5M H_2SO_4 solution and compared with the commercially available 20% Pt/C. Based on the tafel plots, ALD modified Mo_2C support with Pt nanoparticles showed higher HER activity than the commercial 20% Pt/C. Oxygen Reduction Reaction (ORR) activity of both the catalysts were determined by cyclic voltammetry into O_2 saturated 0.5M H_2SO_4 solution. Bare Mo_2C shows very limited stability towards oxidation, but after the surface modification oxidation stability has been increased, which is evidenced by cyclic voltammograms of ORR. The morphology of the ALD modified metal carbide was examined via High Resolution Transmission Electron Microscopy (HRTEM), which proved the island-growth of Pt on TMC with narrow size distribution.

Bibliography

1. Levy, R.B. and Boudart, M., *Platinum-Like Behavior of Tungsten Carbide in Surface Catalysis.* Science, 1973. **181**(4099): p. 547-549.

2. Weidman, M.C., Esposito, D.V., Hsu, Y.-C., and Chen, J.G., *Comparison of Electrochemical Stability of Transition Metal Carbides (Wc, W2c, Mo2c) over a Wide Ph Range.* Journal of Power Sources, 2012. **202**: p. 11-17.

3. Matsumoto, T., Nagashima, Y., Yamazaki, T., and Nakamura, J., *Fuel Cell Anode Composed of Mo[Sub 2]C Catalyst and Carbon Nanotube Electrodes.* Electrochemical and Solid-State Letters, 2006. **9**(3): p. A160.

(377d) Atmospheric-Pressure Atomic Layer Deposition Of Platinum Nanoclusters On Titania Nanoparticles

Goulas, A., Delft University of Technology

van Ommen, J. R., Delft University of Technology

Originally presented on: 11/5/2013 16:30:00 - 16:55:00

Pt-based catalysts are widely used, for example in petrochemical processes, fuel cells and emission control in cars. For optimizing the catalyst performance ? its activity, selectivity, and stability ? while using a minimum amount of scarce and expensive Pt, it is crucial to maintain excellent control over the morphology and distribution of the Pt clusters during their synthesis process. In the past years, atomic layer deposition (ALD) has appeared as an appealing nanofabrication technique for the preparation of precisely tailored heterogeneous catalysts, see e.g. [1].

We will present a novel approach to deposit Pt on nanoparticles in a fluidized-bed ALD reactor at atmospheric pressure. So far, ALD deposition of Pt has only been shown at vacuum. We will demonstrate that the precision characteristics of ALD can be maintained at atmospheric pressure, effectively proving the potential of the process for catalyst fabrication on an industrially relevant scale.

Utilizing (trimethyl)methylcyclopentadienyl platinum (IV) and ozone as precursors we were able to grow Pt islands on TiO_2 (Aeroxide P-25) nanoparticles at 250°C under atmospheric pressure conditions. The nanoparticles were fluidized with nitrogen at a superficial gas velocity of 4.2 cm/s. By applying sub-saturating pulses we achieved ~95% precursor utilization, while the use of a drastic oxidizing agent like ozone ensured that the aimed metal loading could be reached in only a few reaction cycles. Increasing the number of performed cycles resulted in a monotonic increase in the metal loading (Fig. 1), while the obtained particles were highly dispersed (average size of only 1.5nm) and exhibited a narrow particle size distribution (Fig. 2).

The use of relatively mild temperature conditions, the very low impurities inclusion in the materials and the absence of solvent use during the fabrication procedure are rendering ALD an attractive technique for the fabrication and tailoring of materials. We will demonstrate that there is even more space for improvement in the process scale-up by showing coating results obtained in a continuous pneumatic ALD reactor that also operates in atmospheric pressure. Proving that the results obtained in the batch process can be reproduced in the continuous reactor for processing powder is an important development in ALD technology.

[1] King, D.M., Liang, X., Weimer, A.W., Powder Technol. 221 (2012) 13

[2] Goulas, A., van Ommen, J.R., J. Mater. Chem. A 1 (2013) 4647

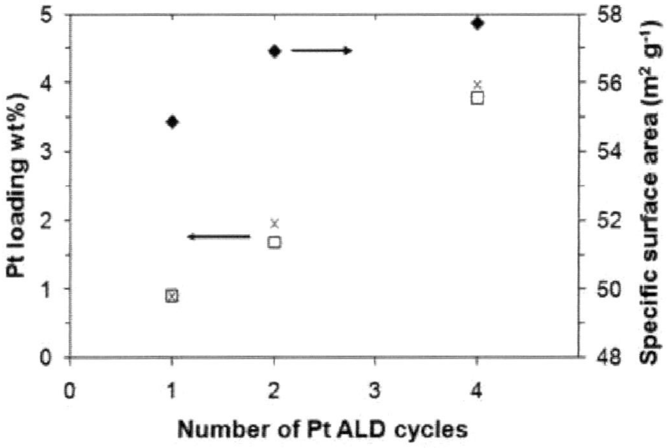

Fig. 1 Pt loading measured (ICP-OES) (o) and predicted for ideal ALD (X), and specific surface area (u) as a function of ALD cycles. [2]

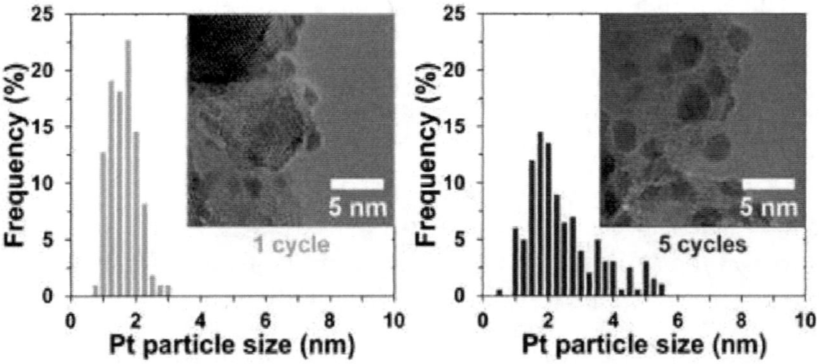

Fig. 2 Pt particle size distributions for nanoclusters deposited with 1 and 5 ALD cycles (inset: TEM images of Pt nanoclusters on the P-25 support nanoparticles). [2]

(377e) Synthesis of Nanostructured Carbon Particle Materials Via Spray Method

Ogi, T.,
Ratna, B.,
Okuyama, K., Hiroshima University
Originally presented on: 11/5/2013 16:55:00 - 17:20:00

Carbon-supported platinum (Pt/C) catalysts with controlled morphologies, namely, “dense-erythrocyte-like (DEL)” and “hollow-porous-microsphere (HPM)” have been synthesized by spray-drying followed by impregnation of Pt nanoparticles. Addition of polystyrene latex (PSL) template particles play a crucial role to determine the carbon morphology, which was confirmed by SEM and TEM images. The preparation of modified catalyst support was completed in only several seconds for DEL and less than 1 h for HPM, which is contrary to currently available methods that require time up to several days to complete the modified catalyst support synthesize. Pt nanoparticles are agglomeration-free, and well-dispersed on the surface of the carbon particles (4 nm). The electrocatalytic activity of the modified Pt/C catalyst particles was comparable to that of a commercially available Pt/C catalyst (TKK). The specific activity of the Pt/C_{HPM} catalyst was 2 times higher than that of Pt/C_{DEL} catalyst, and was up to 2.5 times than that of the commercial one. The mass activity of Pt/CHPM was 239.57 mA/mg Pt and also shows a great enhancement, about 1.55 times higher than the commercial.

(377f) Core-Shell Redox Catalyst for Chemical Looping Reforming of Methane

Shafiefarhood, A., North Carolina State University
Galinsky, N., North Carolina State University
Huang, Y., North Carolina State University
Chen, Y.,
Li, F., North Carolina State University
Originally presented on: 11/5/2013 17:20:00 - 17:45:00

Conversion of methane into syngas, a gaseous mixture composed primarily of carbon monoxide and hydrogen, is a topic of practical relevance for many synthetic fuel and chemical production processes. At present, methane-derived syngas is produced primarily through reforming in the presence of gaseous oxidants such as steam, oxygen, and/or carbon dioxide. Although the reforming based approaches have been successfully utilized at a commercial scale, the efficiencies of the state-of-the-art reforming processes are limited due to the high steam to methane ratio required by the reforming catalysts and/or the needs for energy intensive air separation operations.

Chemical Looping Reforming (CLR) represents an alternative approach for methane reforming. In the CLR scheme, a solid oxygen carrier or "redox catalyst" is used to partially oxidize methane into syngas. The reduced oxygen carrier is then transferred to a subsequent reactor for regeneration with steam and/or air. The cyclic redox operation avoids the needs for air separation. Compared to oxygen carriers in other cyclic redox processes such as chemical looping combustion (CLC), the redox catalyst/oxygen carrier in CLR should exhibit high selectivity towards partial oxidation products.

The current study investigates the performance of a novel "core-shell" redox catalyst in CLR reactions. A number of oxygen carriers composed of a primary metal oxide and mixed ionic-electronic conductor (MIEC), are synthesized, characterized, and tested under redox conditions. Results indicate that the newly developed core-shell redox catalyst is significantly more selective than conventional oxygen carriers for syngas production. It also showed better carbon formation resistance and maintained structural/phase integrity.

(398d) Redistribution of Heavy Metals in Incineration Bottom Ash Through Particle Size Reduction

Yao, Q.,
Samad, N. B.,
Seah, X., Nanyang Technological University
Lau, R., Nanyang Technological University

Incineration is the main treatment process for municipal solid waste in Singapore. The incineration bottom ash (IBA) is used primarily for landfills. Therefore, it is important to study the physical and chemical properties as well as the leaching characteristics of heavy metals in IBA. In this study, both of physical (including size distribution, density and circularity) and chemical (composition and typical heavy metal contents) properties of incineration bottom ash (IBA) are analyzed. Fractional distribution is also used to determine the leaching properties of the heavy metal contents. XRF results indicated that Si, Ca, Al, Fe are the main elements of the bottom ash in Singapore. ICP-AES showed that the ash contained more than 5000 mg/kg of heavy metals. It is found that 17.8% of the heavy metals are leachable while carbonate bound is the main leaching mechanism. The effect of particle size reduction process on the re-distribution of heavy metals is also explored.

(398t) Rheology of Cohesive Granular Materials Across Multiple Dense Flow Regimes

Gu, Y., Rose-Hulman Institute of Technology
Sundaresan, S., Princeton University
Chialvo, S., University of Florida

Understanding flows of granular matter is essential for the design of particulate and gas-solid flow processes, and much research attention has been devoted to studying granular flows of non-cohesive particles. However, cohesive particles are ubiquitous in nature and industrial processes and exhibit behaviors unseen in non-cohesive materials, including jamming at low packing fractions and agglomerate formation. In this work, we study the rheology of cohesive granular material by performing discrete element simulations of homogeneous simple shear flow of assemblies of uniform, cohesive, spherical particles. Cohesion is modeled as a van der Waals force between particles, and frictional and inelastic interactions are also considered. To access multiple flow regimes, a wide array of cohesive strengths, particle volume fractions, and shear rates are simulated. Dense shear flows of non-cohesive materials exhibit three regimes: (i) quasistatic regime, where the stress is independent of shear rate, (ii) inertial regime where the stress varies quadratically with shear rate and (iii) an intermediate regime where the stress manifests power-law dependence with n~1/2 [1]. Cohesion results in bifurcation of the inertial regime into two regimes whose boundaries are determined by a modified Bond number, which is a ratio of the van der Waals force to a characteristic contact force: (a) a new rate-independent regime and (b) an inertial regime. We present a rheological model for cohesive systems that captures all the simulation results.

[1] S. Chialvo, J. Sun, and S. Sundaresan. Phys. Rev. E 85, 021305 (2012).

(398u) Scale-Up of Continuous Rotary Calciners: Heat Transfer and Powder Flow

Emady, H. N., Purdue University
Anderson, K. V., Rutgers University
Glasser, B. J., Rutgers University
Muzzio, F. J., Rutgers, The State University of New Jersey
Borghard, W. G., ExxonMobil
Cuitino, A., Rutgers University

In catalyst manufacturing, continuous rotary calcination is a thermal treatment process in which particles flow through a rotating drum with heated walls. Although this process is widely used, there is a lack of predictive models for material flow and heat transfer, particularly upon scale-up. This work aims at a fundamental understanding of the effects of operation, design, and material properties on residence time and temperature distributions in rotary calcination processes to provide a methodology for scale-up.

For successful calciner performance, the residence time of the particles in the calciner must be longer than the time required for heating and calcination of the particles. This problem can be separated into axial and radial processes. In the axial direction, residence time and dispersion will be important, requiring experiments in full calciners. In the radial direction, heat transfer and dispersion will be important, requiring experiments and discrete element method (DEM) simulations in cross-sections of the calciner. This poster will show results of studies on the competing axial and radial processes and their implications for scale-up of continuous rotary calciners.

(398v) Size and Morphology Control of Polymeric Nanoparticles By Tuning Polymer Components and Process Parameters

Zhou, J., The Hong Kong University of Science & Technology
Chau, Y., The Hong Kong University of Science and Technology

In drug delivery, size and morphology are two key variables that may determine the circulation time, release profile and cellular uptake of biodegradable polymeric drug carriers. However, much remains unknown about how to control self-assembled polymeric particle homogenous in size and morphology. Therefore, the objective of this article is to conduct a study on how block copolymer properties as well as process parameters affect the size and morphology of biodegradable polymeric nanoparticles. Two widely-used biodegradable block copolymers methoxy poly(ethylene glycol)-b-poly(ε-caprolactone) and poly(ethylene glycol)-poly-$_{L}$-lactic acid (mPEG-b-PCL and mPEG-b-PLLA) with different hydrophilic ratios were synthesized and fabricated into nanoparticles via nanoprecipitation and film rehydration method. The size and morphology of nanoparticles were characterized by dynamic light scattering and transmission electron microscopy. It is found that both polymer properties and process physicochemical parameters had an influence on the particle size and polydispersity, but the extent of impact varied greatly. At the same time, the morphology of nanoparticle was not only determined by polymer hydrophilicity as previous research stated, but also dependent on preparation method which is worthy of careful further investigation. It is believed that this article will demonstrate some relationships between polymer properties as well as process physicochemical parameters and nanoparticle features that could be useful in particle engineering.

(398w) Surface Characterization of the Interaction Between Fly Ash Particles and Mercury

Jung, J. E., Stanford University
Wilcox, J., Stanford University
Jew, A., Stanford University
Rupp, E. C., National Energy Technology Laboratory

Fly ash is a primary by-product of the coal combustion process. Despite capture efficiencies of up to 99.95% by particulate control devices, such as electrostatic precipitators or fabric filters, coal-fired power plants in the United States still emit on average 300 tons of fly ash per day. The release of fly ash into the environment is a public and environmental health concern for two reasons. First, the particle size of released fly ash is generally smaller than 2.5 μm ($PM_{2.5}$) and current research suggest that inhalation of small particles can have an adverse health effect. The second concern is because of the presence of toxic trace metals/metalloids, such as mercury (Hg), selenium (Se), and arsenic (As), which may exist as either a component of fly ash, or partition onto fly ash as the flue gas cools. Therefore, it is important to understand the characteristics of fly ash and, specifically, the interaction between trace metals and fly ash components. In this work, various experimental methods were carried out to investigate fly ash interaction with Hg in different particle sizes of fly ash. Bituminous coal ultrafine fly ash particles below 0.1 μm were separated using membrane filtration, and both filtered and non-filtered fly ash samples were exposed to methane combustion flue gas, supplemented with SO_2, NO_x and HCl, that contained Hg in a packed-bed reactor to model Hg sorption. Characterization studies including scanning electron microscopy (SEM) and X-ray diffraction (XRD) were carried out on the fly ash samples to determine changes in their morphologies and mineralogy before and after reaction. Sorption reaction products were characterized using synchrotron-based x-ray fluorescence mapping (XRF) to determine which elements Hg is associated with on the fly ash particulate matter. Mercury hot spots were further analyzed with micro x-ray absorption near edge structure spectroscopy (μ-XANES) to determine the form of Hg contained in these regions. Results indicate that significantly more Hg is associated with the filtered ultrafine fly ash sample when compared to the non-filtered sample. XRF analysis shows that the Hg is partitioned into two major regions in the samples: Fe associated and non-Fe associated. Mercury content in the sample is correlated with Fe, S (reduced), Cl, Br, and Cu, and to a lesser degree with Se and As. There is no discernable association between Hg and the alkali metals, the alkaline earth metals, and the lighter transition metals and the Hg hot spots detected are not associated with the Fe-oxides contained in the fly ash. μ-XANES analysis of the Hg hot spots indicates that these regions are dominated by HgS (cinnabar) crystals. These results indicate that ultrafine fly ash particles could be the primary sorbent of Hg from a gaseous phase.

(398s) Compaction Challenges With Solid Dosage Formulations Containing Vitamin E Tpgs

Sinko, P. D.,

Compaction challenges with solid dosage formulations containing Vitamin E TPGS

Patrick D. Sinko[1], Preetanshu Pandey, Dilbir S. Bindra, Rhye Hamey

[1] Virginia Polytechnic Institute and State University, Materials Science and Engineering

[2] Bristol-Myers Squibb, Drug Product Science and Technology, New Brunswick, NJ USA

Vitamin E TPGS or TPGS (d-alpha tocopheryl polyethylene glycol 1000 succinate) is known to enhance the oral bioavailability of poorly water soluble drugs.[1,2] Given the high percentage of new chemical entities with low aqueous solubility it is may be desirable to incorporate TPGS in the formulation for such cases. This study assesses the processing challenges in incorporating TPGS into oral solid pharmaceutical dosage forms manufactured using a wet granulation process (high-shear and fluid bed). A solution of TPGS was prepared at 20% w/w concentration and added to the formulation as a binder. Tablets with up to 14% w/w TPGS were produced with no significant processing challenges when compressed at low tablet press speeds (dwell times ~ 100 ms). However, at lower dwell times (<20 ms) that are typical of commercial-scale presses, several tablet defects, such as tablet cracking, were observed during compaction.[2] A significant loss in compactability was observed as well. These compaction challenges were overcome through the intra-granular incorporation of Aeroperl® 300 (5-10% w/w), a granulated form of colloidal silicon dioxide. Interestingly, the addition of the same concentration of Aeroperl® 300 to the extra-granular portion did not provide the required compaction enhancement to the formulation. It was shown that the addition of intra-granular Aeroperl® 300 resulted in smaller granule particle size, higher granule porosity and higher compactability than the formulation without Aeroperl® 300.

References

1. Feiyan J, Tatavarti A. Tabletability assessment of conventional formulations containing Vitamin E tocopheryl polyethylene glycol succinate. *Int J Pharm* 2010;389:58-65.
2. Pandey P, Sinko PD, Bindra DS, Hamey R, Vema-Varapu C., Processing challenges with solid dosage formulations containing Vitamin E TPGS. *Pharm Dev Tech* 2013;18(1):296-304.

(398x) Oxygen Carrier Granulation for Chemical Looping Processes

Matsuda, S., National Institute of Advanced Industrial Science and Technology
Hatano, H., National Institute of Advanced Industrial Science and Technology
Kawamura, M., Ohtsuka Ceramics Inc
Komuro, T., Ohtsuka Ceramics Inc.
Ohtsuka, M., Ohtsuka Ceramics Inc.

Chemical looping processes require low cost and higher reactive lattice oxygen carrier, especially for heavy hydrocarbon materials such as coal, biomass and so on. They contain ash and catalytic poison which deactivates lattice oxygen carrier. Spray drying granulation was applied to granulate various kind of particles to support the lattice oxygen carrier. Simultaneous spray drying granulation and post composite method were tested by changing the granulation conditions to reduce the amount of oxygen carriers. Furthermore, various factors were discussed affecting on the reactivity and strength of the granulated particles. Conclusive remarks are as follows. With increasing the calcination temperature, the strength of the granulated particles was increased. However, the specific surface area of them was decreased gradually. On the other hand, the reactivity of them kept almost constant regardless of the decrease of the specific surface area. It was thought that the distribution of the lattice oxygen carrier and voidage in the granulated particles were important.

(398z) Freestanding Zeolitic Imidazolate Frameworks Membrane in Microchannel

Ng, T. N., The Hong Kong University of Science and Technology
Chen, X., The Hong Kong University of Science and Technology
Yeung, K. L., The Hong Kong University of Science and Technology
Sim, V., The Hong Kong University of Science and Technology

Metal-organic frameworks (MOFs) have a wide range of applications including chemical separation, hydrogen storage, sensing, electronics and medicine [1,2]. They incorporate metal clusters and coordinate organic ligands to form networks of pores, giving the material a rich topological and chemical diversity. MOFs, in particular the zeolitic imidazolate frameworks (ZIFs), have a promising future in membrane technology. However, it is still a challenge to prepare high-quality MOF membranes due to poor nucleation, adhesion and intergrowth of the crystals. Although different strategies, including in-situ and seeded growth methods along with others, have been used to prepare supported MOF membranes, very little work has been done to fabricate freestanding MOFs membrane despite the benefits of lower transport resistance and a larger surface area.

Freestanding ZIF-8 micromembranes are fabricated within the confined space of the microfluidic microchannel. This is achieved by establishing two parallel flows of reactants within the channel and inducing nucleation, crystallization and growth at the flow interface. 100 micron thick ZIF-8 was successfully grown along the gap in the microchannel. The freestanding membrane was defect-free and impervious to large dye molecules. Cracks and defects were not evident. The formation of ZIF-8 micromembrane was observed using a confocal fluorescent microscope that enables us to propose a mechanism of formation and growth of ZIF-8 micromembrane in a microfluidic channel.

References:

[1] O. Shekhah, *et al.*, "MOF thin films: existing and future applications," *Chemical Society Reviews,* vol. 40, pp. 1081-1106, 2011.

[1] J. Lee, *et al.*, "Metal-organic framework materials as catalysts," *Chemical Society Reviews,* vol. 38, pp. 1450-1459, 2009

(398aa) Microscopic Models for the Prediction of Liquid Transfer in Granular Flows

Radl, S., Graz University of Technology
Mohan, B., Graz University of Technology
Girardi, M., Princeton University
Sundaresan, S., Princeton University
Khinast, J. G., Graz University of Technology

Previous research in the area of wet granular flows mainly focused on forces connected to liquid bridges. However, there is much less theory concerned with the process of liquid transfer upon particle-particle and particle-wall collisions. To study the liquid transfer upon bridge formation and rupture, the Navier-Stokes equation can be simulated, i.e., direct numerical simulation (DNS) of the three-phase system can be performed. However, DNS of wet granular flow are extremely demanding, and also detailed experiments typically bring only a rather limited insight. Thus, currently available liquid transfer models are based on certain assumptions without rigorous validation.

Here we focus on a comparison of simulation results (using the discrete element method, DEM) for the dispersion of liquid in a range of dense to dilute granular flows. We use various liquid transfer models (e.g., the one proposed by Shi and McCarthy [1], or models developed by our group) in order to highlight key parameters that impact our simulation results. Specifically, we will present results for a simple shear flow, granular flow in a rotating drum (including an experimental validation study), as well as liquid transport in dense and dilute fluidized beds. Our results help in identifying the most appropriate microscopic liquid transfer model for those flow configurations.

References

[1] D. Shi, J.J. McCarthy, Numerical simulation of liquid transfer between particles, Powder Technology. 184 (2008) 64–75.

(398ab) â€œRadicallyâ€ New Deposition of Co Particles

Van Norman, S. A., University of Colorado at Boulder
Falconer, J. L., University of Colorado
Weimer, A. W., University of Colorado at Boulder

Efficient use of precious metal catalysts is essential to enhancing catalytic productivity. Traditional catalyst impregnation proceeds through an oxide phase of the metal followed by reduction. However, with traditional methods the metal is only partially reduced limiting the fraction of active material or the metal sinters under reduction conditions limiting active surface area. The novel deposition of metallic phase Co particles and their properties is discussed.

In this study, metallic Co deposition is performed via a bi-molecular atomic layer deposition (ALD) reaction of cobaltocene and ammonia. Surface adsorbed radicals are formed on a metallic W surface produced by ALD facilitate the reaction. The radical interacts with the metal precursor resulting in chemically bonded Co to the surface. Gas phase reaction is tracked via mass spectrophotometry. Elemental analysis confirms deposition of Co. Metal loading as a function of ALD cycles performed is also explored. The reaction is also shown for oxide surfaces which do not show activity for metallic Co ALD under these conditions.

(398ac) Evaluation of Environmental Materials As Thermal Witness Materials

Vummidi Lakshman, S., New Jersey Institute of Technology
Dreizin, E. L., New Jersey Institute of Technology
Schoenitz, M., New Jersey Institute of Technology

Pyrotechnical events occur on very short time scales, generating temperatures of up to more than 2000 K and quench rapidly. These events are engineered to perform specifically for different applications. To guide the development of new materials used to generate these events, the temperature profile of the combustion clouds needs to be well described.

Measurement of temperature profiles is challenging in these high-stress environments. Electronic sensors may fail due to these stresses, and in addition may not offer the desired time resolution. Optical methods are limited by the optical thickness of the combustion environment, and may provide information of the periphery, or a line of sight average.

Thermal witness materials offer an alternative to real-time measurement methods. Micron sized particles are inserted into the combustion environment, travel along with the expanding gases and undergo irreversible changes during the exposure. The extent of change is determined post exposure. A correlation is made between the extent of change and temperature-time profile the material has experienced.

While several engineered materials are under development as thermal witness materials, this study is aimed at evaluating environmental materials for the purpose of recovering temperature-time information after combustion events. Examples of such environmental materials are minerals with volatile components which are partially lost during exposure. Other materials with non-equilibrium structures, such as glasses, effectively remember thermal conditions when they are quenched-in from their corresponding high-temperature equilibrium state. Both types of materials are sufficiently ubiquitous in both, remote and urban environments to be of interest for temperature forensics.

The examples evaluated include a synthetic jarosite (a complex sulfate hydrate), and silicate glasses.

Jarosites undergo decomposition losing (OH) and SO3. The decomposition process is complex and is governed by a set of serial and parallel reactions. The degree of decomposition depends on the experienced time-temperature profile, and can be determined using thermoanalytical methods. A correlation is made between the residual decomposition and the experienced environment.

The structures of silicate and borate glasses are strongly dependent on the quench rate during glass formation. As an indicator, the glass transition temperature varies linearly with the logarithm of the quench rate. Structural aspects, specifically the degree of connectedness of network forming units [SiO_4] and [BO_3] can be investigated using Raman spectroscopy. This allows one to recognize variations in the temperature, and cooling rate experienced by the material as it cooled from the high-temperature environment.

The use of both types of materials as thermosensors is limited by the resolution available for interrogation: themoanalysis for recovered decomposed minerals, and Raman spectroscopy for glasses. These limitations will be quantified, and general conclusions regarding other related materials that show similar effects will be discussed.

(398y) Low Dew Point Air Production Using Circulating Fluidized Bed

Matsuda, S., National Institute of Advanced Industrial Science and Technology
Hatano, H., National Institute of Advanced Industrial Science and Technology
Uenohara, Y., Tokyo University of Science

New desiccant air conditioning system, Mesh partitioned Counter current flow Moving Bed (MCMB), is proposed to replace conventional air conditioning system. The system can use low temperature exhaust heat for regeneration of adsorbent, thus, it can save consumed energy for dehumidification in supercooling type dehumidifier in which 60 % of energy is expended. Furthermore, the system can be achieved steady and isothermal operation, while it is unsteady and adiabatic operation in commercialized rotor type desiccant air conditioning system. MCMB is also applied to reduce a cost of production of a large amount of very low dew-point air, lower than -60 degree C, required in semiconductor manufacturing and lithium battery processes. In this study, it is measured and analyzed phenomenon of dehumidification section in the MCMB to make clear its properties and enhancing dehumidification performance..

Dehumidifying section of experimental setup has 600 mm long with a rectangular cross-section of 50 mm by 10 mm for particle flow zone and 50 mm by 5 mm for air flow zone. In some runs, 10 mm height is applied for the air flow zone. Air is introduced from the bottom entrance and desiccant particles is introduced from the top entrance, then, counter current contact and steady operation are achieved.

It is observed exponential decrease of the humidity. Because of the dehumidification rate constant increased with increasing φ , it is concluded that rate determining step stay at the mesh. On the other hand, rate determining step is at particle boundary layer for cellulose and silica gel, because dehumidification rate constant is constant. When height of air flow zone is increased to 10 mm, dehumidification rate constant decreased to less than half and rate determining step is considered as the mass diffusion in gas phase. MCMB can produce -52 degree C of dew-point air with 4 % of moisture content of desiccant particles. Especially, 26 degree C of air and particle temperature is extremely higher than that of commercialized dehumidifier. Dew point below -100degree C of dried air will be produced when the moisture content of desiccant is less than 1 %.

This paper makes clear that the rate determining step of three desiccant particles, effect of height in air flow zone on dehumidification rate constant. Furthermore, possibility is demonstrated to produce the -100 degree C of ultra low dew point air.

(398ad) Stabilization of Pt Catalyst With Microporous Layers Obtained By Molecular Layer Deposition

Izar, A., Instituto TecnolÃ³gico y de Estudios Superiores de Monterrey
Hollenbaugh, R., University of Colorado, Boulder
Weimer, A. W., University of Colorado at Boulder

The prevention of sintering is one of the most desired goals in the catalyst industries since its effects are hard to be reversed, leading to huge economic losses. In this paper we analyze the effect of a microporous layer on the sintering of platinum dispersed in silica. This layer was made via molecular layer deposition (MLD) with trimethylaluminum, ethanolamine and maleic anhydride, which after calcination creates a porous layer with an average pore size of 0.8 nm. We studied two cases. In the first case the platinum was dispersed on the silica first, followed by coating with the MLD process. In the second case the platinum was dispersed after the porous layer was created. In both cases the particles were annealed at different temperatures to determine the effect of the coating in the sintering of the platinum. CO oxidation was used to monitor the active life of the catalyst as well.

(398ae) Movement of Cylinder-Shaped Nano-Paticles in a Micro Channel By Individual Particle Tracking

Vo, M., The University of Oklahoma
Papavassiliou, D. V., The University of Oklahoma

The goal of this study is to determine the behavior of nanoparticles in fluid flow by calculating trajectories, orientation and hydrodynamic forces acting on the particles. Initially, a lattice Boltzmann simulation is used to determine the velocity profile in micro-flows with no slip boundary condition at low Reynolds numbers. Flow in microchannels and in microslits with width of 5 microns is considered. Then, cylinder-shaped particles are released at different locations into the flow, representing carbon nanotubes that are moving within the flow field. It is assumed that the presence of the particles does not influence the fluid velocity. The carbon nanotubes experience hydrodynamic forces that include gravity, drag, buoyancy, lift and pressure-gradient forces [1]. Besides, Brownian force is important, because of the nanoscale dimensions of the nanotubes. Additionally, the interaction between the particles and the wall can be accounted by defining a Morse-like potential in regions very close to the wall [2]. The velocity and position of each particle is computed numerically at each time step. Based on hydrodynamic forces, a system of rotation equations is utilized to calculate the angular velocity and orientation of each particle.

The paper will discuss the effect of the aspect ratio of particles on hydrodynamic forces and on the rotation of particles, as well as the relative effect of each force on the particle trajectories and on the collision of the particles with the micro-channel wall. The distribution of the orientation angle of the nanotubes relative to the direction of the flow will also be discussed. Finally, the lattice Boltzman/particle tracking results will be compared to results obtained by dissipative particle dynamics at smaller scales.

This work can find applications in the determination of the trajectories of carbon nanotubes as they are released in the subsurface, where they can find their way to the water table, or in cases where nanoparticles are used to enhance oil recovery [3] or as tracers in haydraulic fracturing fluids.

ACKNOWLEDGEMENTS

The financial support of the Advanced Energy Consortium (AEC BEG08-022) and the computational support of XSEDE (CTS090017) are acknowledged.

REFERENCES

1. Maxey, M.R., and J.J. Riley, "Equation of motion for a small rigid sphere in a non-uniform flow", *Phys. Fluids*, **26**, 883-889, 1983

2. Kolmakov G., Ravanur, R., Tangirala, R., Emrick, T., Russell, T.R., Crossby A.J., and A.C. Balazs, "Using nanoparticle-filled capsules for site-specific healing of damaged substrates: Creating a repair and go system", *ACS Nano,* **4**(2), 1115-1122, 2010.
3. Kong, X. , and M. M. Ohadi, "Applications of Micro and Nano Technologies in the Oil and Gas Industry - Overview of the Recent Progress", *Society of Petroleum Engineers*, paper # 138241, 2010.

(398e) Flow Regimes Transition and On-Lone Monitoring of Gas-Solid Fluidized Bed

Zaid, F. M., Missouri University of Science and Technology
Al Dahhan, M. H., Missouri University of Science and Technology

The fluidized bed reactors can operate in different flow regimes. The reactor performance can change significantly as a result of flow regime change. It is very important to know how to identify the boundaries of the different flow regimes since reactor volume productivity, mixing, mass and heat transfer are affected by the prevailing flow regime. In this work implement a technique based on gamma ray densitometry (GRD) to identify and monitor on-line flow regime and pattern in 0.14 m ID Plexiglas fluidized bed. The flow regime and its transition will be first investigated using an optical probe and pressure transducer. Statistical analyses have been performed on the time series measurement of these techniques for identifying the flow regimes and their transitions Pressure transducer was measured at the wall while optical probe measured inside the bed and Gamma- ray densitometry for monitored on-line at different axial positions.

Based on the analysis notable differences were observed in flow regimes transition along the bed axial and radial locations particularly between the sparger (Z/D=0.286) and the fully developed flow (Z/D= 1.7) regions. There are slight variations between the flow regime transitions obtained using pressure transducer at the wall and those obtained inside the bed using optical probe and Gamma- ray densitometry at the studied height bed. Such information is needed for preparing a flow regime map and scale-up. Detailed data and findings will be discussed and presented in this talk.

(398ag) Anomalous Aqueous Dispersion of Hydrophobic Hedgehog Particles

Bahng, J. H., University of Michigan
Yeom, B., University of Michigan
Wang, Y., University of Michigan
Tung, S. O., University of Michigan
Kim, J. Y., University of Michigan
Kotov, N., University of Michigan

Forcing hydrophobic particles to be dispersed in water is a common requirement for numerous applications from drug delivery to oil recovery. Such dispersions are typically achieved by using surfactants or polymers that "camouflage" hydrophobic surfaces with hydrophilic groups. Aqueous dispersion of hydrophobic particles without the use of any surfactants has both technological and academic significance. In this study, we demonstrate that aqueous dispersion of hydrophobic particles without any chemical canopy is indeed possible with reengineering of the interfacial topography. We sculptured the surface of a polystyrene microsphere to feature hydrophobic and rigid ZnO nano-spikes. The Hedgehog particles (HHPs), reflective of its morphology, forms a three-phase interfacial shell; radially distal portion occupied with water and the proximal portion trapped with air both of which are interdigitated with ZnO nano-spikes. We believe the aqueous stability of hydrophobic HHPs is due to its comparable density to the medium and increased charge repulsion due to high surface area. Furthermore, we demonstrate biomedical applications where we anchored the hydrophobic HHPs onto the cell membrane which opens up the possibilities of unconventional ornaments of biological cues onto the cells and of increased complexities in cell-cell signaling.

(398ah) Investigation of Permeation of Dispersed Particles Through a Pore and Transmembrane Pressure Behavior in Dead-End Constant-Flux Microfiltration Using Direct Numerical Simulation

Fuse, H., Kobe university, Center for Membrane and Film Technology
Ishigami, T., Kobe university, Center for Membrane and Film Technology
Kagawa, Y., Kobe university, Center for Membrane and Film Technology
Asao, S.,
Matsuyama, H., Kobe university, Center for Membrane and Film Technology

Introduction

Microfiltration (MF) plays an important role in water treatment processes. Constant-flux filtration, by which end users can obtain a constant desired supply of filtered water, is often used in water treatment processes by MF. In the case of the constant-flux filtration, the transmembrane pressure represents the filtration performance of the MF membrane. It is important in water treatment processes to understand the filtration dynamics of dispersed particles, because these govern transmembrane pressure behavior corresponding directly to the pumping power required to supply the feedwater. During dead-end MF, particles could permeate the pores in some cases where the particle size is smaller than the membrane pore size. It is well-known that the transmembrane pressure increases due to the membrane fouling. However, it is not as well understood that the transmembrane pressure also increases by the hydrodynamic interactions between fluid and particles, without fouling, as the suspension permeates through the membrane. This is because it is very difficult to experimentally study the flow characteristics and particle motions within the pore that influence hydrodynamic interactions. The aim of this study was to investigate the effects of fluid and particle motions on transmembrane pressure behavior during dead-end, constant-flux filtration. The effects of pore size on the transmembrane pressure were investigated under constant-flux microfiltration. On the basis of these investigations, we clarify the effect of hydrodynamic interactions between the fluid and particles on the transmembrane pressure.

Numerical method

The governing equations of fluid flow are the equation of continuity and the Navier-Stokes equation. The interaction force between the particle and fluid is described as the momentum exchange at the fluid-particle interface. The momentum exchange is solved by the immersed boundary method developed by Kajishima *et al*[1] without specification of the boundary condition around the particles. The motion of each particle was tracked with a Lagrange scheme by solving the equations of linear momentum and angular momentum. Hydrodynamic, electrostatic, van der Waals, and contact forces are considered as forces acting on a particle.

The computational domain is assumed for the simulation of dead-end, constant-flux filtration. The size of the domain was L_x = 20.0d, L_y = 6.0d and L_z= 6.0d, where d is the diameter of 1.0 μm particles. A membrane with a column straight pore was mounted in the center of the streamwise direction of the computational domain, with the membrane thickness set at 10 μm. The centers of particles were initially placed at x = 1.0 μm and then flowed in with the same translational velocity as the inlet uniform velocity of the fluid. To investigate the effect of pore size on the transmembrane pressure, the pore diameter D was changed in three steps of 2.0, 3.0, and 5.0 μm.

Results and Discussion

In all cases, particles move with the fluid flow and the fluid velocity around the particles then decreases when particles enter the pore. On the other hand, the central velocities in all cases do not decrease significantly as particles penetrate and pass through the pore. This results indicates that the effects of particle penetration on hydrodynamic resistance do not differ significantly with pore size. We investigated the time variations of normalized transmembrane pressure, $\Delta p/\Delta p_0$. Δp is normalized by the transmembrane pressure at steady state without particles. The calculated result revealed that, the mean values of $\Delta p/\Delta p_0$ were similar in all cases. In addition, the number of particles that permeate per unit time are also similar in all cases because the volume fractions of feedwater and the water flux rate are the same in the present conditions. These tendencies are consistent with the velocity field as described above. On the other hand, the transmembrane pressure behaviors show periodicity in all cases. The peaks correspond to the times when particles are positioned at the inlet and outlet of the pore. Particles constantly enter from the inlet of the computational domain in the present conditions, giving rise to the transmembrane pressure periodicity. The peak values in the case of D = 2.0 μm are much higher than those in other cases because the velocity gradient at the inlet and outlet of the pore becomes sharp with decreasing pore size. This sharp velocity gradient contributes a high drag force between the fluid and particles.

Reference

[1]Kajishima, T. et al., Turbulence structure of particle-laden flow in a vertical plane channel due to vortex shedding. *JSME Int. J., Ser, B* 2001, **44**, 526

(398ai) Importance of Temperature On the Gel Electrophoresis of a pH-Regulated, Zwitterionic Sphere

Hsu, J. P., National Taiwan University
Huang, C. H.,
Hsu, C., National Taiwan University
Lee, D. J., National Taiwan University

Gel electrophoresis has been adopted widely in soft matter science to characterize/separate entities such as DNAs and proteins. Compared with that in aqueous electrophoresis, the mobility of samples in gel electrophoresis is slower and, therefore, easier to be observed. Due to the presence of Joule effect, temperature can be a significant factor in electrophoresis. Because all the physicochemical properties of the dispersion medium and those of the dispersed entities vary with temperature, its influence on the electrophoretic behavior of an entity can be significant. In this study, we consider the electrophoresis of a pH-regulated, zwitterionic spherical particle in a gel medium under various levels of temperature. This type of particle is capable of simulating a wide class of entities including biocolloids such as DNAs, proteins, and microorganisms, and inorganic colloids such as SiO_2 and TiO_2, where dissociation/association reactions occur on their surface and the associated charged conditions depend upon nearby environment. Choosing an SiO_2 particle in a dilute agarose gel as an example, a detailed numerical simulation is conducted to investigate its electrophoretic behavior under various conditions. The parameters examined include temperature, pH, and the bulk salt concentration. To simulate the usual experimental conditions, we assume that the temperature and the solution pH vary from 283 K to 323 K and 5 to 8.5, respectively.

(398f) Jamming of Elastic Particles in Granular Systems

Swaminathan, S., Purdue University
Wassgren, C. R., Purdue University

Particle jamming in the gap between a rotating blade and container frequently occurs in particle handling equipment. Jamming can be desired, as in case of particle mills or detrimental, for example in mixers and granulators. In this study, we present an analytical model to predict the onset of jamming for elastic particles. It was found that the onset of jamming can be calculated from the particle properties and surface interaction properties. Further, we present experimental results to compare to the model.

(398af) Diffusiophoresis of a Charged Sphere in a Necked Nanopore

Lo, T. W.,
Hsu, C., National Taiwan University
Liu, K. L.,
Hsu, J. P., National Taiwan University
Tseng, S., Tamkang University

In an attempt to improve the performance of single-molecule nanopore based biosensors, we study theoretically the diffusiophoresis of a particle along the axis of a charged, necked nanopore, and the mechanisms involved. Numerical simulation is conducted by varying the geometry of the nanopore and its charged status, the particle position, and the background salt concentration to examine the particle behavior under various conditions, and to gather necessary information for future device design. We show that the strength of the local concentration gradient, the charge density of the nanopore, and the presence of the nanopore wall yield complicated and interesting diffusiophoretic behaviors. For instance, if both the particle and the nanopore are negatively charged, the particle mobility decreases monotonically as it approaches the throat region, shows a local minimum in its throat region, and a local maximum in the outlet region. In the inlet and outlet regions of the nanopore, if the nanopore is negatively (positively) charged, the mobility is larger (smaller) than that if it is uncharged. These arise from the effect of ion concentration polarization. In nanopore based DNA sequencing, raising the DNA capture rate by the nanpore and reducing its translocation velocity inside the nanopore are crucial challenging problems. These problems can at least be partially solved by adopting our design.

(398ak) Production of Nano/Micro Particles of Antiretroviral Drugs Using Supercritical Technology

Bevilaqua, G.,
Rosa, P. T. V., State University of Campinas

An important demand in the pharmaceutical industry is the production of bioactive particles with well-defined morphology and reduced size in order to increase the bioavailability of the drug in the body. This makes it possible to reduce the need for daily drug intake with consequent reduction of side effects to the patient. One can also increase the circulation time of drug in the body, so as to increase the efficiency of the therapy. Since the particle size distribution of the drug may affect its pharmacological action, it is desirable that it has the lowest size distribution.

Various techniques are used to achieve these objectives in the formation of particles, such as spray-drying, freeze-drying and granulation, among others, which require drastic conditions of operation and may not allow control of the size of the drug particles formed. As an alternative to the limitations imposed by conventional techniques, supercritical technologies have been used in micronization of bioactive particles, producing particles of defined morphology and small size distribution. These technologies utilize a fluid in supercritical conditions, usually carbon dioxide, because of their low toxicity and high diffusion coefficient. Its use is fairly stated in micronization of labile substances, since its critical point is considered mild. The supercritical technologies have the advantage of easy control of particle characteristics with small variations in operating conditions such as pressure and temperature. In addition, the formed particles tend to have low residual solvent quantity than those formed by traditional techniques and a lower number of separation processes in order to obtain the final product.

Various techniques for the production of particles using supercritical technology where carbon dioxide can act as a solvent, as in the Rapid Expansion of Supercritical Solutions (RESS) or anti-solvent techniques such as Supercritical AntiSolvent (SAS) and Solution-Enhanced Dispersion by Supercritical Fluids (SEDS). The application of the technique depends on the drug solubility in the supercritical fluid.

The SAS technique, used in this work, is used in micronization of drugs with low solubility in supercritical media. Thus, the bioactive substances are solubilized in an organic solvent soluble in the supercritical fluid. The fluid is injected into a vessel and expands, while the organic solution containing the drug is also injected into the same vessel. Carbon dioxide removes the organic solvent, promoting supersaturation of the drug in the middle and causing precipitation. The SAS technique has proved effective not only in micronization of drugs, but also in the synthesis of drug delivery systems and encapsulation of bioactive compounds in various substances.

There are several factors that affect the size of the particles obtained, their size distribution and morphology. In this work will be tested different operating conditions in micronization of antiretroviral drugs such as temperature, pressure, concentration and solution flow drug/solvent, among others. The organic solvent used to solubilize the drug is acetone.

It is expected thereby producing nano and micrometric particles through the SAS technique to improve the bioavailability of drugs in the body and to study the influence of various operational parameters on the characteristics of the produced particles.

(398c) Recent Advances in Microreactor-Based Calorimetry for Monitoring Rapid and Strong Exothermic Processes

Loebbecke, S., Fraunhofer Institute for Chemical Technology (ICT)
Antes, J., Fraunhofer Institute for Chemical Technology ICT

Powerful process analytical techniques which provide a high level of information while simultaneously keeping experimentation times short are essential for the process design of energetic materials syntheses. For example, calorimetry is a universal analytical method for investigating thermal effects resulting from chemical reactions and/or changes in physical states. Calorimetric measurements make it possible not only to make direct statements about thermodynamics but also about kinetics and physical constitutions. Knowing the heat flows that occur during a reaction therefore provides crucial information for the successful and safe design of processes producing energetic materials.

In this context, the combination of calorimetric measurement technologies with continuously operated microreactor processes offers an extremely promising methodical approach. The technical advantages offered by microreaction technology make it particularly suited for use in processes with an increased hazard potential – such as removing strong reaction heat, preventing unwanted side or decomposition reactions or handling explosive and other unstable products and intermediates in small hold-ups under short residence times.

Here we report on the development of various small-volume, continuously operated calorimeters on the basis of microreactor chips which permit rapid screening of thermokinetic key data of chemical processes. At the heart of the calorimeters are sensor arrays based on miniaturized Seebeck elements for the localized, quantitative characterization of heat flows. The sensor arrays consist of up to 40 individual sensors, which can collect data concerning the reaction heat generated in a microreactor with a high degree of temporal and spatial resolution. Due to the large surface-to-volume ratio of microreactors and the resulting small time constants of just a few seconds, it is also possible to perform isothermal calorimetric measurements for rapid and strong exothermic reactions. The test volume is extremely small ($< 100\ \mu L$), which enables a fast reaction screening.

The combination of continuous microreaction technology and efficient heat flow sensors thus opens the way to rapid and extremely efficient screening of chemical reaction parameters and allow acquisition of safety data. During screening experiments, the influence of individual process parameters (e. g. concentration, stoichiometry, use of alternative reactants, temperature, residence time, etc.) can be followed directly by observing the altered heat signal – both in a qualitative and quantitative manner. The measurement of the heat flows is in real-time; time-consuming calibrations of the heat transfer are not required. Even targeted and quantitative analysis of the energetic potential of critical process conditions (worst case scenarios) can be conducted safely.

(398al) Modeling Dust Explosions

Spijker, C., Universitiy of Mining Leoben

Kern, H., University of Mining Leoben

Raupenstrauch, H., Montanuniversität Leoben

Held, K., Montanuniversität Leoben

Introduction

Dust explosions are a major safety topic, on every plant, that handles flammable granular or pulverized substances. Commonly dusts are characterized by two scalars, the maximum explosion pressure and the maximum rate of pressure rise. With that method it is not possible to know how the flame front and the pressure act in the geometry. When a description of the behavior of a dust explosion in the geometry is needed, a CFD (Computational Fluid Dynamics) model, which can describe the physics of a dust explosion, has to be used

For the modeling of gas explosion, two types of combustion models are common. On the one hand chemical and mixture limited reaction models like the eddy dissipation concept, on the other hand flame speed approaches. To model the dust explosion with a chemical and mixture limited reaction model, the pyrolysis of the dust has to be characterized. The calculation of the pyrolysis in the dust particles needs a kinetic and enormous computing recourses.

If a flame speed approach is used, the durst can be characterized by the laminar flame speed and the burnable fuel fraction as function of the dust concentration and the heat of combustion. The flame speed approach is also relatively cheap in terms of computing resources.

So a CFD code to model dust explosions, which is based on a flame speed approach was developed. This solver is based on the XiFOAM solver of the software package OpenFOAM.

Description of the model

Basic concept

The key parameter is the turbulent flame speed, which is highly depended to the turbulence and also depends on the laminar flame speed. The laminar flame speed is a function of the dust concentration. These values are based on measurements with a Siwek 20-liter vessel. With the pressure time curve from the experiment, the laminar flame speed can be calculated, shown by Skjold [1]. It is also possible to calculate the burnable fuel fraction by energy balancing. Figure 1 shows the laminar flame speed and burnable fuel fraction of lycopodium as function of the concentration.

The influence of the turbulence on the flame speed follows the description of Dahoe [2].

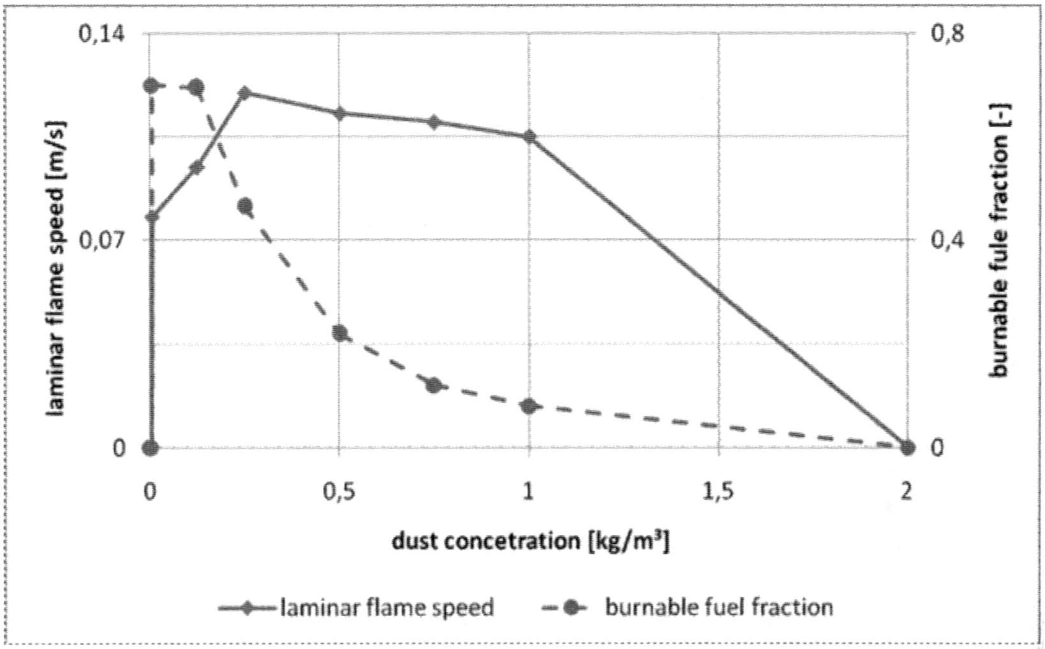

Figure 1: laminar flame speed and burnable fuel fraction of lycopodium as function of the concentration

Particle and gas motion

To describe the movement of the dust particles an Eulerian approach was chosen because of the high particle count. In bigger geometries a Lagrangen approach is too expensive in terms of computing resources.

The exchange coefficient for momentum *K* is based on the work of Syamlal [3] and the drag coefficient, to calculate is described by the Schiller Neuman approach [4]. The granular pressure *ps* and the granular temperature are computed by an analytical published by Syamlal [5].

Equation 1 describes the momentum equation of the gaseous phase and equation 2 the momentum equation of the particle phase.

$$\frac{\partial\left(\varepsilon\rho\vec{u}\right)}{\partial t}+div\left(\varepsilon\rho\vec{u}\vec{u}\right)-div\left(\varepsilon\tau_{g}\right)-\varepsilon\rho\vec{g}+K\left(\vec{u}-\vec{v}\right)=-grad\left(p\right)\varepsilon \tag{1}$$

$$\frac{\partial\left(D\vec{v}\right)}{\partial t}+div\left(D\vec{v}\vec{v}\right)-div\left(\tau_{p}\right)-D\vec{g}+K\left(\vec{v}-\vec{u}\right)=-grad\left(p\right)\left(1-\varepsilon\right)-grad\left(ps\right) \tag{2}$$

Equation 3 and 4 show the continuity equation. The source term on the right side, describes the mass transfer between the phases through the combustion process.

$$\frac{\partial\left(\varepsilon\rho\right)}{\partial t}+div\left(\varepsilon\rho\vec{u}\right)=\rho_{unburned}U_{lam}Bsh\left|grad\left(C\right)\right|B_{Aus}D \tag{3}$$

$$\frac{\partial\left(D\right)}{\partial t}+div\left(D\vec{u}\right)=-\rho_{unburned}U_{lam}Bsh\left|grad\left(C\right)\right|B_{Aus}D \tag{4}$$

A comparison between the used Eulerian and the Fluent DPM model is shown in figure 3. In a simple 2D geometry, shown in figure 2 particles where injected. The curves in figure 3 show the dust concentration on the Y axis after different times.

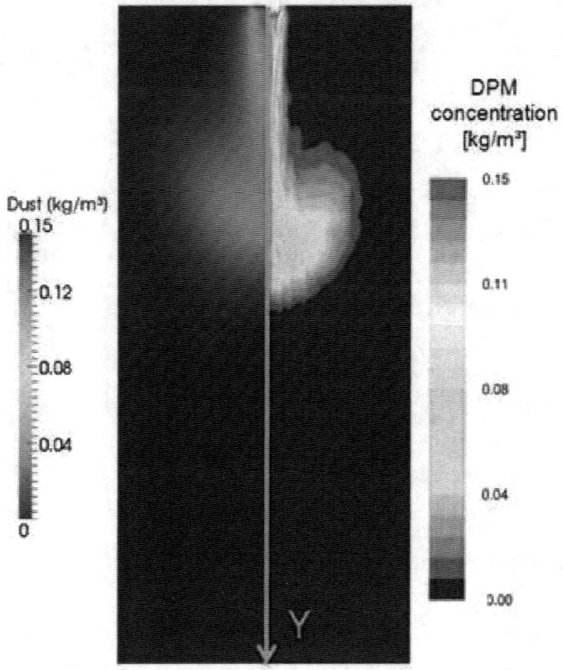

Figure 2: Geometry for comparison between the used Eulerian and the Fluent DPM model

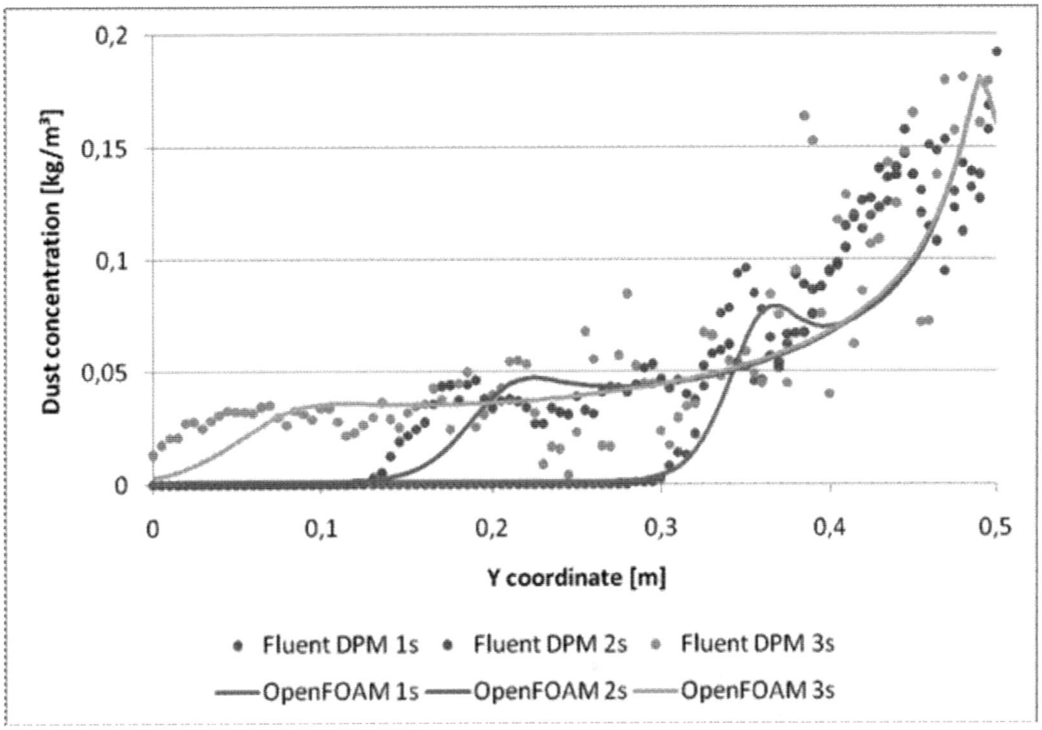

Figure 3: Comparison between the used Eulerian and the Fluent DPM model

Turbulence modeling

Early simulations had shown that the in OpenFOAM implemented RANS fail. The reason for that is the high relative velocity in front of the flame front. The high relative velocity generates turbulent kinetic energy behind the particles. To model this effect an additional source tem, based on the consistent approach of Crowe [6] was implemented in the k equation of the standard k-e model. Equation 5 shows the modified k equation.

$$\frac{\partial(\rho k)}{\partial t}+div(\rho k\dot{u})-div\left[\mu+\frac{\mu_t}{\sigma_k}grad(k)\right]+P_k+=\frac{D}{\tau_p}|\vec{u}-\vec{v}|^2-\rho eps \qquad (5)$$

Combustion model

Like in every flame speed approach the state of combustion is defined by a progress variable, in this case called C. If C has a value of 1 the dust-air mixture is unburned. For a burned mixture C has a value of 0. The source tem of conservation equation for C, which is shown in equation 6, includes the laminar flame speed, the variable Bsh, which describes the ratio between the laminar and the turbulent flame speed and the magnitude of the gradient of C. That source term is also used in other equations, to describe changes based on combustion processes.

$$\frac{\partial(\rho C\varepsilon)}{\partial t}+div(\vec{u}\,\rho C\varepsilon)-div[a_{eff}\,grad(C)]=\rho_{unburned}U_{lam}Bsh|grad(C)| \qquad (6)$$

Energy Equations

To be able to consider the high relative velocities in front of the flame front, for each phase a separate energy equation was implemented. The energy equation for the gaseous phase, shown in equation 7, has two source terms. The first one describes the energy of combustion based on the heat of combustion, dust concentration, the burnable fuel fraction and the combustion source term. The second term describes the energy exchange to the particle phase, based on the heat exchange coefficient of Ranz-Marshall [4].

The energy equation for the particle phase, shown in equation 8 has just a storage function. The source term just describes the energy transfer to the gaseous phase.

$$\frac{\partial(\varepsilon\rho T_g)}{\partial t} + div(\varepsilon\rho\bar{u}T_g) - div[a_{Eff}\, grad(T_g)] = \rho_{unburned} U_{lam} Bsh|grad(C)| B_{Aus} H_{Vbr} D + \alpha_{g,p}(T_p - T_g) \quad (7)$$

$$\frac{\partial(DT_p)}{\partial t} + div(D\bar{u}T_p) = \alpha_{g,p}(T_g - T_p) \quad (8)$$

Species Equation

The density of the gas is a very important parameter for expansion. It is based on pressure, temperature and the molar mass. To describe the molar mass the variable b was introduced. This variable describes the off gas- air mixture. The conservation equation for b is shown in equation 9. The source term is based on a coefficient, dust concentration, the burnable fuel fraction and in the combustion source term.

$$\frac{\partial(\varepsilon\rho b)}{\partial t} + div(\varepsilon\rho\dot{u}b) - div[d_t\, grad(b)] = -\rho_{unburned} U_{lam} Bsh|grad(C)| B_{Aus} Q_{Vbr} D \quad (9)$$

Results and comparison to the experiment

To evaluate the model, in house experiments from Kern [7] were used. Figure 4 shows the experimental set up. A screw conveyer delivers a defined amount on dust, in that case lycopodium, in to a vertical duct.When the dust gets a constant defined concentration, the mixture is ignited by an electric spark.

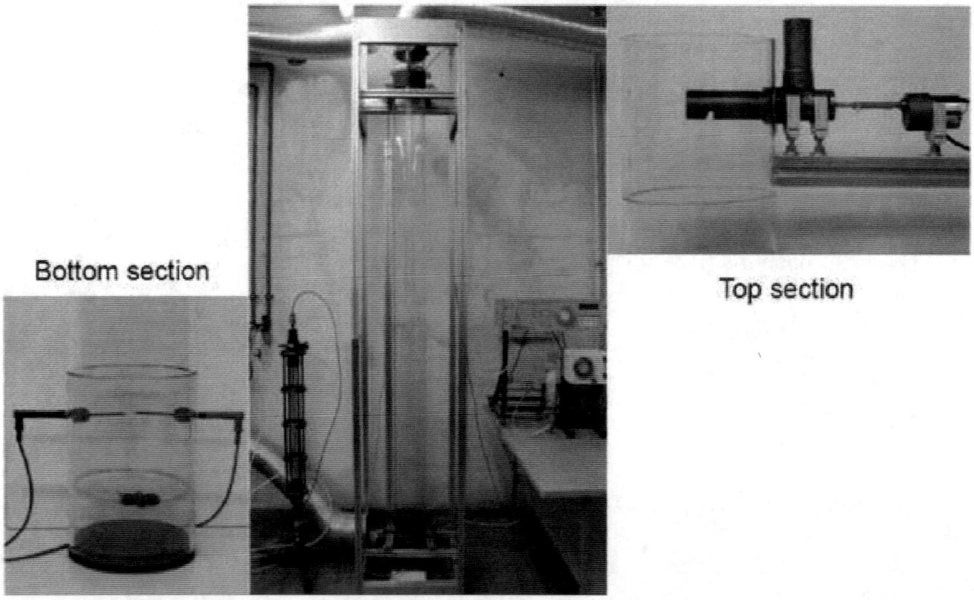

Figure 4: Experimental set up

In figure 5 the comparison of the flame front between experiment and the model is shown. For a concentration of 300g/m³ lycopodium. The model fits well to the experiment in the first 200ms. After that the results gets different. The reason is that the momentum out of the duct influences turbulence after 200ms. In the experiment a flame arrester at the top of the duct and the off gas system has to be used. This parts are not implemented in the model.

In the future, experiments with the duct at another location are planned, where no flame suppression barrier and the off gas system is needed.

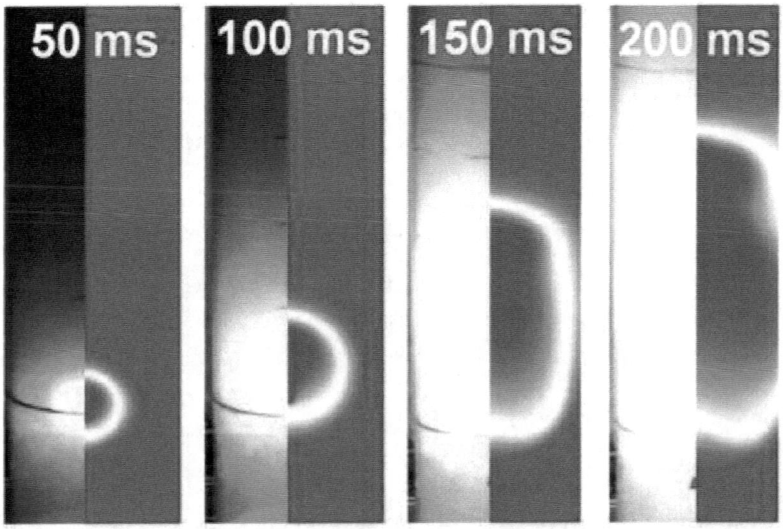

Figure 5: comparison between experiment and model

<>List of abbreviations

a_{Eff}	Effective thermal diffusivity	t	Time
b	Progress variable off gas- air mixture	T_g	Temperature gaseous phase
B_{aus}	Burnable fulefraction	T_p	Temperature particle phase
Bsh	Ratio between laminar and turbulent flamespeed	$\vec{u}$	Velocity vector gaseous phase
C	Progress variable flamefront	U_{lam}	Laminar flame speed
D	Dust concentration	$\vec{v}$	Velocity vector particle phase

eps	Dissipation of the turbulent kinetic energie	$a_{g,p}$	Heat exchange coefficient
$\vec{g}$	Gravity vector	ε	Volume fraction gaseous phase
H_{Vbr}	Heat of combustion	τ_g	Stress tensor gaseous phase
K	Momentum exchange coefficient	τ_p	Stress tensor particle phase
p	Pressure	ρ	Density gaseous phase
ps	Granular pressure	ρ *unburned*	Density gaseous phase unburned
Q_{Vbr}	Chemical coefficient		

<>

<>Literature

[1] T. Skjold, B.J. Arntzen, O.R. Hansen, O.J. Taraldset, I.E. Storvik, R.K. Eckhoff, Simulating Dust Explosions with the First Version of DESC, Process Safety and Environmental Protection, Volume 83, Issue 2, March 2005,151?160

[2] A.E., Dahoe, 2000. Dust explosions: a study of flame propagation, PhD

thesis, Delft University of Technology, Delft, Holland

[3] M. Symlal, J. O'Brain, (1989), ?Computer Simulation of Bubbles in a Fluidized Bed?, AIChE Symp. Series 85, 22-31

[4] FLUENT (2010), ?FLUENT (2003), ?Ansys Fluent Theory Guide?, Fluent Inc

[5] M. Symlal, J. O'Brain, W. Rogers, (1993), ?MFIX Documentation?, National Technical Information Service, Springfield, US.

[6] C. T. Crowe, On models for turbulence modulation in fluid-particle, International Journal of Multiphase Flow 26 (2000) 719-727

[7] H. Kern, K. Held, H. Raupenstrauch, Investigations on the influence of the oxygen concentration on the flame propagation in lycopodium/air mixtures, ACHEMA 2012

Frankfurt, June 21st 2012

(398g) Near Infrared Spectroscopic Method Development Based On Effective Simple Models for Continuous Quantitative Monitoring of Powder Mixing and Final Dosage Product in a Pharmaceutical Process

Colon, Y., University of Puerto Rico

Romañach, R. J., University of Puerto Rico at Mayaguez

NIR Spectroscopic Method Development Based on Effective Simple Models for Continuous Quantitative Monitoring of Powder Mixing and Final Dosage Product in a Pharmaceutical Process

This research addresses the need to make near infrared (NIR) calibration models more robust. NIR spectroscopy has been effectively used for chemical quantification and to determine various physical properties of pharmaceutical samples [1-4]. Multivariate calibration methods have proven to be an excellent tool for extracting quantitative and physical information from NIR spectra. In spite of this progress there are still examples of NIR calibration models developed with lab prepared calibration samples with very poor prediction ability of manufacturing process samples [5, 6]. These research efforts provide evidence that there is still the need to identify how process variation affects the effectiveness, robustness and reliability of NIR calibration models. Within an established quality system and for a particular manufacturing process, one would expect an inverse relationship between the level of process understanding and the risk of producing a poor quality product [7]. Therefore, the proposed work will provide knowledge on the specific variation of the NIR spectra resulting from the effects of the individual manufacturing unit operations.

References

1. Barajas, M.J., et al., *Near-Infrared Spectroscopic Method for Real-Time Monitoring of Pharmaceutical Powders During Voiding.* Applied Spectroscopy, 2007. **61**(5): p. 490-496.

2. Ropero, J., et al., *Near-infrared Spectroscopy for the In-line Characterization of Powder Voiding Part I: Development of the Methodology.* Journal of Pharmaceutical Innovation, 2009. **4**(4): p. 187-197.

3. Blanco, M. and M. Alcalá, *Simultaneous quantification of five active principles in a pharamacutical preparation: Development and validation of near infrared spectroscopic method.* European Journal of Pharamaceutical Sciences, 2006. **27**: p. 280-286.

4. Ropero, J., et al., *Near-Infrared Chemical Imaging Slope as a New Method to Study Tablet Compaction and Tablet Relaxation.* Appl. Spectrosc., 2011. **65**(4): p. 459-465.

5. Blanco, M., M. Bautista, and M. Alcalá, *Preparing calibration sets for use in pharmaceutical analysis by NIR spectroscopy.* Journal of Pharmaceutical Sciences, 2008. **97**(3): p. 1236-1245.

6. Blanco, M., et al., *Influence of the procedure used to prepare the calibration sample set on the* performance of near infrared spectroscopy in quantitative pharmaceutical analyses. Analyst, 2001. 126(7): p. 1129-34

7. FDA, Guidance for industry: PAT ? A framework for innovative pharmaceutical development, manufacturing and quality assurance; September 2004

(398a) Effective Separation Chamber Length of Uniflow Cylones for Optimal Separation Efficiencies
Kraxner, M., MCI

Effective Separation Chamber Length of Uniflow Cylones for optimal Separation Efficiencies

M. Kraxner[1], F. Lauterbach[1], M. Pillei[1], T. Kofler[1], J. Bolda[2], B. Huurdeman[2], U. Muschelknautz[1]

[1]*MCI – The Entrepreneurial School, Environmental, Process & Energy Engineering, Maximilianstraße 2, A-6020 Innsbruck, AUSTRIA*

[2]*MANN+HUMMEL GmbH, Brunckstraße 15, D-67346 Speyer, GERMANY*

Uniflow cyclones used for the separation of particles from gases are characterized, compared to standard reverse flow cyclones, by its compactness and its low energy consumption. The compact construction gives plant operators the possibility of easy implementation in piping systems. The length of the separation chamber affects the particle collection efficiency considerably. The separation procedure of particles in a uniflow cyclone is physically identical to a sedimentation process in a centrifugal field. Therefore it can be assumed that elongating the separation chamber length increases the separation efficiency. On the other hand *Gauthier et al.* showed in the early 1990 that the length of the separation chamber should be kept short ($L/D \leq 4$), to achieve highest collection efficiencies [1].

The aim of this work is to shed more light on the observation of an optimum separation chamber length. For this purpose 2D/3C-PIV measurements have been carried out to evaluate the change of the gas velocities and thus of the swirl rate as a function of the separation chamber length. The PIV laser measurements of the flow field and dedusting tests, which were performed parallel, show that the length of the separation chamber should be within $1 \leq L/D \leq 2$ to generate highest separation efficiencies. The solids-loaded gas flow in the outer layer of the separation chamber suffers high friction between at the wall and leads to a rapid decrease of the centrifugal forces.

[1] Gauthier, T.A., Briens, C.I., Bergougnou, M.A., Galtier, P., Uniflow cyclone efficiency study, Powder Technology, Nr.62: 217-225, 1990.

(398b) Optimization of the Inlet Vane Design in Uniflow Cyclones
Pillei, M.,
Kofler, T.,

Optimization of the Inlet Vane Design in Uniflow Cyclones

M. Pillei[1], M. Kraxner[1], T. Kofler[1], U. Muschelknautz[1]

[1]MCI – The Entrepreneurial School, Environmental, Process & Energy Engineering, Maximilianstraße 2, A-6020 Innsbruck, AUSTRIA

Uniflow cyclones represent a seperator class which is advantageous in some applications compared to standard reverse-flow cyclones. Economic benefits may from the compact design, the low pressure drop and the easy implementation in existing piping systems.

The main aim of this investigation is to examine the influence of shortening the guide vanes of the swirl generator on collection efficiency and pressure drop. Therefore the length of the inlet vane guides is decreased systematically. Preliminary results already presented at the AIChE 2012 Annual Meeting (639f, Influence of the Inlet Vane Geometry on the Uniflow Cyclones Performance) show an optimization potential. This study extends these results and confirms the expected positive effect. For this purpose particle image velocimetry (2C-PIV) and CFD simulations are applied. The results from these measurements show that the alternating shortening of the guide vanes improves the collection efficiency. To minimize experimental efforts a numerical sensitivity study has been carried and the rated range has been identified. The 2C-PIV is used to validate the CFD simulation data. Moreover experiments with solid-loaded flows indicate similar tendencies.

The results show that the shortening of the guide vanes leads to a detachment of the flow at the backside of the vanes leading to higher velocities in the vane channel. As a result the collection efficiency increases at an unchanged constant pressure drop.

(398aj) Numerical Simulation of Interaction of Solid Particles With Elastic Bodies By a Lagrangian-Lagrangian Approach

Mizutani, S., The University of Tokyo
Sakai, M., School of Engineering, the University of Tokyo

Interaction between solid particles with elastic bodies often occurs in industrial products, for example powder conveying using rubber tubes or blade-cleaning process. In such processes, elastic bodies, or rubberlike material often changes their shape extremely large. It is difficult to acquire the internal information of elastic bodies in such processes by experiments, and the methodology of contact problem of elastic bodies with large deformation has been not established because Finite Element Method, the conventional structural analysis method, has a difficulty of simulating both of contact problems and large deformation analysis. Therefore development of a method that can deal with such phenomena is expected in industrial area.

Accordingly, we developed the coupled analysis method using Finite Deformation theory based Particle Method (FD-PM), which was developed on the assumption of large deformation and Discrete Element Method (DEM) as the model to represent contact between solid particles with elastic bodies. The extra algorithm is not required to combine these two methods because FD-M and DEM are both Lagrangian methods. Numerical simulations were executed for validation. The deformation of the elastic bodies was measured and the results of the simulations corresponded with experiments at the same scale.

(398h) Drag Force Correlation for Assemblies of Non-Spherical Particles in Gas-Fluidized Bed

Chen, Y., ETH Zurich
Third, J., ETH Zurich
Muller, C. R., ETH Zurich

Gas-solid systems are of significant industrial importance and are the basis of various reactor concepts, e.g. gas-fluidized beds or rotary kilns. However, with regards to numerical modeling of those systems, very little is known about the force that exerts on the individual particles, when fluid flows past an assembly of particles, the drag force. Thus, a good understanding of particle-fluid interactions is essential to predict more accurately the dynamics of these systems.

Using a multi-scale modeling approach, the closure relationship which describes the effective particle-particle interaction can be obtained from discrete particle simulations, whereas the closure relationship describing the effective fluid-particle interaction can be obtained from empirical relations or direct numerical simulation (DNS). Currently, the drag force correlations of Ergun (1952) or Wen &Yu (1966) are commonly used. However, these correlations are often applied for conditions in which their validity is questionable. Direct numerical simulations using e.g. the lattice Boltzmann method (LBM) allows us to measure fluid-particle interactions and, thus, formulate accurately drag force correlations.

Using LBM, Koch (Koch and Hill, 2001) developed a drag force correlation for an array of spheres. Van der Hoef and Beetstra (van der Hoef et al., 2004; Beetstra et al., 2007) extended the work of Koch and Hill (2001) establishing a new drag force correlation for mono- and bi-disperse arrays of spheres. The drag force correlations derived from lattice Boltzmann simulations showed a better agreement with experimental measurements than the traditional Ergun and Wen &Yu correlations.

However, due to a lack of drag force correlations for *assemblies* of *non-spherical* particles, numerical simulations of gas-fluidized beds have been restricted mostly to beds containing spherical particles. Indeed, even for *single* non-spherical particles comparatively little work has been reported. For example, Tran-Cong et al. (2004) measured the drag force coefficients for *isolated* non-spherical particles constructed from several identical spheres. On the other hand, Hölzer et al. (2007, 2009), using lattice Boltzmann simulations, correlated the drag force acting on *single* non-spherical particles with particle orientation and Reynolds number. However, these correlations cannot be applied directly to simulate gas-fluidized beds since they do not account for the influence of the solids volume fraction on the drag coefficient. Here we report the development of a drag force correlation for non-spherical particle assemblies constructed from identical spheres using LBM. The effect of particle shape and Reynolds numbers on the drag force is studied in detailed. The new drag force correlation developed here is suitable for Euler-Euler and Euler-Lagrangian simulations of gas-fluidized beds comprised of non-spherical particles.

Reference

[1] S. Ergun. Fluid flow through packed columns. Chem. Eng. Progr., 48:89-94, 1952.

[2] C.Y. Wen and Y.H. Yu. Mechanics of fluidization. Chem. Eng. Progr. Symp. Series, 62:100-111, 1956.

[3] D.L. Koch, R.J. Hill. Inertial effects in suspension and porous media flow. Annu. Rev. Fluid Mech., 33, 619, 2001.

[4] M.A. Van der Hoef, R. Beetstra, J.A.M. Kuipers. Lattice-Boltzmann simulations of low Reynolds number flow past mono- and bi-disperse arrays of spheres: results for the permeability and drag force. J. Fluid Mech., 528: 233-54, 2004.

[5] R. Beetstra, M.A. Van der Hoef, J.A.M. Kuipers. Drag force of intermediate Reynolds number flow past mono- and bi-disperse arrays of spheres. AIChE J., 53: 489-501, 2007.

[6] S. Tran-Cong, M. Gay, E.E. Michaelides. Drag coefficients of irregularly shaped particles. Powder Technol., 139: 21-32, 2004.

[7] A. Hölzer, M. Sommerfeld. New simple correlation formula for the drag coefficient of non-spherical particles. Powder Technol., 184: 361-365, 2007.

[8] A. Hölzer, M. Sommerfeld. Lattice Boltzmann simulations to determine drag, lift and torque acting on non-spherical particles. Comput. Fluids, 38: 572-589, 2009.

(398i) Investigation of Multiphase Multicomponent Aerosol Spray From Pmdis Through Commercial Spacers

Sarkar, S., University of Connecticut
Chaudhuri, B., University of Connecticut
Kakhi, M., US Food & Drug Administration
Peri, P., US Food & Drug Administration

The purpose of the current study is to develop a numerical model which can characterize flow and the resulting particle metrics (surface deposition, PSD etc.) when a commercial suspension based pMDI is sprayed into a spacer. The effect of different variables like volumetric rate of coaxially flowing air, spacer geometry and delay are investigated in order to develop labeling guidelines for patient use for similar products. Flow behavior of the multiphase multicomponent aerosolic spray, in static air, and within the spacers (Aerochamber Plus and Optichamber Advantage) in the presence of coaxial air were simulated using phenomenological 2 equation RANS models . Computational geometries of the inhaler and spacers were realized and meshed (tet dominant) using ANSYS Workbench tools. The standard k- ω model with Lagrangian tracking available within ANSYS Fluent 14.0 was used to simulate spray emanating out of the Proventil inhaler in a 2 way coupling mode with concurrent coaxial air flow at specified volumetric flow rates of 28.3 and 11 LPM flowing through the entire domain. The cone angle of the spray used as a model input was experimentally measured by high speed video imaging. The predictions made by the model were validated against known experimental results. The effect of delay (0 and 5 seconds) was also investigated for both the spacers.

The CFD model predictions favorably compared with experimental data when velocity and size data were compared at two different locations when the inhaler was sprayed in static air. The results for inhaler-spacer systems show an increase of drug escaping the spacers on increasing the volumetric flow rate in accordance with experimental determinations. The recirculation observed near obstructions in axial path of the spray within the spacer is determined to be critical in enhancing spray retention within the spacer. Introduction of a delay is found to trap more drug in the spacer. The MMAD of the drug that escapes the spacer is smaller than what is injected for all cases, while the coarser particles are trapped within the spacer surfaces. In conclusion, a validated CFD model was developed for characterizing inhaler-spacer interactions which can be used for similar systems to predict desired particle metrics.

(398k) Mechanically-Alloyed Alâ"™Ti Powders With Customized Particle Sizes

Aly, Y., New Jersey Institute of Technology
Schoenitz, M., New Jersey Institute of Technology
Dreizin, E. L., New Jersey Institute of Technology

Aluminum is a high combustion enthalpy metal capable of boosting the energy density of energetic formulations used in propellants, explosives, and pyrotechnics. A number of approaches have been investigated that shorten aluminum ignition delay, increase burn rate, and decrease the tendency of aluminum droplets to agglomerate. Aluminum-based materials, including mechanically alloyed powders, are of interest as alternatives to aluminum in these formulations. Mechanically alloyed Al·Ti powders, in particular, have previously been shown to have higher burn rates compared to similarly sized Al powders. Upon their heating, an intermetallic Al_3Ti forms causing a heat release accelerating their ignition. A practical drawback of this material was that it could only be prepared as a powder with relatively coarse particle sizes, making it impossible to replace fine aluminum powders commonly used in the energetic formulations. This work is aimed at preparing Al·Ti powders with customized particle compositions and sizes. Recent efforts showed a successful reduction in the particle sizes for Al·Mg powders using a staged mechanical milling approach. The first stage was aimed to prepare the mechanically alloyed powder and the second stage was used to adjust its particle size distribution. A small amount of elemental iodine (I_2) was added during the final milling stage, helping to reduce the particle size distribution of the final product. A conceptually similar approach is explored here for Al·Ti alloys. The milling protocol is modified using combinations of process control agents and staged milling to prepare equiaxial, micron-scale particles suitable for laboratory evaluations of their oxidation, ignition, and combustion characteristics. Particle size distributions are measured using low-angle laser light scattering. Electron microscopy and x-ray diffraction are used to examine particle morphology and phase makeup, respectively. Combustion of aerosolized powder clouds is studied using a constant volume explosion apparatus. Combustion of individual particles is studied using a laser ignition setup. Oxidation and decomposition are studied using thermo-analytical measurements. For all materials, ignition and combustion characteristics are compared to those of pure Al. Additionally, ignition and oxidation characteristics are compared to determine which events lead to particle ignition. Compositions with improved performance (i.e., shorter ignition delays, higher burn rates, and faster pressurization rates) compared to pure Al are identified.

(398l) Plasmonic Nanoshell Colloids With Reversible Magnetic Field Induced Optical Anisotropy: Synthesis, Characterization and Simulation

Klupp Taylor, R. N., University of Erlangen-Nuremberg
Zhuromskyy, O., University of Erlangen-Nuremberg

Small particles possessing a localized plasmon resonance in the visible and near-infrared spectral regions are promising for applications ranging from (bio)chemical sensors and drug delivery systems to photovoltaic devices and optical metamaterials. Recently there has been a growing interest in hierarchical structures and coatings based on such plasmonic building blocks which, due to electrodynamic interactions, display markedly different optical properties compared to those of isolated particles. Most structures produced using colloidal nanoparticles rely on the irreversible free or templated deposition or assembly in order to obtain plasmonic interactions. However, the addition of a superparamagnetic component to a plasmon resonant particle is an approach which permits the reversible adjustment of interactions and thereby the optical properties. Although the literature contains some reports of superparamagnetic and plasmon resonant structures, the influence of magnetic field on their optical behaviour is rarely considered. In this presentation we will describe the synthesis, characterisation and electrodynamic simulation of dispersions of magneto-optical nanocomposite particles produced using colloidal techniques. We utilize core particles with different superparamagnetic iron oxide loadings produced by self-assembly and emulsification techniques. These are then coated with gold or silver in order to form superparamagnetic nanoshells which possess visible and near-infrared resonances, the wavelengths of which depend on the core and shell dimensions. Using a modified spectroscopic ellipsometer we show that an applied magnetic field of moderate strength is sufficient to produce reversible and anisotropic linear optical response in dilute suspensions of such multifunctional particles. In particular, this response is significantly enhanced near the dipole resonance of the metal shell. Electrodynamic simulations based on a combination of the Lorenz-Mie, T-matrix and Maxwell Garnett effective medium approaches confirm that optical interactions between particles in chain-like configurations parallel with the direction of magnetic field are responsible for the field-induced optical changes observed experimentally. Finally, we attempt to use simulations to design synthetically realisable structures with even larger field-induced responses.

(398m) Numerical Optimization Meets Wet Chemistry in the Design of Novel Pigments

Klupp Taylor, R. N., University of Erlangen-Nuremberg
Pflug, L.,
Semmler, J.,
Seifrt, F.,
Leugering, G.,
Zhuromskyy, O., University of Erlangen-Nuremberg
Peschel, U.,
Peukert, W., University of Erlangen-Nuremberg

Small particles are exploited for their optical properties in a vast array of applications such as paint, inks, textiles and cosmetics. In many of these application areas, features of the particles (size, shape and topology) can be on the nanoscale, a fact which is often fundamentally important to their desirable properties. With the broad range of relevant product areas, future development of new optical materials will require an intimate understanding of this microstructure-function relationship in order to improve cost efficiency, environmental safety and sustainability. At the same time, there needs to be a shift from materials development through empirical synthetic experiments towards a "process chain" linking theoretical design to the practical realization of application-optimized particles. Making this possible are significant advances, on the one hand in computational tools which allow the inversion of optical simulation for the discovery of non-intuitive structures, and on the other, synthetic capabilities which go beyond the formation of single phase spherical or ellipsoidal particles.

Many aspects of this new approach to colorant development have been explored in the framework of an interdisiplinary project carried out in the framework of the Erlangen-based Cluster of Excellence "Engineering of Advanced Materials". The aim has been, on the one hand, to develop systematic and efficient theoretical screening tools for the exploration of the physically possible parameter space for nanostructured particles. On the other hand, we have considered the practical realisation of nanostructured optical particles, in several cases having motifs directly inspired by the theoretical work. In view of potential applications, focus on scalable and tunable processes has been a particular feature of the project.

This poster will exemplify the work of the project described above for the case of particles with multifunctional visible and IR optical extinction properties, a requirement for heat reflecting windows or "cool" roof-tiles. We demonstrate how mathematical optimization techniques can take over where intuitive screening using Lorenz-Mie approaches reaches practical limitations. Furthermore, we show that the designed particle structures can be fabricated and the resulting optical properties are consistent with the original target function. This work shows how mathetical approaches could in the near future revolutionize the pigment design process.

(398j) Contact Force Measurement for Liquid-Solid Systems At Different Liquid Saturation States

Lievano, D. M., University of Pittsburgh
Velankar, S., University of Pittsburgh
McCarthy, J. J., University of Pittsburgh

Solid-liquid interactions in the processing of solid particles is of great importance to applications from particle drying to powder granulation. Cohesive and viscous forces that result from the formation of liquid bridges between wet particles can greatly alter the behavior and flow of an otherwise identical dry system. While the effect of pendular bridges in particles systems has been an area of active study, resulting in research on methods to induce separation or mixing, the effect of different (higher) levels of liquid saturation have received dramatically less attention. This work explores the liquid-solid interaction forces of a simple system using one particle type and one fluid. A stationary container of a small number of particles (from 2-4) is filled with different volumes of liquid corresponding to different saturation levels (funicular, pendular, capillary, droplet). A particle is then forced into contact with the system creating a solid-liquid interface. The liquid-induced forces are measured at the solid-liquid boundary. These results will be compared to existing theoretical/modeling predictions.

(398o) Comparison of Pyrolysis Behaviors of Large and Pulverized Coal Particles

Yang, C.,
Li, S., Institute of Process Engineering
Song, W., Institute of Process Engineering, Chinese Academy of Sciences
Lin, W., Institute of Process Engineering, Chinese Academy of Sciences

Pyrolysis is an alternative to conventional liquefaction. The liquid products from coal pyrolysis contain high value chemicals such as BTX and PCX, and can also be upgraded into a substitute for transportation fuel. Particle size is an important factor influencing pyrolysis behaviors of coal. Large coal particles(mm-sized) and pulverized coal (μm-sized) may exhibit different pyrolysis behaviors. The heat and mass transfer inside coal particles can significantly affect product distribution and composition from pyrolysis. Large coal particles tend to fragmented into small pieces at high temperatures due to thermal stress and inner pressure. Thus, a better understanding of particle size effect on pyrolysis behaviors of coal is crucial to optimize pyrolysis processes. In this work, the experimental results on pyrolysis behaviors of mm-sized coal particles and μm-sized coal particles in a fixed bed reactor were presented. The effects of particle size, heating rate and pyrolysis temperature on pyrolysis product distribution and composition were examined in an attempt to explore the relationship between particle size effect and pyrolysis product composition.

(398n) Experimental and Multi-Scale Modelling to Understand Electrostatic Behavior of Pharmaceutical Powders

Naik, S. S., University of Connecticut
Gupta, V., University of Connecticut
Sarkar, S., University of Connecticut
Hancock, B. C., Pfizer Inc.
Abramov, Y., Pfizer
Yu, W., Pfizer
Chaudhuri, B., University of Connecticut

In pharmaceutical formulation processes, particle charging is often a nuisance and can cause problems in the manufacture of products, such as affecting powder flow, and reducing fill and dose uniformity. For a fundamental understanding of the powder triboelectrification, it is essential to study charge transfer under well-defined conditions. Hence all experiments were conducted in a humidity controlled glove box at 20%rh. To understand tribocharging materials, different surface viz. PVC, aluminum, Teflon, PMMA were used along with two pharmaceutical excipients and two API's. The experiments were performed in V-blender at a prescribed speed of 13 rpm. A triboelectric series of pharmaceutical materials was generated. For the pharmaceutical materials the work function was determined using MOPAC. MOPAC is a general-purpose semi-empirical molecular orbital package for the study of solid state and molecular structures and reactions providing band gap and ionization potential of molecules. In addition a discrete element method (DEM), including charge transfer and screened coulombic forces was developed for the V-blender system. There are different competing mechanisms involved in tribocharging. The two most important ones are electron transfer and ion transfer. Electron transfer has been well established for metal-metal and to some extent for insulator-metal contacts. There is still an ongoing debate for insulator-insulator charging whether the charge carriers are electrons or ions. As per work function hypothesis or established tribo-electric series powders should charge negatively against Alumimium, PMMA surfaces and positively against Teflon surfaces and PVC surfaces. According to the experiments in V-blender both the excipients (lactose and MCC) and API's charged negatively against PMMA and Aluminum and positively against Teflon and PVC surfaces.

(398p) Carbon Combustion Synthesis of Rare Earth Ferrites

Dannangoda, C.,
Sauchimov, A.,
Kumekov, S.,
Martirosyan, K. S., University of Texas at Brownsville

We report the synthesis of submicron powders of rare earth ferrites based on $LnFe_2O_4$, where Ln is Y, Er, Yb by using Carbon Combustion Synthesis of Oxides (CCSO). The CCSO is a modified form of self-propagating high temperature synthesis that uses carbon as a fuel instead of a pure metal. The reactant mixture contains in addition to oxides and/or carbonates about 6-12 wt. % a carbon powder. Following local ignition, the exothermic reaction between carbon and oxygen, (=-393.5 kJ/mol), provides the heat for the solid reactions and heats the adjacent reactant layer. This generates a self-sustaining temperature (up to 1000° C) wave that propagates through the reactant mixture at a velocity of up to 2.5 mm/s. The major parameters affecting the process are the carbon concentration in the reactant mixture and the ability of the oxygen infiltration to the reaction zone. A stable self-propagating reaction front can be obtained only at carbon and oxygen concentrations exceeding a critical value. The reaction front temperature and average velocity increased as the carbon concentration in the reaction mixture increased approaching eventually asymptotic values. The concentration of the carbon in the reactant mixture enabled control of the moving front temperature and average velocity as well as the particle size and surface area.

(398r) Characterization of Lithium Cobaltate Nanoparticles Produced By Ccso

Mamyrbaeva, Y.,
Hobosyan, M. A., University of Texas at Brownsville
Martirosyan, K. S., University of Texas at Brownsville
Kumekov, S., Kazakh National Technical University named after K.I. Satpaev

The crystalline $LiCoO_2$ nanoparticles were produced by Carbon Combustion Synthesis of Oxide (CCSO). The thermo-gravimetric analysis was used to identify interaction features in the system $LiNO_3$ - Co_3O_4 - Carbon to produce $LiCoO_2$. The electrochemical testing of the lithium cobalt oxide-based nano-composite were prepared consisting of the active powder material (93.5 wt. %), carbon black (4 wt. % as the conductor) and polyvinylidenefluoride (PVDF, <1µm, Sigma Aldrich) dissolved in N-methyl pyrrolidinone as the binder (2,5 wt.%). After all components were mixed thoroughly, the cast was coated on an aluminum foil substrate; the coating was followed by pressing and drying at 120 °C for 12 h under a vacuum. Laboratory made pouch-type battery was assembled in glove box using separator membrane with special type of polyethylene: thickness 16~25 µm, Surface Density 10~14 g/m2, Porosity 36 - 44 %, Pore Size 0.03 µm, Charge-discharge cycles were carried out in the potential range of 2.7 - 4.2 V. The analyser discharges the battery at constant current down to 2.7 V, then after 1 minute at constant current the battery was charged up to 4.2 V. The cathode materials prepared from as synthesized materials shows good performance for lithium-ion batteries. The CCSO is giving submicron and nano-sized particles which are showing excellent consistency of specific capacity and retaining over 90 % of initial capacity after 30 cycles. The as-prepared $LiCoO_2$ has over 200 mAh/g capacity which is more than 66% of theoretical maximum value for $LiCoO_2$. The CCSO is a promising method to produce ultrafine $LiCoO_2$ cathode materials with improved electrochemical performance.

(398q) Bubble Volume Fraction Using Electrical Capacitance Tomography and High Speed Photography

Sardeshpande, M. V., National Chemical Laboratory, Pune-8

Ranade, V. V., National Chemical Laboratory, Pune-8

Bubble volume fraction using Electrical Capacitance Tomography and High Speed Photography

By

Madhavi V. Sardeshpande and Vivek V. Ranade

Industrial Flow modeling group, National Chemical Laboratory, Pune ? 411 008, INDIA

mv.sardeshpande@ncl.res.in, vv.ranade@ncl.res.in

Abstract

A two-phase flow is one of the most common flows in nature as well as in industrial applications; it covers gas-solid, liquid-liquid, solid-liquid and gas-liquid flows. Among these, the gas-liquid flows can be encountered in wide variety of industrial applications including boilers, distillation towers, chemical reactors, oil pipelines, nuclear reactors, etc. The measurement of two-phase flow parameters such as flow regime and void fraction is considerably important and plays an important role in operational safety, process control and reliability of variety of processes (Dong, F. *et al.,* 2003). In process industries, void fraction is one of the most important parameter to characterize the hydrodynamic behavior of two phase dispersion system in a bubble column. The void fraction is a dimensionless quantity and is often termed as ?holdup or fraction? in two-phase flows. It is defined as the ratio of the volume of that phase to the total volume of the pipe (Corneliussen, S. e*t al.,* 2005) or can be defined as the fraction occupied by the gas phase in the total volume of a two or three phase mixture in a bubble column (Tang, C., 2006). Electrical Capacitance Tomography (ECT) technique is new and upcoming technology in the area of measurement of phase hold up distribution. ECT technique gives qualitative data for phase hold up distribution not the quantitative information. Therefore, it was thought desirable to validate this new technique with conventional photography technique for void fraction measurement. Therefore, an attempt was made here to design and construct an experimental test rig of laboratory scale bubble column for two-phase air-water bubble flow by using in-house facilities. This laboratory scale equipment is used to generate the phenomenon of void fraction in an ECT sensor (i.e. co-current bubble column) and its subsequent measurements. Series of experiments will be performed for the analysis of physical parameters of two-phase flow.

Experimental set up and techniques

Bubble column experimental set up was made up of Perspex of ID-48mm and height 1.2m. Two exactly similar set up assembled on test rig where (i) one is for ECT sensor/electrodes mounted on Perspex section (0.4m in length) (ii) similar transparent Perspex section fabricated for flow visualization (for high speed photography). Both the columns were connected to each other by ½" diameter of SS tube. The ECT sensor consists of a ring of 12-electrodes (which are separated from each other by small gap) on its outer periphery. It is connected with the data acquisition system which is then connected to computer by the in- built ECT software. A sieve plate sparger with multiple equally spaced holes has been installed at the bottom of the column with total 13 and 21 (1 mm diameter each) holes on its surface to generate uniform bubbles. A schematic of an experimental test setup is shown in Figure 1.

Electrical Capacitance Tomography

Tomography by definition refers to the process of exploring various characteristics of multi-component systems using measurement of change in internal properties of fluid and solid; depending on electrical properties of the components that is being analyzed and background conditions during measurement. Electrical capacitance tomography makes the use of capacitance measurements, section by section, using pairs of electrodes from an 8 or 12 electrode ring assembly. It provides subsequent reconstruction of the obtained data in the form of concentration profile images by various image reconstruction techniques. It can be used to investigate the internal dynamic behavior by presenting cross-sectional distribution of materials, i.e. tomographic images. The overall voidage of the contents of ECT sensor can be calculated by using the normalized pixel values in the reconstructed ECT image (Donthi, S.S., 2004).

High speed photography

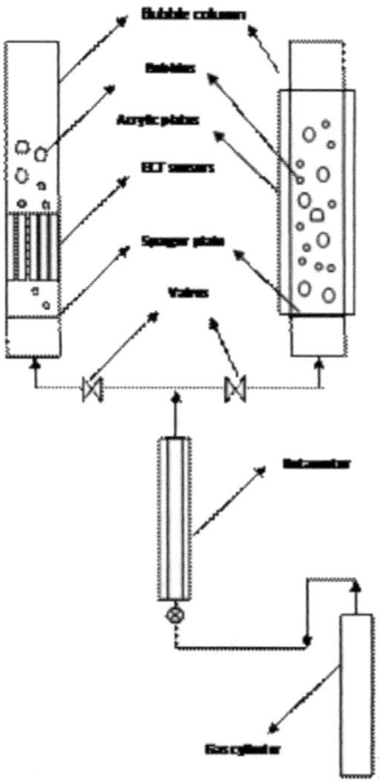

Figure 1: A schematic of an experimental test setup

High speed NR4-S1 camera was used for the current experiments with resolution 1024×1024 @1000fps and pixel size not more than 14μm. Bubble volume fraction will be measured for different sieve plate sparger designs as well as different gas flow rates. Based on the knowledge of volumetric void fraction, this current study also calculate the void fraction by using the photographic technique

The comparative study of tomography and photography would be useful to get quantitative information regarding void fraction in two phase flow. It will also be useful for two phase flow boiling applications where boiling tubes are opaque and there ECT technique will play a promising technology in order to get bubble volume fraction/vapor quality.

***Key words**: two phase flow, ECT, void fraction, flow regimes*

References:

Corneliussen, S., J.P. Couput, E. Dahl, E. Dykesteen, K.E. Frøysa, E. Malde, H. Moestue, P. Moksnes, L. Scheers and H. Tunheim, 2005. Handbook of Multiphase Metering. The Norwegian Society for Oil and Gas Measurement.

Dong, F., Z.X. Jiang, X.T. Qiao and L.A. Xu, 2003. Application of Electrical Resistance Tomography to Two-Phase Pipe Flow Parameters Measurement. Flow Measurement and Instrumentation, 14: 183-192.

Donthi, S.S., 2004. Capacitance Based Tomography for Industrial Applications. M.Tech. Credit Seminar Report, Electronic Systems Group, EE Department IIT Bombay, pp: 1-18.

Tang, C. and T.J. Heindel, 2006. Estimating Gas Holdup via Pressure Difference Measurements in a Cocurrent Bubble Column. International Journal of Multiphase Flow, 32: 850-863.

Ismail I, Shafquet A and Karsiti M. N. Void Fraction Estimation by Using Electrical Capacitance Tomography and Differential Pressure in an Air-Water Co-Current Bubble Column Australian Journal of Basic and Applied Sciences, 5(11): 1533-1541, 2011

(417a) Interference-Based Characterization of Microspheres: Radius, Local Curvature and Roughness Determination

Contreras-Naranjo, J. C., Texas A&M University
Chang, J., Texas A&M University
Ugaz, V. M., Texas A&M University
Originally presented on: 11/6/2013 8:30:00 - 8:49:00

Accurate measurement of microspheres size and roughness is fundamentally important in manufacturing settings; therefore, the use of characterization techniques capable of simultaneously measure micrometer-sized particles and nanometer-scale features offers a significant advantage. Interference-based techniques are known for their setup simplicity and practical implementation while offering nanometer/microsecond-scale resolution. However, the accurate use of interference for the study of microscopic objects has been difficult to achieve because current methods for the analysis of interferograms are strictly valid only when dealing with planar-like geometries.

Here, we describe work aimed at overcoming these barriers. Interferograms obtained from microspheres observed using reflection interference contrast microscopy (RICM) are analyzed with novel methods that allow accurate measurements of the microspheres size. In addition, RICM offers a unique non-invasive "view from below" perspective that provides valuable information about the nanometer-scale topography of the microspheres in close proximity to a transparent substrate, making it possible to measure roughness and local deviations from the overall particle radius. Two parameters possibly related to the surface roughness of microspheres are considered: first, a finite separation distance between the particles and the substrate measured by RICM when the surfaces are in contact and, second, fringe visibility in RICM interferograms.

(417b) Investigation of Packing and Flow Property Influenced By Particle Size and Shape

Xi, Z., Tel-Tek
Ding, S., Tel-Tek
Ratnayake, C., Tel-Tek
Originally presented on: 11/6/2013 8:49:00 - 9:08:00

A study of the influence of particle size and shape of carbon anthracite on the flow and packing behavior has been made. The candidate particle was first grinded by a lab-scale pin mill in order to manufacture different particle sizes and shapes. Image analyzer was employed for determination of particle shape size and shape.

(417c) Determination of Structure, Porosity and API Distribution in Granules By the Computed Micro-Tomography

Kaspar, O., Institute of Chemical Technology, Prague
Tokarova, V., Institute of Chemical Technology, Prague
Oka, S., Rutgers, The State University of New Jersey
Ramachandran, R., Rutgers University
Stepanek, F., Institute of Chemical Technology, Prague
Originally presented on: 11/6/2013 9:08:00 - 9:27:00

A mechanistic understanding of the correlation between the properties of the primary powder and granules can lead to enhanced and efficient process development in tablet manufacturing via high-shear wet granulation (HSWG), which is currently the most common route undertaken. Moreover, Quality-by-Design (QbD) in pharmaceutical manufacturing desires the identification of an optimal design space of critical quality attributes (CQAs) as a function of critical process variables and material properties. This work deals with the application of micro-tomography (CT) technique and image analysis to study granules prepared by the high shear granulator. The influence of granulation process parameters (binder content, wet massing time and vessel RPM) on particle structure (porosity, API distribution) has been investigated. The formulation comprises of acetaminophen and microcrystalline cellulose (30% APAP, w/w). Water is used as a binder. CT measurement allows us to study localization of the API (acetaminophen) in the matrix of granule (cellulose) and evaluate the diffusion mechanism of API from granules into surroundings for different process variables. This approach could be very beneficial in optimization of granulation process and properties of the final product.

(417d) Artificial Thermalization of Non-Brownian Microparticles for the Fabrication of Close-Packed Colloidal Crystals

Lash, M. H., Rutgers University
Little, S. R., University of Pittsburgh
McCarthy, J. J., University of Pittsburgh
Originally presented on: 11/6/2013 9:27:00 - 9:46:00

Colloidal crystals have been explored in the literature for applications in molecular electronics, photonics, sensors, and drug delivery. However, much of the research on colloidal crystals has been focused on nano-sized particles with limited attention directed towards building blocks with dimensions ranging from tens to hundreds of microns. This can be attributed, in part, to the fact that large-scale particles are less prone to assemble in an organized fashion due to the relative absence of thermalizing forces. Nevertheless, ordered arrays of large particles are of interest both as a basis for the production of hierarchically structured materials as well as for tissue engineering scaffolds due to large pore sizes. In this work, ultrasonic agitation is being explored as a means of artificially "thermalizing" these particles in order to overcome kinetic barriers to packing in the creation of close packed, highly ordered, crystalline structures from large microparticles (18-750um). Using this process, we have been able to create and characterize both two- and three-dimensional structures by adjusting properties of the system and observing the changes using a variety of microscopy techniques. We are currently investigating the significance of various substrate and solution properties such as surface tension, viscosity, particle concentration and particle composition both experimentally and computationally. The interactions between the constituent particles within these colloidal formations can be estimated and analyzed, ideally providing insight into, and control over the formation of desired crystal structures.

(417e) Internal Structure Characterization of Porous W Particles Formed By ALD On Sacrificial Polymers

Van Norman, S. A., University of Colorado at Boulder
Tringe, J. W., Lawrence Livermore National Lab
Falconer, J. L., University of Colorado
Weimer, A. W., University of Colorado at Boulder
Originally presented on: 11/6/2013 9:46:00 - 10:05:00

Engineered porous particles have a wide range of applications, such as gas sensing and catalyst supports. Porous support materials of tungsten/alumina are fabricated via atomic layer deposition (ALD) on sacrificial templates of hydroxylated methacylic polymer. Conformal films are formed on the sacrificial material by two separate sequential self-limiting surface half reactions of $Al(CH_3)_3$ and H_2O for alumina and similarly for other materials. Reactions were performed at temperatures below material phase transition temperatures; once the desired film thickness was achieved the polymer template was volatilized at temperatures above 400°C.

In this study, controlled physical characteristics of particles are explored. Physical properties of particles are characterized before and after material deposition and after removal of the sacrificial polymer. Properties include particle size, surface area, pore volume, pore size and porosity as well as chemical composition are presented. Internal structure of the porous particles as well as particle packing is explored via 3D X-ray tomography.

(417f) On-Line Characterization of Nanoparticles During Gas-Phase Synthesis

Grohn, A. J., Swiss Federal Institute of Technology (ETH Zurich)
Pratsinis, S. E., Swiss Federal Institute of Technology (ETH Zurich)
Wegner, K., ETH Zurich

Originally presented on: 11/6/2013 10:05:00 - 10:24:00

Real-time aerosol characterization is important for monitoring continuous nanoparticle producing processes as primary and agglomerate particle size determine product properties significantly. Sampling these aerosols, however, is all but trivial. Particle concentrations can be as high as 10^18 per kg of gas, requiring an enormous dilution to quench particle growth processes. The sampling probe has to withstand temperatures up to ~1500 K and possibly corrosive environments. Therefore, mostly ex-situ methods have been used to determine particle size and morphology in such conditions.

Here, an in-situ method for nearly real-time determination of average agglomerate mass, volume, mobility and structure along with the constituent primary particle size is employed. It is performance-tested in a flame spray aerosol reactor for production of zirconia nanoparticles (Gröhn et al., 2012). A sampling probe for continuous extraction and immediate dilution of the hot and highly concentrated aerosol was designed and constructed. Online characterization of the diluted aerosol is achieved by combining a differential mobility analyzer (DMA) with an aerosol particle mass analyzer (APM) and applying a power-law correlation to derive average primary particle size as proposed by Eggersdorfer et al. (2012).

Primary particle growth was shown to be completed shortly above the burner as a constant primary particle diameter was observed for all radial and downstream axial positions. The average agglomerate size increased with axial distance from the burner. However, larger agglomerates were observed at the fringes of the aerosol plume attributed to prolonged residence time due to lower gas velocity there. Results were compared against off-line particle size characterization by nitrogen adsorption and thermophoretic sampling/transmission electron microscopy. The method could also be applied successfully in continuous pilot-scale manufacture of zirconia nanoparticles, providing instantaneous quality control.

References:

Eggersdorfer, M.L., Gröhn, A.J., Sorensen, C.M., McMurry, P.H. and Pratsinis, S.E. (2012), J. Colloid. Interface Sci. 387, 12 - 23.

Gröhn, A.J., Pratsinis, S.E., and Wegner, K. (2012), Chem. Eng. J. 191, 491 - 502.

(417g) Cereal Fiber Based Encapsulation of Selected Polyphenols and Their Solid State Characterization

Pai, D. A., Institute of Chemical & Engineering Sciences
Ng, J. W.,
Ng, W. K., Institute of Chemical and Engineering Sciences
Tan, R., Institute of Chemical and Engineering Sciences, A*STAR (Agency for Science, Technology and Research)
Originally presented on: 11/6/2013 10:24:00 - 10:43:00

Polyphenols are naturally occurring potent antioxidants which have been shown to have anti-inflammatory and cardiovascular protective effects but have processing and formulation challenges since they are sensitive acidic pH and are bitter in taste. In order to overcome these challenges, they are often encapsulated, mainly using technologies such as spray drying. Although spray drying is an extensively used process in the food industry, the materials used to encapsulate an active ingredient are limited to gum Arabic, maltodextrins and modified starches. The overall goal of this research work is to evaluate potential of a novel prebiotic water soluble cereal fiber to encapsulate the polyphenols Quercetin (Qn) and Naringin (Nn) using spray drying. The cereal fiber chosen for these experiments shows faster and higher solubility in cold water and is more economical compared to Gum Arabic, which is the most widely used encapsulation material.

The cereal fiber was found to show extremely high solubility with a maximum of 62% (w/v) dissolving in water. The solutions were found to Newtonian in nature and a 20% solution was only found to be twice as viscous as water, whereas a 10% Gum Arabic solution was found to be almost five times as viscous as water. These low viscosities enable higher fiber loading for spray drying. The moisture sorption behavior of the cereal fiber was studied using dynamic vapor sorption between relative humidities of 0-95% using two cycles of adsorption-desorption. The fiber was found to be highly hygroscopic beyond RH of 60% with a maximum adsorption of 50% moisture (w/w) at a RH of 95% while forming an irreversible gel-like structure.

The influence of polyphenol type, cereal fiber viscosity and spray dryer inlet temperature on the morphology and size distribution of spray dried particles was studied. Suspensions were prepared with 3% Qn/Nn in either a 20% or 40% cereal fiber solution. Inlet temperatures of 155 and 180 ° C were evaluated for spray drying at a constant pump flow rate of 4 mL/min. For Naringin encapsulation, most of the resulting particles were found to be smooth, spherical and fully amorphous, indicating conversion of crystalline Naringin to amorphous form as a result of spray drying. However, for the encapsulation of Quercetin particles, X-Ray Diffraction spectrum consistently showed small proportion of crystalline peaks indicating incomplete transformation of crystalline Quercetin to amorphous form. The peaks were found to not overlap with those of pure Qn, suggesting formation of a different crystalline form upon spray drying. This could possibly be due to incomplete solubilization of Qn prior to spray drying, owing to its poor water solubility, leaving a proportion of particles in crystalline state, which transformed during spray drying. SEM data showed that a mix of particles with spherical and ellipsoidal morphology, but with a smooth surface, indicating their encapsulation by CFG. SEM results showed larger and smooth particles compared to the original needle shaped Qn/Nn crystals, indicating effective encapsulation of polyphenols by the cereal fiber.

The particle size determined by Laser diffraction was found to vary from 0.15-30 μm with a median particle size of 5-7 μm for encapsulated Qn and Nn particles with an exception: Nn spray dried in a 20% of the cereal fiber solution at 155 ° C, with a median particle size of 11 μm. The spray dryer inlet temperature and %CFG variations were found to have no significant effect on the median particle size ($p<0.05$).

Measurement of Hausner and compressibility ratios indicated poor to very, very poor flow behavior for all particles according to USP classification. The encapsulated Naringin particles showed a maximum moisture sorption of 21% (w/w) which was a significant reduction compared to the pure fiber (50% w/w).

Hence, in conclusion, it was qualitatively shown that the cereal fiber under study effectively encapsulated both Naringin and Quercetin. The encapsulated Naringin particles showed a decreased hygroscopic tendency compared to the cereal fiber, decreasing its hygroscopicity. However, traces of crystallinity were found for encapsulated quercetin particles which pose the risk of re-conversion to crystalline form during storage/processing. This risk needs to be mitigated by changing the solvent for spray drying or reducing the content of Qn in the slurry. Encapsulation also resulted in poor particle flow behavior which needs to be improved depending on the final product of intended use. Future work will be dedicated to quantify the amount of polyphenols encapsulated, understand the dissolution kinetics of spray dried powders in simulated gastric and intestinal fluids and sensory evaluation to evaluate taste masking potential of the cereal fiber compared to gum arabic.

(417h) Zeta Potential: A Neglected Parameter in the Characterization of Nanoparticulate Suspensions
Fairhurst, D.,
Originally presented on: 11/6/2013 10:43:00 - 11:02:00

Zeta Potential: A Neglected Parameter in the Characterization of Nanoparticulate Suspensions

David Fairhurst PhD and Patrick O'Hagan PhD

While particle size (distribution) has been long recognized as an important metric, the most fundamental and critical aspect, especially of nanoparticulate suspensions, is the molecular interaction between the suspended particles and the medium at the interface. It is such an interaction, along with the particle–particle interactions, that determines the dynamic properties of the dispersion and strongly affect the performance of the suspension.

The zeta potential (ZP) is a parameter (symbol ζ), which is related to the surface charge (*vide* the interfacial chemistry), a property that all materials possess, or acquire, when suspended in a fluid. The sign and magnitude of the ZP affect process control, quality control, and product specification; at the simplest level, the ZP can help maintain a more consistent product and at a complex level, it can improve product quality and performance. At the very least its measurement answers the question: Is the electrical charge on the material particle positive or negative? Such information is often sufficient to suggest further steps in formulation or processing. The next higher level of inquiry has to do with quality control: Has the product sufficient electrostatic repulsion (because of the magnitude of the surface charge) to maintain its stability?

The ZP represents, in the practical sense, the *effectiveness* of the particle surface charge in solution. While it may derive, initially, from the fundamental number of surface sites, (how many, what type, etc.), more importantly, from a *practical* application, are the solution conditions themselves, because they control the resulting final sign and magnitude of ζ. The consequence of this to the process of dispersing particles is critical. And importantly, it is useless to quote a measured ζ value without specifying the suspension conditions under which the measurement was made.

In this paper we will discuss the effect of ZP on suspension properties and present a variety of applications of ZP measurements to illustrate its practical broad utility and commercial significance.

(425a) Numerical Modeling Of Particle Agglomeration By Dynamic Liquid Bridge

Nakamura, H., Osaka Prefecture University
Watano, S., Osaka Prefecture University
Kan, H.,
Yamakmto, Y.,
Originally presented on: 11/6/2013 8:30:00 - 8:52:00

In fluidized bed granulation, a binder solution is sprayed to fluidized particles and the particles are agglomerated by liquid bridge of the sprayed binder, resulting in particle size enlargement by the agglomeration. This liquid bridge formed between particles is not static and can be easily compressed, deformed, and ruptured, because the particles inside the granulators are always moving. Therefore, particle agglomeration by such a “dynamic” liquid bridge should be modeled to understand micro scale particle agglomeration mechanism inside the granulators. In this study, modeling of the dynamic liquid bridge between two particles was performed by a numerical approach.

Numerical modeling of the dynamic liquid bridge between two particles was conducted using a coupling simulation of a discrete element method (DEM) with a computational fluid dynamics (CFD). In this simulation method, a particle motion was solved by the DEM, and a liquid–gas two-phase fluid flow consisting of a liquid bridge was simulated by the CFD. To solve governing equation of the liquid–gas two-phase fluid flow, the constrained interpolation profile (CIP) method was used. A surface force acting at an interface between liquid-gas-solid phases, i.e., contact line, was calculated by a continuous surface force (CSF) model so that a wetting behavior of a droplet on a particle was simulated. A capillary force acting on the particles was calculated by integrating the surface forces of the contact line. A viscous force acting on the particles was simulated using a immersed boundary (IB) method.

A wetting behavior of a droplet on flat and curved surfaces with different contact angles (different hydrophilicity/hydrophobicity) was initially simulated. Repealing and spreading of the droplet was well simulated. A break up behavior of a liquid bridge between two spheres was then simulated, and the simulation result showed good agreement with an experimental result. Finally, particle coalescence and rebound mediated by the dynamic liquid bridge were simulated at various conditions, and key parameters determining the particle agglomeration were investigated. Consequently, numerical modeling of the dynamic liquid bridge between particles was successfully performed. This numerical approach can provide significant insights into a micro scale mechanism of particle agglomeration in fluidized bed granulation as well as the other wet granulation processes.

(425b) Validation of Filtered Two-Fluid Models Using NETL-PSRI Bubbling Fluidized Bed Challenge Problem

Ozarkar, S., ANSYS Inc.
Yan, X., Department of Chemical and Biological Engineering, Princeton University,
Wang, S.,
Milioli, C.,
Milioli, F.,
Sundaresan, S., Princeton University
Originally presented on: 11/6/2013 8:52:00 - 9:14:00

Filtered two-fluid models offer a promising practical alternative to microscopic two-fluid model for the simulation of gas-particle flows in industrial systems generally characterized by heterogeneous structures ranging from micro-scale to macro-scale. In our prior studies [1,2] we performed highly resolved simulations of gas-particle flows using microscopic two-fluid model and extracted constitutive models for filtered drag coefficient and filtered particle phase stresses in terms of filtered particle volume fraction and filter size as markers that classify inhomogeneity of sub-filter scale flow structures. In our recent study [3] we further investigated various markers and identified filtered fluid-particle slip velocity as an important additional marker for filtered drag coefficient. In that study new constitutive models were proposed. Filtered drag coefficient was expressed in terms of filtered particle volume fraction, filter size and filtered slip velocity while pressure and viscosity of both phases were expressed in terms of filtered particle volume fraction, filter size and filtered scalar shear rates.

In present study, we focus on validation of Milioli et al. filtered models [3] using NETL-PSRI bubbling fluidized challenge problem [4]. Simulations were performed using different levels of relatively coarse grid resolutions and differential pressure gradient (also indicator of bed expansion) data from experiments was used for comparison. It was found that Milioli et al. filtered models predicted correct bed expansion and differential pressure gradient predictions compared well with experiments while microscopic two-fluid model and Igci et al. filtered models [1] predicted unphysical bed expansion thereby clearly demonstrating the relative importance of filtered drag and its dependence on filtered slip velocity. Milioli et al. filtered models were also tested under different flow conditions including initial static bed depth and gas superficial velocity. Overall, model predictions were found to be in good agreement with experiments.

1. Y. Igci and S. Sundaresan, Ind. Eng. Chem. Res., 50, 13190–13201 (2011).

2. Y. Igci and S. Sundaresan, AIChE J., 57, 2691-2707 (2011).

3. C.C. Milioli, F.E. Milioli, W. Holloway, K. Agrawal, S. Sundaresan, Accepted to *AIChE Journal* (2012).

4. NETL-PSRI Bubbling Fluidized Bed Challenge Problem 2010, https://mfix.netl.doe.gov/challenge/index_2010.php.

(425c) Entrainment Review - Review of Existing Correlations

Chew, J. W., Nanyang Technological University
Hrenya, C. M., University of Colorado at Boulder
Cocco, R., Particulate Solid Research, Inc. (PSRI)
Originally presented on: 11/6/2013 9:14:00 - 9:36:00

Despite the abundance of entrainment correlations made available in the past few decades[1-6], the discrepancy between empirical prediction and experimental data is up to a hundredfold in some cases[6]. Hence, a comprehensive review of the available correlations is warranted, with the goal of extracting the underlying physics giving rise to the disparities, and thereby providing insights towards enhanced understanding of elutriation. Using lognormal Geldart Group A particle size distribution (PSD) as a base case, comparisons of the correlations reveal a difference of up to 21 orders of magnitude difference among the total entrainment flux predicted for a wide PSD. Further probing indicates qualitative discordance on the shapes of the elutriated PSDs, the most notable of which is the non-physical anomaly whereby larger particles are predominantly elutriated. Furthermore, whether a particular correlation was developed for a specific Geldart Group or for monodisperse and/or polydisperse systems seems inconsequential, since correlations largely overlap in the same range. These observations underscore the need for more physically-based models to enable a more accurate prediction of the elutriation phenomenon.

References

1. Leva M, Wen CY. Elutriation. In: Davidson JF, Harrison D, eds. *Fluidization*. New York, N.Y.: Academic Press Inc.; 1971:627-650.

2. Geldart D. Elutriation. In: Davidson JF, Clift R, Harrison D, eds. *Fluidization*. London: Academic Press; 1985:383-411.

3. Kunii D, Levenspiel O. *Fluidization Engineering*. 2nd ed. Newton, MA: Butterworth-Heinemann; 1991.

4. Fan LS, Zhu C. *Principles of Gas-Solid Flows*. New York, NY: Cambridge University Press; 1998.

5. Rhodes MJ. *Introduction to Particle Technology*. 1st ed. West Sussex, England: John Wiley & Sons Ltd.; 1998.

6. Tasirin SM, Geldart D. Entrainment of FCC from fluidized beds - a new correlation for the elutriation rate constants K-i infinity*. *Powder Technology*. Mar 1998;95(3):240-247.

(425e) Understanding the Origin of Pressure Oscillations in Fluidized Beds

Boyce, C. M., University of Cambridge

Holland, D. J., University of Cambridge

Scott, S. A., University of Cambridge

Dennis, J. S., University of Cambridge

Originally presented on: 11/6/2013 9:58:00 - 10:20:00

The origin of pressure oscillations in fluidized beds has stood for a long time as a major question in the understanding of two-phase granular flow. Pressure oscillations are important to understand because they can easily be measured on fluidized beds of all sizes and their relation to the passage of void regions has led to their use in defining different flow regimes of fluidization [1]. While it has long been understood that pressure oscillations correlate well with bubble passage through fluidized beds, there has been contention as to whether pressure oscillations are tied to bubble formation [2], coalescence [3], or eruption [4]. This paper compares detailed results from a recently developed computational model [5] with experimental results [4] to shed light on the origin of pressure oscillations in bubbling fluidized beds.

A bubbling fluidized bed with a diameter of D_{bed} = 50 mm, filled with d_p = 1.2 mm diameter particles to various tapped bed heights, H_0, was modelled computationally for direct comparison to combined magnetic resonance (MR) imaging and pressure sensor experiments conducted by Müller *et al.* [4] on the same bed. A 3D cylindrical discrete element model with computational fluid dynamics (DEM-CFD) described and validated for simulating bubbling fluidized beds elsewhere [4] was used to model the fluidized bed. The bed was fluidized at excess gas velocity $(U-U_{mf})$ = 0.263 m/s for direct comparison of model and experiment.

Figure 1 shows simulation results of the bed with initial height H_0 = 40 mm over 275 ms of steady bubbling. The first two rows show voidage (ε) and differential pressure drop in the vertical direction ($-dp/dz$) maps of the bed every 25 ms. The third row gives the pressure drop across the bed over the time period and the fourth row gives the 1D voidage traces at the bottom and top of the bed, capturing bubble formation and eruption, respectively, with spikes in voidage. The results show that bubbles form and erupt at the same frequency, since bubbles form, rise and erupt without interacting with bubbles above or below them in this short bed. Pressure drop occurs most rapidly in regions dense with particles due to the drag force on these particles. Almost no pressure drop occurs in the vertical region spanning from the bottom to the top of a bubble, since the lack of particles in the bubble provides a free path through which fluid can pass. The pressure drop across the entire bed is at a minimum when a single, fully-formed bubble is rising through the bed because the bubble provides a long path of low resistance for fluid. The pressure drop reaches a maximum just as a bubble erupts because the free path for fluid disappears and the fluid now has to pass through a tall region dense with particles.

Figure 2 gives the same sequence of simulation results of the same bed, except with initial height H_0 = 50 mm, and voidage and differential pressure drop maps every 50 ms over a period of 550 ms. Approximately every 260 ms, a large bubble forms and begins to rise, spanning almost the entire diameter of the bed. Soon after, a smaller bubble forms in its wake and rises more quickly behind the first large bubble. The first bubble erupts, followed by the second in quick succession. The pressure drop across the entire bed is at a minimum the entire time the first bubble rises, but begins to increase dramatically as the first bubble erupts since the free path for fluid created by this large bubble suddenly disappears. The pressure drop reaches a maximum just as the second bubble erupts, since the next large bubble has not formed yet and the bed is left with a large height of dense particles, through which fluid must pass.

Table 1 gives the frequency of oscillations in pressure drop, bubble formation and bubble eruption for various bed heights, determined by taking the Fourier transform of the respective traces shown in the third and fourth rows of Figures 1 and 2 over 2.1 seconds of steady bubbling. In the 40 mm tall bed, bubble formation, eruption and pressure oscillations all occur at the same frequency, since the bubbles do not interact, and thus the same particle arrangements which create easy and difficult paths for fluid passage occur for every bubble formed. In the 50 mm tall bed, the frequency of bubble formation is twice that of bubble eruption and pressure oscillations. Even though the bubbles do not fully coalesce, they strongly interact, causing the arrangement of particles around the first bubble to be vastly different from the second. Since, due to drag, arrangements sparse in particles allow for little pressure drop over a vertical distance and arrangements dense in particles create large pressure drops, maxima in pressure drop occur only once for every two bubbles formed. While the bubbles do not erupt at exactly the same time, the second bubble erupts in the wake of the first eruption, causing the frequency of the oscillations of the voidage trace at the top of the bed to be half of that at the bottom.

The results for the four bed heights in Table 1 demonstrate $H_0 = 50$ mm to be a critical height at which two consecutive bubbles begin to interact, and thus the frequency of bubble eruption and pressure oscillations suddenly half, deviating significantly from the frequency of bubble formation. This result matches well with experiments conducted on the same bed [4], which showed a sudden decrease in bubble eruption and pressure oscillation frequency at $H_0 = 50$ mm, as shown in Table 1. Finding the same frequency of bubble eruption and pressure oscillation beds with H_0 ranging from 20-161 mm, but vastly different frequencies of bubble formation in beds 50 mm or taller, Müller *et al.* [4] postulate that pressure oscillations in fluidized beds are derived from bubble eruption. The detailed results provided by the DEM-CFD simulations show that maxima in pressure drop across narrow fluidized beds occur just after bubble eruption due to the large height of dense particles left behind by the bubbles. Müller *et al.* [4] also found that the transition point in pressure fluctuations occurred at a bed height at which the theoretical value for effective bubble diameter, D_e, became large enough that the bed entered the slugging regime, defined as $D_e/D_{bed} > 0.6$. This hypothesis is confirmed by DEM-CFD results, shown in Table 1, which saw a drastic increase in average D_e, with the bed entering the slugging regime between $H_0 = 40$ mm and $H_0 = 50$ mm.

This paper will further investigate the origin of pressure oscillations in fluidized beds by using the DEM-CFD model to look at pressure oscillations in beds ranging from 30-156 mm in tapped bed height and further comparing to the results of Müller *et al.* [4]. Investigating this larger range of heights will provide understanding on how bubble coalescence and a variety of types of slugging affect pressure oscillations.

References:

[1] Bi, H.T. (2007). *Chem. Eng. Sci.*, **62**, 3473-3493.

[2] Fan, L.T., Ho, T., Hiraoka, S. and Walawender, W.P. (1981). *AIChE J.*, **27**, 388-396.

[3] Littman, H. and Homolka, G. (1973). *Chem. Eng. Sci.*, **28**, 2231-2243.

[4] Müller, C.R., Davidson, J.F., Dennis, J.S., Fennell, P.F., Gladden, L.F., Hayhurst, A.N., Mantle, M.D., Rees, A.C., Sederman, A.J. (2007). *Powder Technol.*, **177**, 87-98.

[5] Boyce, C.M., Holland, D.J., Scott, S.A., Dennis, J.S. (submitted 2013). *Phys. Rev. E.*

Figures:

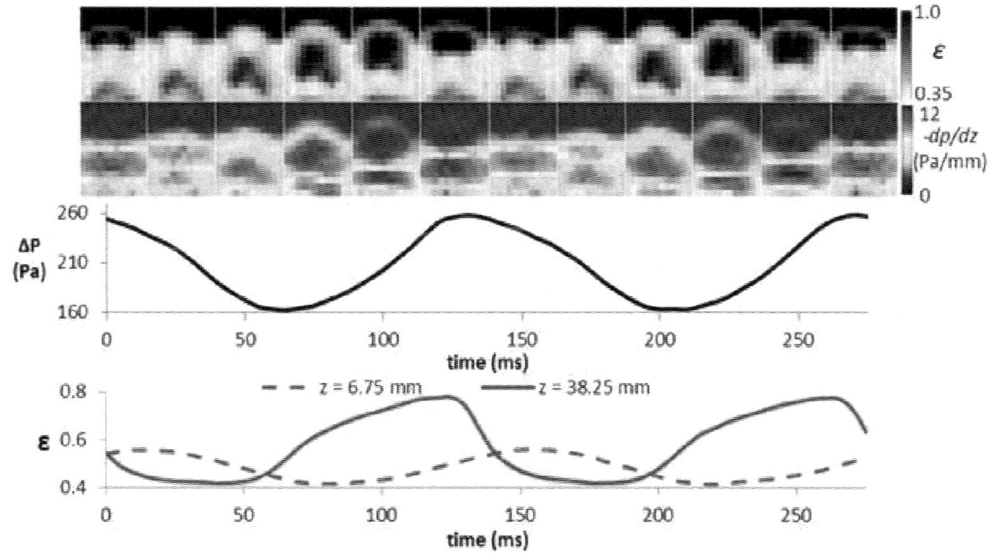

Figure 1. Analysis of pressure oscillations in a fluidized bed with H_0 = 40 mm and D_{bed} = 50 mm. Over 275 ms this figure gives: voidage maps (first row) and differential pressure drop in the vertical direction maps (second row) every 25 ms, as well as traces of pressure drop across the entire bed (third row) and 1-D voidage at the top and bottom of the bed (fourth row).

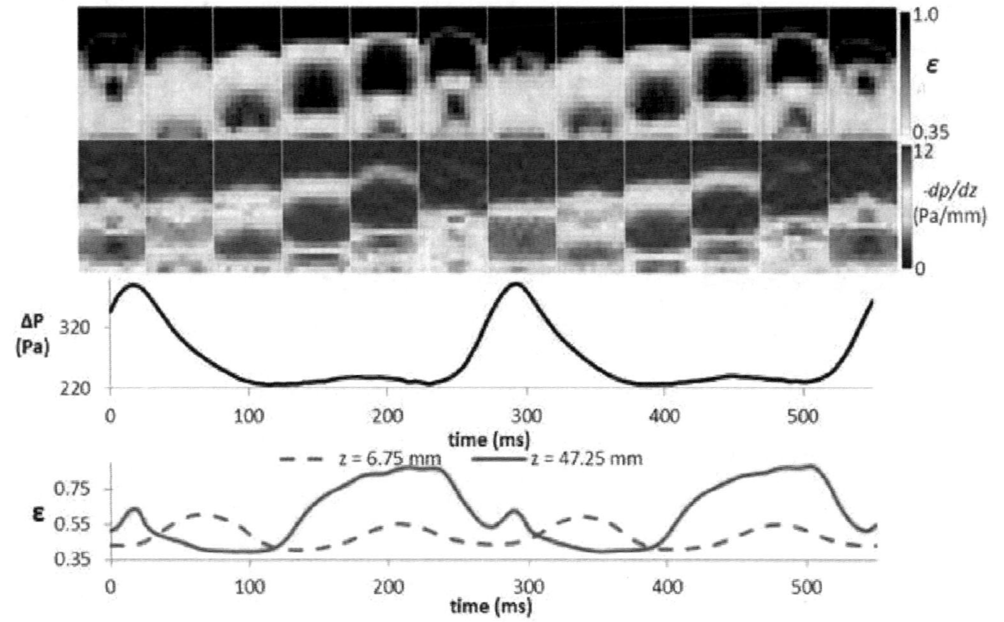

Figure 2. Analysis of pressure oscillations in a fluidized bed with H_0 = 50 mm and D_{bed} = 50 mm. Over 550 ms this figure gives: voidage maps (first row) and differential pressure drop in the vertical direction maps (second row) every 50 ms, as well as traces of pressure drop across the entire bed (third row) and 1-D voidage at the top and bottom of the bed (fourth row).

Tables:

Table 1. Frequencies of bubble formation (f_{form}), bubble eruption (f_{erupt}) and oscillations in pressure drop ($f_{\Delta P}$), as well as the ratio of effective bubble diameter to bed diameter (D_e/D_{bed}) in a 50 mm diameter bubbling fluidized bed with varying tapped bed heights (H_0). The results are given for a DEM-CFD model as compared to experiments [4] using a combination of magnetic resonance (MR) imaging and a pressure sensor.

	DEM-CFD Results				Experimental Results	
H_0 (mm)	f_{form} (Hz)	f_{erupt} (Hz)	$f_{\Delta P}$ (Hz)	D_e/D_{bed}	f_{erupt} (Hz)	$f_{\Delta P}$ (Hz)
30	6.96	6.96	6.96	0.41	6.4	6.0
40	7.12	7.14	7.15	0.44	5.6	5.5
50	7.42	3.73	3.72	0.66	2.9	3.0
60	7.57	3.75	3.79	0.65	2.9	3.1

(425f) Experimental Research and Analytical Modelling of a Gas-Solid Vortex Reactor (GSVR)

Kovacevic, J., Ghent University

Heynderickx, G. J., Ghent University

Marin, G. B., Ghent University

Originally presented on: 11/6/2013 10:20:00 - 10:42:00

Experimental research and analytical modelling of a Gas-Solid Vortex Reactor (GSVR)

J. Z. Kovacevic, G. J. Heynderickx*,G. B. Marin

Laboratory for Chemical Technology, Ghent University, Krijgslaan 281 (S5) 9000 Gent, Belgium;

Tel +32 926 44 516 Fax +32 926 44 999; email: Geraldine.Heynderickx@UGent.be

Process Intensification requires the development of new reactor technologies. The Gas-Solid Vortex Reactor (GSVR) also known as the Rotating Fluidized Bed Reactor in Static Geometry (RFB-SG), provides intensification opportunities by substituting gravitational by centrifugal force. The most common limitations of operating a Static Fluidized Bed (FB) in the gravitational field are thus overcome. The high slip velocities (~1 to 10 m/s), the high solids fractions (~0.3 to 0.6), the short residence times (~50 ms), as shown by Ashcraft [2,3], allow the GSVR to promote high throughput operations, uniform gas-solid contacting and enhanced heat and mass transfer, resulting in an intensified process on particle scale as well as on reactor scale. As an over-all result, the dimensions of the reactor can be drastically reduced. Process Intensification can be realized for different conditions: operating conditions as well as geometrical configuration.

A Fluidized Bed reactor technology in which the gravitational force field is replaced by a centrifugal force field has been considered since the seventies mainly with a rotating reactor. Extensive experimental research on the subject has not been performed, the main reason being the problematic scale-up of a rotating reactor. In the present work the centrifugal field is realized in a static reactor by a tangential introduction of the fluid phase as described in detail by Ekatpure et al [4]. The presented study focuses on the influence of solids density, particle diameter and gas velocities on reactor hydrodynamics. 2D Particle Image Velocimetry (PIV) is used as a measuring technique to determine the particle velocities. Captured images are post processed and a particle velocity profile is obtained as shown in the added figures. At the same time the pressure profile in the rotating bed is measured using multiple pressure probes. By combining velocity and pressure profiles, the bed behaviour and influence of solids density, particle diameter and gas velocities are examined, and used to develop an analytical model for the GSVR based on the Chen[5] partial fluidization model.

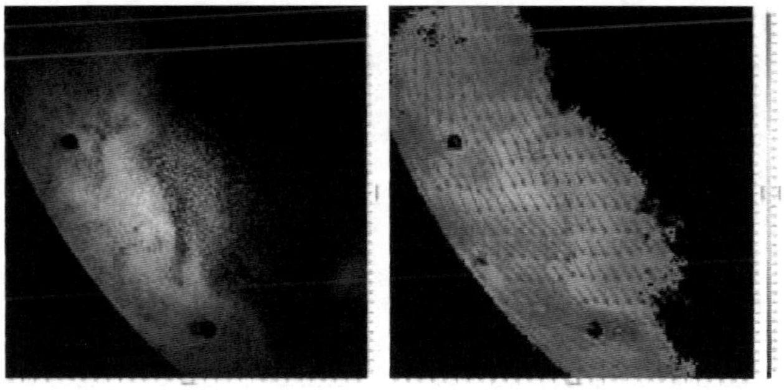

Typical 2D solids velocity profile in the GSVR based on PIV measurements

References

1. De Wilde, J. and A. de Broqueville, *Rotating fluidized beds in a static geometry: Experimental proof of concept.* AIChE Journal, 2007. **53**(4): p. 793-810.

2. Ashcraft, R.W., G.J. Heynderickx, and G.B. Marin, *Modeling fast biomass pyrolysis in a gas?solid vortex reactor.* Chemical Engineering Journal, 2012. **207?208**:p. 195-208.

3. Robert W. Ashcraft, Jelena Kovacevic, Geraldine J. Heynderickx, and Guy B. Marin, et al., *Assessment of a Gas?Solid Vortex Reactor for SO2/NOx Adsorption from Flue Gas.* Industrial & Engineering Chemistry Research, 2013. **52**(2): p. 861-875.

4. Rahul P. Ekatpure, Vaishali U. Suryawanshi, Geraldine J. Heynderickx, Axel de Broqueville, Guy B. Marin, *Experimental investigation of a gas?solid rotating bed reactor with static geometry.* Chemical Engineering and Processing: Process Intensification, 2011. **50**(1): p. 77-84.

5. Chen, Y.M., *Fundamentals of a Centrifugal Fluidized Bed* AIChE Journal, 1987. **33**(5): p. 722-728.

Acknowledgment

This work was supported by the ?Long Term Structural Methusalem Funding? by the Flemish government. The work was also partially supported by a European Research Council Advanced Grant, ERC-MADPII, project number 41L02311. ADDIN EN.REFLIST

(425g) A Model for the Size of Polar and Apolar Nanoparticle Agglomerates in a Fluidized Bed

de Martín, L., Delft University of Technology

van Ommen, J. R., Delft University of Technology

Originally presented on: 11/6/2013 10:42:00 - 11:04:00

Fluidization is emerging as a promising but challenging technique for processing of nanoparticles, for instance, to produce coated nanoparticle. Nanoparticles do not fluidize individually but form agglomerates due to strong interparticle forces, dominant at the nano-scale [1]. The type of fluidization and transport phenomena inside the agglomerates strongly depend on the agglomerates structure and size, two variables in turn related to the forces present between particles and agglomerates.

The size of the fluidized agglomerates is commonly estimated with a balance between attractive interparticle forces, such as van der Waals, electrostatic and capillary force and separating forces, such as gravity and drag force. When dealing with hydrophobic (apolar) nanoparticles, capillary is neglected and van der Waals is the only interparticle attractive force usually considered. In the case of hydrophilic (polar) nanoparticles, it is common to add a capillary force to the interparticle forces if the fluidization is carried out with ambient air. If the fluidizing gas is dry, the approximation is similar to that with apolar nanoparticles and only van der Waals interactions are included. Tahmasebpoor *et al.* [2] showed that attractive forces between dry and polar particles is considerably larger than between dry and apolar nanoparticles. This was attributed to the formation of hydrogen bonds between the surfaces of the polar nanoparticles.

In this work, we propose a new model to estimate the average agglomerate size in dry fluidized beds by taking into account the type of surface, polar or apolar. We do this by including a term in the inter-particle forces that takes into consideration the possibility of formation of hydrogen bonds.

In our model, complex agglomerates are built by porous and spherical primary agglomerates [3]. The attraction between primary agglomerates is described according to Rabinovichxs model [4], considering that the nanoparticles act as asperities in the primary agglomerates. The Hamaker coefficient has been corrected with the two-body summation approach to take into account the porosity of the primary agglomerates. The contribution of the hydrogen bond has been modeled taking into account the geometry of the contact point between agglomerates. The only separating force considered is gravity. The average agglomerate size is therefore the size that makes the Bond number of the agglomerates equal to unity.

The resulting average agglomerate size is a function of d_p, ρ_p, D_{f1}, D_{f2}, d^*, A_H, N and α, where d_p and ρ_p are the size and density of the nanoparticles, respectively. D_{f1}, D_{f2} are the fractal dimensions of the primary and complex agglomerates, respectively. d^* is the average size of the primary agglomerates. A_H is the Hamaker coefficient of the nanoparticles material. N is an empirical constant related to the connectivity of the primary agglomerates, being 1.38 for all the nanopowders. α is an empirical constant that takes into account the formation of hydrogen bond between nanoparticles. α is 0 for apolar nanoparticles and ≈1.67e-3 N/m for polar nanoparticles.

The proposed model successfully approximates the size of most of the nanopowders reported in literature, both in conventional and in centrifuged beds. It has no systematic error as a function of the particle size, density or Hamaker coefficient. The average prediction error of the proposed model is ≈20 %, while for the model proposed by Valverde and Castellanos [5], it is ≈45 %.

1. Ommen, J.R., J. Valverde,
and R. Pfeffer, Journal of Nanoparticle Research, 2012. **14**(3): p. 1-29.

2. Tahmasebpoor, M., L. de
Martín, M. Talebi, N. Mostoufi, and J.R. van Ommen, Physical Chemistry Chemical Physics, 2013. DOI: 10.1039/C3CP43687J.

3. Yao, W., G. Guangsheng, W.
Fei, and W. Jun, Powder Technology, 2002. **124**(1?2): p. 152-159.

4. Rabinovich, Y.I., J.J.
Adler, A. Ata, R.K. Singh, and B.M. Moudgil, Journal of Colloid and Interface Science, 2000. **232**(1): p. 10-16.

5. Valverde, J.M. and
A. Castellanos, Chemical Engineering Journal, 2008. **140**(1?3): p. 296-304.

(425h) Investigation of Electrostatic Reduction in Gas-Solid Fluidized Bed By In-Situ Corona Charge Eliminator

Dong, K., Zhejiang University
Zhou, Y., Zhejiang University
Zhang, Q., Zhejiang University
Huang, Z., Zhejiang University
Wang, J., Zhejiang University
Yang, Y., Zhejiang University
Originally presented on: 11/6/2013 11:04:00 - 11:26:00

When particles are fluidized, they acquire electrostatic charges due to impacts and frictions. The accumulation of excess electrostatic charges in fluidized bed reactors (FBRs) can cause problems such as particles agglomeration, wall sheets and even an unscheduled shutdown. Reduction of the electrostatic charge accumulation can prevent these issues. This study develops a technique called In-situ Corona Charge Eliminator (ICCE), based on the mechanisms of gas corona discharging and particle corona charging, for reducing electrostatic charge in FBRs. The experiment was carried out in a 420 mm internal diameter plexiglass fluidized bed, with LLDPE particles and dry air as the fluidized particles and gas respectively. The electrostatics in the fluidized bed at various superficial gas velocities (0.2~0.7 m/s) were first characterized in multiaspect, such as electrostatic potential, particle charge-to-mass ratio and electric field strength, by means of electrostatic probe, faraday cup and field meter respectively. The results show that the axial electrostatic potential in the bubble column, particle specific charge and electric field strength in the dilute phase increased with increasing gas velocity. At small gas velocities, the particles in the fluidized bed were bipolar charged, with negative charged particles at the bottom and positive charged particles at the top. However, polarity reversal was observed for the bottom particles at large gas velocities. Moreover, six corona electrodes were installed on the inner wall of fluidized bed and the electrostatic charge level was found to be significantly reduced by ICCE. The polarity of the particle charge reversed under the action of ICCE at low gas velocities due to limited charge accumulation on particles caused by impacts and frictions. With increasing gas velocity, the charge accumulation on particles increased obviously due to strengthened impacts and frictions. Therefore, electrostatic charge was partially reduced at large gas velocities by the same setting of ICCE as used at lower gas velocity. The results indicate that ICCE is an effective and promising technique for reducing electrostatic charges in FBRs.

(431a) Study of Flow and Mixing in High Shear Mixers Using Numerical Techniques

Dubey, A., Tridiagonal Solutions Inc.
Muzzio, F. J., Rutgers, The State University of New Jersey
Vedapuri, D., Tridiagonal Solutions Inc.
Originally presented on: 11/6/2013 8:30:00 - 8:50:00

A variety of blending equipment is available for use in the pharmaceutical industry, such as V-blenders, double-cone blenders, and continuous mixers. Ribbon and ploughshare mixers are a high shear type, where the action of one or more blades produces convective motion in the granular bed. Their names reflect their respective blade designs – helical ribbon type in one and plough shaped in the other. These are often found to be suitable for mixing highly cohesive materials such as active pharmaceutical ingredients (APIs). During the blending process, ribbon blenders are generally observed to create a lesser degree of size-segregation compared to tumbling blenders. They are, however, prone to having dead zones and the high shear can cause over-lubrication problems in case a lubricant like magnesium stearate is used. This study focuses on a large scale discrete element method (DEM) based simulation of the effect of process parameters on blending performance of these two blenders. The effects of fill level, loading order of cohesive/finer material (API or lubricant) relative to the excipient, and blade speed are investigated. The nature of powders is simulated using slump tests and DEM simulation parameters are fine-tuned to simulate material flow behavior. Multi-core/multi-processor parallel computing is used to scale the DEM simulations to more realistic number of particles. The findings are compared with available experimental data [1,2]. The results show the applicability of such mechanistic models can be extended further up from small lab-scale equipment to pilot-scale and above.

[1] Laurent B.F.C and Cleary P.W. Comparative study by PEPT and DEM for flow and mixing in a ploughshare mixer. Powder Technology, 228 (2012) 171-186.

[2]Muzzio, F.J., Llusa M., Goodridge C.L., Duong N., Shen E. Evaluating the mixing performance of a ribbon blender. Powder Technology, 186 (2008) 247-254.

(431b) Micro-Mixing Characterization of Continuous Mixing Processes Using Near Infrared Chemical Imaging

Osorio, J. G., Rutgers, The State University of New Jersey
Hernandez, E.,
Romanach, R., University of Puerto Rico, Mayaguez
Muzzio, F. J., Rutgers, The State University of New Jersey
Originally presented on: 11/6/2013 8:50:00 - 9:10:00

Continuous powder mixing has been a subject of much interest in the past few years. The mixing efficiency of these processes is commonly characterized by estimating the blend homogeneity from relevant samples. The concentration of the critical component (i.e. active pharmaceutical ingredient, API) is measured from these samples and the blend homogeneity is quantified by estimating the relative standard deviation (RSD). This bulk homogeneity (macro-mixing) does not yield much information on the dispersion or degree of agglomeration (micro-mixing) of the API in the blend. In this study, powder blends obtained from a continuous mixing process are characterized using near infrared chemical imaging (NIR-CI). This allows the micro-mixing quantification of the critical ingredient in the final blends. An experimental design was used to vary the concentration of API in the blend, the total flow rate through the system, and the impeller rotation rate of a continuous mixer (GCM 250, Gericke). These led to changes in cohesion of the blend, mean residence time and hold-up [1]. The blend used was composed of semi-fine acetaminophen (APAP, ~45 μm) and Avicel PH 200 (~220 μm). Samples were taken at the outlet of the mixer for analysis and a Spectral Dimension SyNIRgy chemical imaging spectrometer (Malvern) was used to analyze the powder blend samples.

The chemical images showed the areas in which APAP could be in aggregates. The chemical images were treated and converted to binary images by choosing a specific threshold using the distribution of pixels of each image. Binary images were used to quantify aggregate size statistics. The statistical analysis from the binary images yielded several qualitative and quantitative results. For example, the aggregate equivalent area measured was correlated to the rotation rate for several APAP concentrations. These correlations were comparable to the previous results obtained in the bulk characterization of this continuous mixer [1]. Previous results showed a maximum number of blade passes and minimum RSDs at intermediate rotation rates in which the powder experiences the most shear and total strain. Both of these results are in accordance with blends having a minimum aggregate equivalent area at intermediate impeller rotation rates. More correlations between the micro-mixing state of the final blends as a function of the mixing parameters and concentration of the API will be presented. The use of this technique can help in the development of pharmaceutical blend formulations and the optimization of continuous powder mixing processes.

[1] A.U. Vanarase, F.J. Muzzio, Effect of operating conditions and design parameters in a continuous powder mixer, Powder Technol 208 (2011) 26-36.

(431c) Continuous Tumble Mixing Characterization Using DEM Simulations

Florian, M., University of Puerto Rico - Mayaguez Campus
Velazquez, C., University of Puerto Rico - Mayaguez Campus
Mendez, R., University of Puerto Rico - Mayaguez Campus
Originally presented on: 11/6/2013 9:10:00 - 9:30:00

Powder mixing is an essential operation in the manufacture of many pharmaceutical products. Currently, the pharmaceutical industries are performing this operation in batch mode. The FDA have been working in furtherance the application of the continuous processes as an alternative to batch mixing processes that in principle allows for easier on-line control and optimization of mixing performance. In this study, a new continuous tumble mixer was developed using as a model the batch drum mixers to study granular mixing. The principal results obtained using different flow rates, API concentrations, mixer speeds and feeding angles demonstrated that the tumble mixer is capable of achieving a good mixing based on the RSD values.

Based on these results, a combination of different parameters has been study using Discrete Element Methods (DEM) simulations to analyze the effect of the material properties and equipment parameters on the final uniformity. The blender simulated has exactly the same dimensions of the experimental system, and the experimental design included simulations at 50 and 70 RPM with and without cohesion. The principal parameters used in these simulations include two particle generations, 1 mm particle diameter, and flow rate.

To measure the final uniformity; the effect of cohesion, flow and revolutions per minute have been evaluated using the RSD values. To better understand the powder phenomena inside the mixer, the velocity profile, residence time distribution and Froude Number were used and different areas were selected to study the particle homogeneity. The mixing dynamic was evaluated creating sampling volumes in each exit to measure concentration. Simulation results showed good mixing with RSD values below 3%, similar behavior compared to experimental part, and that the powder behavior inside the mixer change with the flow rate, cohesiveness, and the mixer speed.

(431d) Implementation of a Control Loop for a Continuous Powder Mixing Process

Zhao, X., Universtie de Toulouse, centre RAPSODEE, ecole des Mines d'Albi-Carmaux
Gatumel, C., Universite de Toulouse, centre RAPSODEE, ecole des Mines d'Albi-Carmaux
Dirion, J. L., Universite de Toulouse, centre RAPSODEE, ecole des Mines d'Albi-Carmaux
Berthiaux, H., Universite de Toulouse, centre RAPSODEE, ecole des Mines d'Albi-Carmaux
Cabassud, M., Laboratoire de Genie Chimique, UMR 5503 CNRS/INPT/UPS,
Originally presented on: 11/6/2013 9:30:00 - 9:50:00

Powder mixing is an essential operation in various industrial fields, such as pharmaceutics, agro-food, cements, etc. While important advances on modeling and simulation of powder flow have been made over the past years, the understanding of powder behavior inside a mixer is still too poor to be implemented in a model that may be used in real-time in the frame of a process control loop. Whatever the mixture considered, the efficiency of a continuous powder mixer strongly depends on various operating variables, such as flow rates and stirrer's rotational speed. Performing the regulation of these is therefore essential to guarantee the homogeneity of the mixture produced.

This work is part of an overall study on the elaboration of a process control loop to minimize the impact of current perturbations, such as changing the inlet dosage mode, and reach the required homogeneity at the outlet of the vessel. For this, we have used a Gericke GCM500 continuous mixer, for which a method of on-line image analysis had previously been developed to derive the homogeneity of mixtures in real time. Different modes of regulation for the rotational speed of the stirrer have been studied, based on either the mean content or the coefficient of variation. Case studies are defined and presented to evaluate the impact of the regulation modes on the mixer's efficiency during strong perturbation of the inflow rate and inlet composition.

(431f) Enhanced Mixing of Particulate Matter in Multi Orifice Silos

Doshi, P., National Chemical Laboratory

Orpe, A. V., National Chemical Laboratory

Kunte, A., National Chemical Laboratory

Originally presented on: 11/6/2013 9:55:00 - 10:15:00

Enhanced Mixing of particulate matter in multi orifice silos

Silos and hoppers have a ubiquitous existence in several industries which handle material in a powdered form. In some instances silos/hoppers act as a feeder to a particular process while in some other instances silos are used for processing material (e.g. nuclear pebble bed reactors). One of the key factors which affects the efficient design of a silo is the flow pattern and mixing of the material which drains down the silo and exits from the orifice. Typically, a silo (flat bottomed) or a hopper (conical bottomed) consists of a single orifice/exit either located centrally or at the periphery, which is wide enough to prevent jamming of the flowing particles. Such systems have been widely studied using experiments and numerical simulations to understand the flow and mixing behaviour of the particles.

In this work we provide newer insights about particle dynamics occurring in a flat bottomed silo having more than one orifice. We employ discrete element method (DEM) simulations to study the flow and mixing of the spherical particles draining down the silo. The method employs a Hertzian contact model and is simulated by the large-scale atomic/molecular massively parallel simulator (LAMMPS) developed by Sandia National Laboratories. In parallel, experiments are carried out using model granular particles, whose motion (near the wall) is captured using high speed camera imaging. Position of every particle is tracked over several images to determine their mean and fluctuating properties. The mixing index is quantified by counting different types of particles exiting from different orifices.

For the silos operated with "n" orifices each being wide enough, we observed that the overall flow behaviour, measured in terms of the component of mean velocity profiles, is equivalent to the "n" silos with a single exit orifice acting in parallel with little or no interaction between different regions of flow. However, very interesting behaviour is observed when width of every orifice is equal to that which can induce jamming. This causes spontaneous jamming at different orifices at different times, leading to co-existence of regions of flow and no flow which interact with each other dynamically.

To explore this behaviour systematically, we created a protocol in the simulations whereby the orifices are artificially closed and opened in a prescribed sequence and for a predetermined time period. We observe that such systematic closing/opening of the orifices caused the particles initially far apart from each other, to mix whilst coming out of the orifices. Figure 1 shows two snapshots taken from a typical simulation. The mixing efficiency is obtained by measuring the concentration of particles coming out of different orifices relative to their initial position higher up the silo. The observed behaviour in simulations is being validated by experiments conducted using a model granular system. The early results clearly show that the multi orifice silo operated using the abovementioned strategy leads to an improvement in particle mixing as compared to silos with a single orifice.

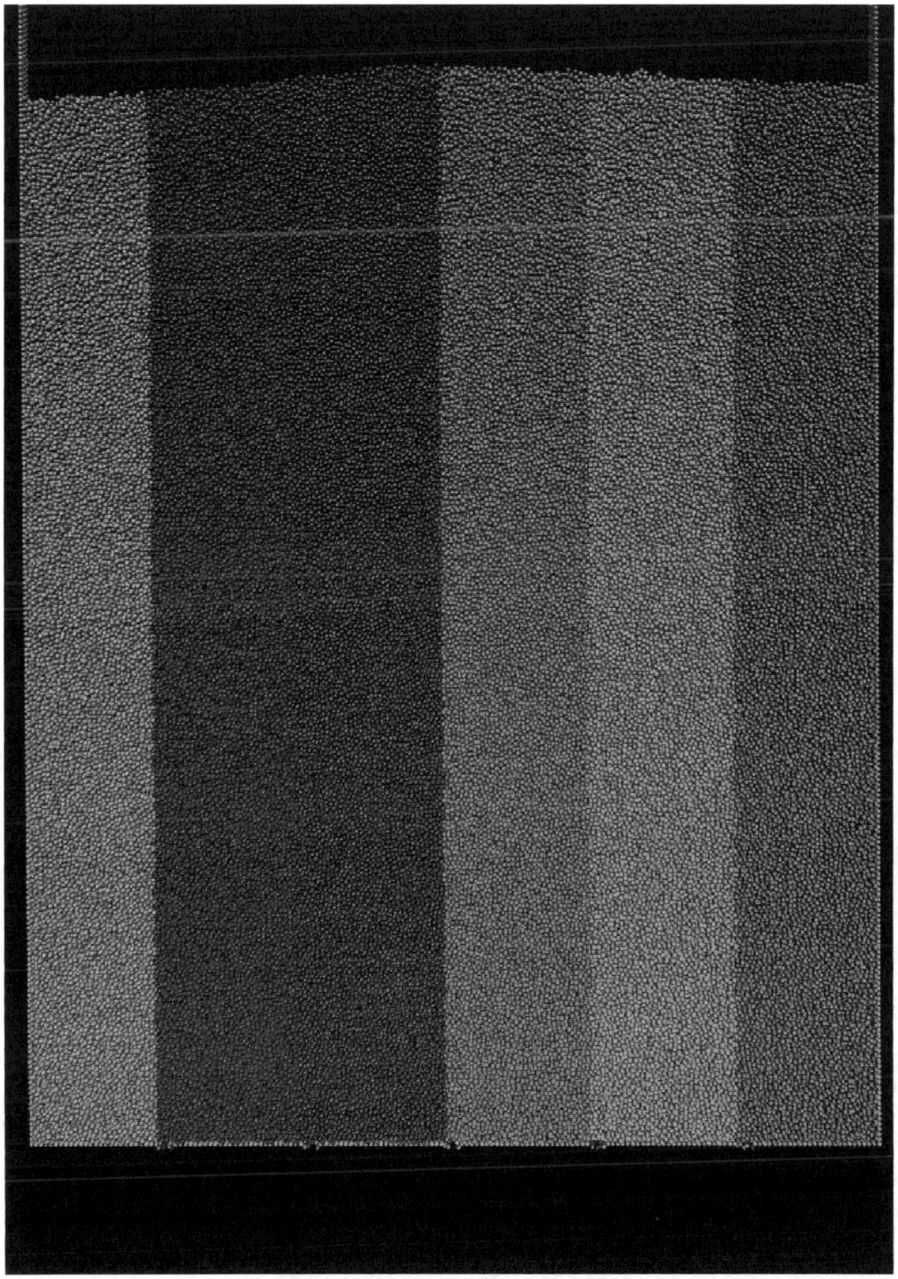

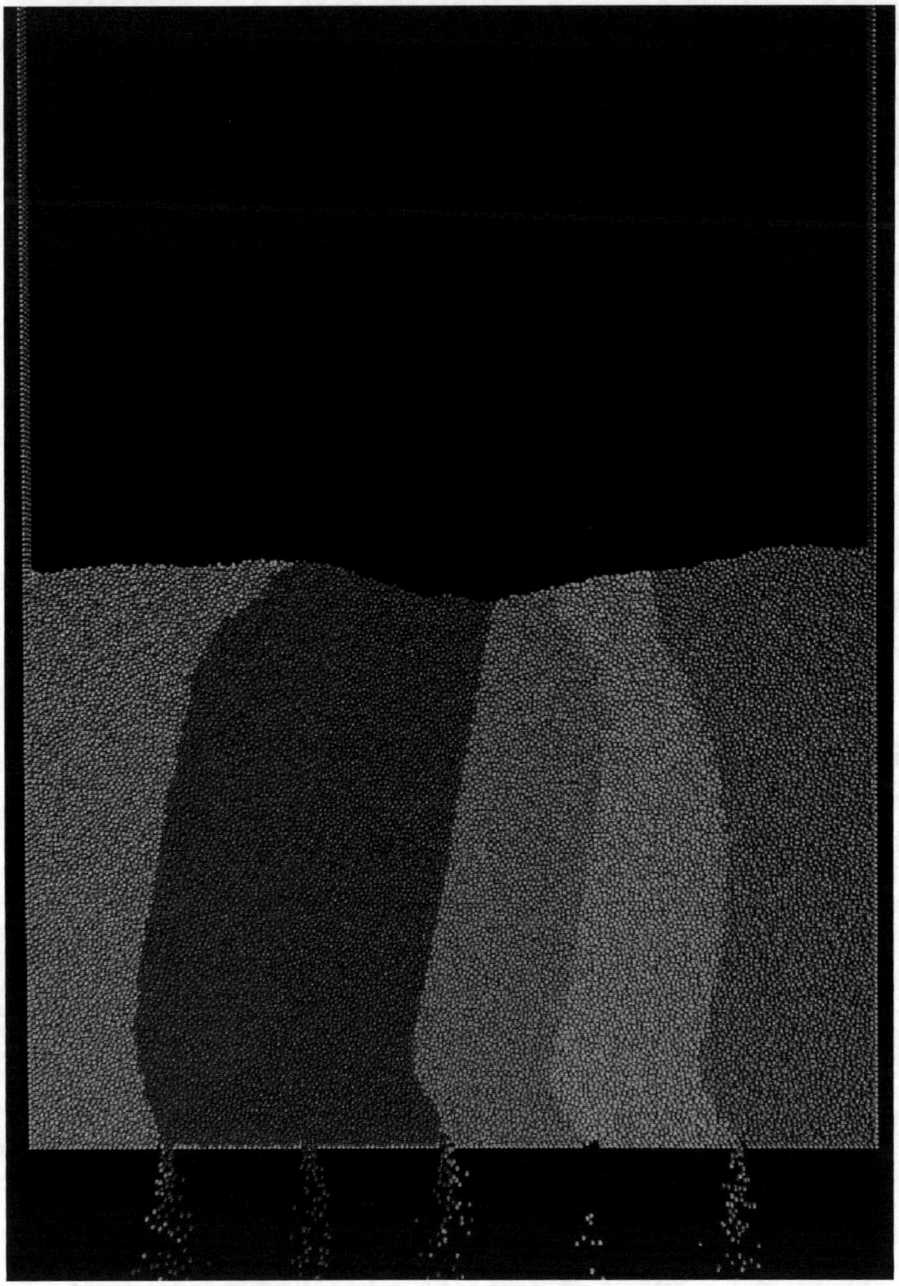

(a) **(b)**

Figure 1. Snapshot of silo drainage: (a) at initial time, and (b) at a subsequent time when silo is close to half empty. The silo is filled with identical particles but colored differently for ease of visualization. Notice the flow of brown colored particles from three different orifices shown in b.

(431g) Optimizing Screw Mixer Effectiveness Using Composition and Statistical Analysis Techniques

Kingston, T. A., Iowa State University
Heindel, T. J., Iowa State University
Originally presented on: 11/6/2013 10:15:00 - 10:35:00

The mixing of granular materials has a significant influence on the yield and/or quality of the desired products in numerous industrial processes including energy generation, concrete manufacturing, food processing, and pharmaceutical production. The operating conditions of the mixing equipment require careful selection because of their impact on the homogeneity of the mixture, especially when handling materials which vary in size, shape, and/or density. In this study, the mixing effectiveness of a double screw mixer blending red oak chips and glass beads was analyzed under numerous operating conditions. The mixer exit stream was sectioned into four equal regions and samples were collected across the exit stream to determine the mixture composition variance, which indicates the spatial homogeneity (or lack thereof) of the granular mixture. The effect of four factors on the composition variance was investigated: (i) screw rotation speed at levels of $\omega = 20$, 40, and 60 rpm, (ii) dimensionless screw pitch at levels of $p/D = 0.75$, 1.25, and 1.75, (iii) screw rotation orientation at levels of co-rotating, countering-rotating up-pumping, and counter-rotating down-pumping, and (iv) material injection configuration at levels of the red oak and glass beads being injected into port one and two and port two and one, respectively. A full factorial design of experiments was used taking on all possible combinations of the prescribed factors, and repeatability was incorporated by performing three tests at each of the 54 operating conditions. A model equation was applied using an analysis of variance (ANOVA) statistical procedure. Of the 15 main-effect and interaction terms, eight were found to be statistically significant, and a reduced model equation was applied using the significant terms. Most notably, two three-way interaction terms consisting of the interaction between the screw rotation speed, dimensionless screw pitch, and screw rotation orientation, and between the screw rotation speed, screw rotation orientation, and material injection configuration were significant. The mixture composition variance was minimized (i.e., highest mixing effectiveness) at a 60 rpm screw rotation speed, dimensionless screw pitch of 1.75, counter-rotating down-pumping screw rotation orientation, and the red oak chips and glass beads injected into port one and two, respectively.

(431h) Characterizing the Mixing Efficiency of the Planetary Powder Mixer Triaxe® Through An Extended Power Consumption Study

Gatumel, C., Université de Toulouse, centre RAPSODEE, ecole des Mines d'Albi-Carmaux

Berthiaux, H., Université de Toulouse, centre RAPSODEE, ecole des Mines d'Albi-Carmaux

André, C., UC Lille

Delaplace, G., UC Lille, Hautes Etudes Ingénieurs (H.E.I.), Laboratoire de Génie des Procédés

Originally presented on: 11/6/2013 10:35:00 - 10:55:00

While mixing processes have been largely employed in industry since a long time, a need of comprehensive studies of these processes is recently growing. Indeed modern materials are frequently performed in very tight intentions, especially in high added-values industries. It is usual to find formulations containing several tenths of ingredients, but for which end-use properties finally depend on the efficiency of a single mixing step.

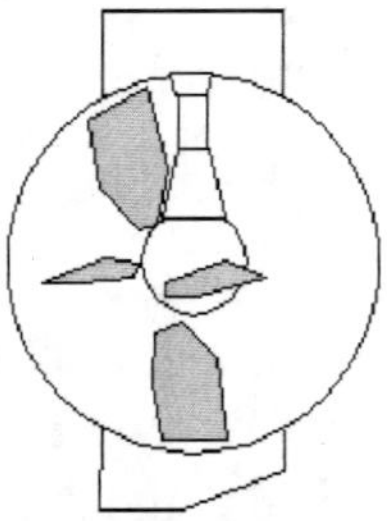

Fig 1: Triaxe® mixing system

Planetary mixers take a large place among the proposal of equipment manufacturers. The combination of several motions ensures interesting mixing conditions and allows the mixer to be adapted to a wide range of raw material properties. The Triaxe® mixing system combines a gyration motion and a rotation motion of an impeller across a spherical tank, as shown in fig. 1. Previous studies from our group enhanced the versatility of this mixer, being able to achieve mixing of free-flowing or cohesive materials in a short mixing time. For a given product, mixing time and power consumption depend on gyrational and rotational speeds of the impeller. A characteristic speed, corresponding to the maximum tip speed of the impeller for given gyration and rotation conditions, is defined and its versatility is demonstrated. It is concluded from mixing and power consumption experiments that this speed governs agitation and mixing of free-flowing powders. In addition, some correlations involving dimensionless Power numbers and Froude numbers deduced from this characteristic speed are highlighted. This is enlarged to account for powders of very different flow characteristics, as well as their mixtures.

(436a) Hydrogen Production From Water

McDaniel, M. D.,
Ekerdt, J. G., University of Texas at Austin

Originally presented on: 11/6/2013 8:30:00 - 9:00:00

Silicon-based photocathodes show promising potential for photoelectrochemical (PEC) water splitting for renewable production of hydrogen. Trade-offs between stability and efficiency have limited the success of Si-based photoelectrodes. To overcome this issue, we utilize a thin layer (≈ 2 nm) of strontium titanate, $SrTiO_3$ (STO), which is grown epitaxially on p-doped Si (001) by molecular beam epitaxy. The STO layer greatly enhances the stability of the Si-based photocathode. In addition, the high quality STO layer and clean, low defect interface between STO(001) and Si(001) provides minimal trap states for charge carriers. Moreover, there is no conduction band offset between STO and Si. This allows the photo-generated electrons to easily move from the Si substrate to the STO surface, where hydrogen ions are reduced to form gaseous hydrogen. Based on our electrochemical measurements, we find that the STO-Si photocathode is both stable and exhibits high conversion efficiency. Studies with continuous Ti/Pt films for H_2 evolution yield a current density above 15 mA/cm^2; patterned metal layers are expected to give higher current densities. The results indicate that Si-based photocathodes for PEC water splitting can be improved using a thin STO layer, providing a viable method for clean hydrogen production.

(436b) Core-Shell Photoelectrochemical Electrodes for Water Splitting

Peng, Q., Duke University
Kalanyan, B., North Carolina State University
Hoertz, P.,
Miller, A.,
Kim, D. H., North Carolina State University
Hanson, K., UNC-Chapel Hill
Alibabaei, L., UNC-Chapel Hill
Liu, J., Duke University
Meyer, T., UNC-Chapel Hill
Parsons, G. N., North Carolina State University
Glass, J. T., Duke University
Originally presented on: 11/6/2013 9:00:00 - 9:18:00

Photoelectrochemical (PEC) processes offer one of most promising ways to convert solar energy into fuels (e.g. H_2, and CH_3OH), especially H_2. In order to synthesize high efficient and stable PEC water splitting electrodes, one must optimize the light absorption, charge separation, and charge transfer processes of the electrode simultaneously. Core-shell nanostructures with transparent conductive cores and semiconductor shells show great promise for optimizing PEC performance of photoactive materials, such as TiO_2. However, limited research has been performed in this field.

We will present the result of using transparent conductive network consisting of low cost, solution-processed antimony doped tin oxide nanoparticles (nanoATO) and tin doped indium oxide nanoparticles (nanoITO) for enabling core-shell TiO_2 PEC electrodes. In detail, we will present the synthesis of TiO_2 PEC electrodes by coating solution-processed nanoATO on FTO glass with TiO_2 through atomic layer deposition (ALD). The conductive, porous nanoATO film-supported TiO_2 electrodes, yielded a highest photocurrent density of 0.58 mA/cm^2 under AM 1.5G simulated sunlight of 100 mW/cm^2. This is approximately 3× the maximum photocurrent density of planar TiO_2 PEC electrodes on FTO glass. Transient photocurrent measurements showed that nanoATO films reduce charge recombination by accelerating transport of photoelectrons through the less defined conductive porous nanoATO network. Further, we will show that the effect of interfacial reaction during the TiO_2 ($TiCl_4/H_2O$) ALD on transparent nanoATO and nanoITO on their PEC performance. The scalable solution-processed porous transparent nanoATO and nanoITO films are promising as a framework to replace other conductive scaffolds (Si nanowire forest, CNTs, and $TiSi_2$ nanonets) for PEC electrodes. The results are not only valuable for PEC, but also informative for the related DSSC research.

(436c) Design of Photocatalysts Active in Infrared and Visible Range

Lee, D. C., Korea Advanced Institute of Science and Technology (KAIST)

Originally presented on: 11/6/2013 9:18:00 - 9:36:00

Infrared- and visible-active semiconductors are promising candidates for light-harvesting materials. When the semiconductor nanocrystals are within quantum confinement regime, they exhibit size tunable bandgap, enabling light-harvesting even more controllable. Furthermore, the dynamics of excited states can also be controlled through a design of heterostructure semiconductor nanocrystals. We synthesized PbSe/CdSe/CdS core/shell/shell nanocrystals with varying structural design, e.g., spheres, pyramid and tetrapod. These nanocrystals show varying electron-hole exchange interactions and radiative recombination lifetimes depending on their dimensions. We also prepared semiconductor/metal nanojunction and used the nanomaterials as photocatalysts for reduction of various dyes by growing metal tips at the corners of semiconductor nanocrystals. Interestingly, the morphology change leads to the difference in the tip growth pattern. In addition, the growth of metal particles affects the Fermi level of the entire hybrid composites, leading to the change in their photocatalytic activity. In the presentation, the implications of the heterostructure quantum dots and their hybrid composites with metal nanoparticles will be discussed in the context of energy applications.

(436d) Efficient Photoelectrochemical Water Splitting With Si-Based Metal-Insulator-Semiconductor Photoelectrodes

Esposito, D. V., University of Delaware
Lee, Y., National Instittue of Standards and Technology
Talin, A. A., Sandia National Laboratory
Moffat, T. P., National Institute of Standards and Technology
Originally presented on: 11/6/2013 9:36:00 - 9:54:00

Photoelectrochemical (PEC) water splitting is an attractive pathway for renewable production of hydrogen, but the efficiency and stability of semiconducting photoelectrodes must be improved. One promising approach to achieve both high efficiency and good electrochemical stability is the metal-insulator-semiconductor (MIS) photoelectrode architecture.[1,2] The MIS design consists of catalytic metal structures, or collectors, deposited on a 1-3 nm thick oxide-covered semiconductor. Of great importance to this design is the thin insulating oxide layer, which must protect the semiconductor from the potentially corrosive electrolyte while simultaneously mediating minority carrier tunneling between the semiconductor and collectors. In recently published work,[2] we have demonstrated significant improvement in the performance of Si-based MIS photocathodes through interfacial engineering. These planar photocathodes exhibit excellent stability and appreciable photoelectrode conversion efficiencies, up to 2.9%, but further improvement in efficiency must be achieved in order to make this process economically viable.

In this work, nano- and micro-structured Si-based MIS photoelectrodes fabricated by reactive ion etching and vapor-liquid-solid chemical vapor deposition are explored as a means to further increase the efficiency of Si-based photoelectrodes. The use of 3D-structured electrodes offer several potential advantages for achieving higher efficiency, including increased surface area for catalysis, improved collection of photo-generated carriers, and decreased optical reflection losses. However, the use of 3D photoelectrodes also introduces detrimental effects that create trade-offs between optical, catalytic, and electronic properties. In this talk, design principles for optimizing the performance of 3D-structured photoelectrodes are described and a comparison of the performance of planar and 3D-structured MIS photoelectrodes is presented. The incorporation of MIS photoelectrodes into stand-alone PEC water splitting devices will also be discussed. Calculations based on the method of detailed balance show that PEC devices utilizing Si-based MIS photoelectrodes are capable of achieving solar-to-hydrogen conversion efficiencies greater than the 2015 efficiency target of 15% set by the Department of Energy.[3]

References

[1.] H.J. Lewerenz, et al., *Electrochem. Acta*, **2011**, 56, 10726.

[2.] D.V. Esposito, I. Levin, T.P. Moffat, A.A. Talin, *Nature Materials*, **2013**, (in press).

[3.] Fuel Cells Technologies Office Multi-Year Research, Development and Demonstration Plan. US Department of Energy Efficiency and Renewable Energy. Available from: http://www1.eere.energy.gov/hydrogenandfuelcells/mypp/ (2012)

(436e) Graphene Oxide Photocatalysts for Water Splitting and Its Upconverted Photoluminescence
Yeh, T. F.,
Teng, H.,
Originally presented on: 11/6/2013 9:54:00 - 10:12:00

Graphene-Oxide Photocatalysts for Water Splitting and Its Upconverted Photoluminescence

Te-Fu Yeh[1] and Hsisheng Teng[1,2,*],

[1]*Department of Chemical Engineering and Research Center for Energy Technology and Strategy, National Cheng Kung University, Tainan 70101, Taiwan*

[2]*Center for Micro/Nano Science and Technology, National Cheng Kung University, Tainan 70101, Taiwan*

**E-mail: hteng@mail.ncku.edu.tw*

Abstract

Graphene oxide (GO) is a p-type semiconductor with a suitable band positions for photocatalytic watersplitting. GO sheets have a hydrophilic 2D carbon structure that allows for extensive chemical modification. This study presents strategies of tuning the conductivity type of GO from p-type to n-type for simultaneous evolutions of H_2 and O_2 under visible light irradiation. Further converting the n-type graphene oxide sheets to quantum dots results in stoichiometric evolution of H_2 and O_2 at 2-to-1 ratio. The quantum dots also serve as an energy-converting medium with upconverted photoluminescence to promote the activity of metal-oxide photocatalyts.

Introduction

Polymeric semiconducting molecules, which have a large contact area with water, attract significant attention as alternatives to metal-containing photocatalysts. Graphene oxide (GO) derived from graphite oxidation is a molecule-like semiconductor [1]. GO can be extensively dispersed in water to molecular scale and has a large exposed area. Graphene is a zero gap semiconductor whereas the GO gap increases with the oxidation level [2,3]. The bonding with electron-withdrawing oxygen functionalities results in the p-type conductivity of GO. N-type conductivity appears when graphene covalently bonds to electron-donating nitrogen functionalities [4-6].

In addition to chemical doping, we exploit the quantum-confinement effect to modulate electronic properties. Unlike two-dimensional graphene, the zero-dimensional graphene quantum dots (GQDs) have a finite band gap depending on their size [7]. The confined p-electron in sp^2 domain gives the quantized discrete levels [8]. With these features, GQDs exhibit upconvered photoluminescence (PL) [9].

This study presents the potential of GO as a photocatalyst for water splitting. P-type GO can catalyze H_2 evolution with the presence of electron donors. Ammonia-treated GO (NGO) exhibits n-type conductivity and promising photocatalytic activity for generating H_2 and O_2 simultaneously, but with the ratio below the stoichiometric value (that is, 2). Miniaturized N-doped graphene oxide quantum dots (NGO QDs) function as photocatalysts for stoichiometric H_2 and O_2 generation from water cleavage under visible light irradiation. We also demonstrate the upconversion ability of NGO QDs.

Experimental

GO was prepared using a natural graphite powder through a modified Hummers' method. NGO was obtained through nitridation of the as-prepared GO performed at room temperature using a flow of NH_3 gas. NGO QDs were prepared from oxidation of nitrogen-doped graphene. Photocatalytic reactions were conducted at approximately 25 Â¢XC in a gas-enclosed inner irradiation system or a side irradiation system.

Result and Discussion

We demonstrated the photocatalytic activity of GO in the gas-enclosed circulation system with mercury-lamp irradiation. Continuous H_2 evolution over metal-free GO was observed for reactions conducted in pure water and an aqueous methanol solution [1]. The total evolution of H_2 from aqueous methanol solution far exceeds the amount of H_2 obtained from pure water. GO cannot catalyze O_2 generation even with the presence of the electron scavenger Ag^+ ion.

We modified the as-synthesized GO photocatalyst by ammonia treatment to form NGO, which exhibited n-type conductivity. The change of conductivity type was contributed by the electron-donating property of the introduced amino groups. NGO effectively catalyzes O_2 evolution under mercury-lamp irradiation [6]. The n-type conductivity of NGO attracts hole migration toward the interface region for O_2 evolution. NGO gave rise to simultaneous evolutions of H_2 and O_2 from the side irradiated system under visible light irradiation [6]. The molar ratio of the evolved H_2 and O_2 was smaller than the stoichiometric ratio 2-to-1.

Quantum confinement effect elevates the conduction band edge, promoting the electron injection for H_2 evolution from water. We found stoichiometric evolutions of H_2 and O_2 from pure water over NGO QDs with visible-light irradiation. Figure 1 shows the PL spectra of NGO QDs excited by long-wavelength light (660-900 nm), with the upconverted emission located in the range of 450-530 nm. The upconverted PL emission of NGO QDs excited $NaTaO_3$ to catalyze H_2 evolution from an aqueous methanol solution under visible light irradiation.

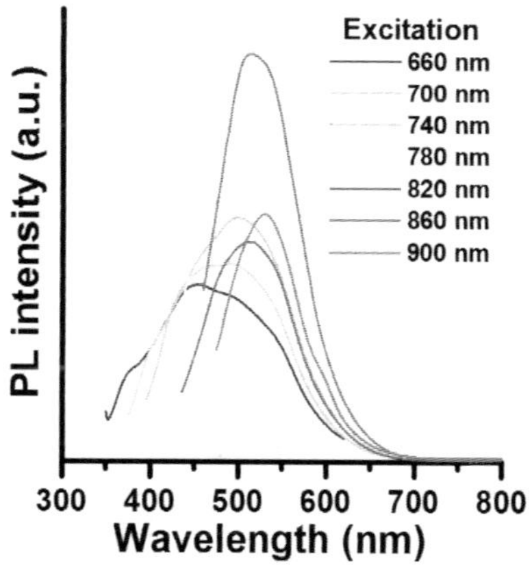

Figure 1. Upconverted PL from NGO QDs. The excitation wavelength ranges from 660 nm to 900 nm. The upconverted emission located in the range of 450-530 nm.

Conclusions

The p-type GO catalyzed H_2 evolution from an aqueous methanol solution under mercury-lamp illumination. The introduction of nitrogen functionalities converted the p-type GO to an n-type NGO, which was capable of catalyzing simultaneous H_2 and O_2 evolutions from water under visible light. The NGO QD catalyst achieved overall water splitting ($H_2/O_2 = 2$) under visible light irradiation. The upconversion ability of NGO QDs made wide-gap photocatalysts active under visible-light illumination.

Acknowledgements

This research is supported by the National Science Council of Taiwan (101-2221-E-006-243-MY3 and 101-2221-E-006-225-MY3) and the "Aim for the Top-Tier University and Elite Research Center Development Plan" of National Cheng Kung University.

References

[1] T.F. Yeh, J.M. Syu, C. Cheng, T.H. Chang, H. Teng, *Adv. Funct. Mater.* **2010**, 20, 2255.

[2] T.F. Yeh, F.F. Chan, C.T. Hsieh, H. Teng, *J. Phys. Chem. C* **2011**, 115, 22587.

[3] T.F. Yeh, J. Cihlář, C.Y. Chang, C. Cheng, H. Teng, *Materials Today* **2013**, 16, 78.

[4] X. Wang, X. Li, L. Zhang, Y. Yoon, P. K. Weber, H. Wang, J. Guo, H. Dai, *Science* **2009**, 324, 768.

[5] C. C. Hu, H. Teng, *J. Phys. Chem. C* **2010**, 114, 20100.

[6] T.F. Yeh, S.J. Chen, C.S. Yeh, H. Teng, *J. Phys. Chem. C* **2013**, 117, 6516.

[7] L. A. Ponomarenko, F. Schedin, M. I. Katsnelson, R. Yang, E. W. Hill, K. S. Novoselov, A. K. Geim, *Science* **2008**, 320, 356.

[8] G. Eda, Y. Y. Lin, C. Mattevi, H. Yamaguchi, H. A. Chen, I. S. Chen, C. W. Chen, M. Chhowalla, *Adv. Mater.* **2010**, 22, 505.

[9] H. Li, X. He, Z. Kang, H. Huang, Y. Liu, J. Liu, S. Lian, C. H. A. Tsang, X. Yang, S. T. Lee, *Angew. Chem. Int. Ed.* **2010**, 49, 4430.

(436f) The Study of Cu2ZnSnS4 Nanocrystal/TiO2 Nanorod Arrays Heterojuction Photoelectrochemical Cell for Hydrogen Generation

Ho, T. Y.,
Chen, L. Y.,
Originally presented on: 11/6/2013 10:12:00 - 10:30:00

In this study, we reported TiO_2 nanorods arrays (TiO_2 NRAs) synthesized by hydrothermal method with controllable densities, diameters and lengths for photoelectrochemical hydrogen generation. The densities and diameters of the TiO_2 NRAs could be easily controlled by the concentration of titanium precursor (titanium isopropoxide, TTIP) and the amount of the HCl. With different reaction time, the lengths of TiO_2 NRAs could be increased from 2 μm to 9 μm. The photocurrent density and conversion efficient were increased with the lengths of TiO_2 NRAs. The maxima current density and conversion efficiency are 2.5mA/cm^2 at 1.23V vs RNHE and 0.95% with 4-5 μm in length under 100mA/cm^2 Xenon lamp illumination with AM1.5 filter, respectively. To improve the conversion efficiencies, various sensitizers, such as dye or quantum dot, are usually used to cover the absorption in the visible light range. Herein, Cu_2ZnSnS_4 (CZTS), which owns direct band gap semiconductor with 1.5eV and suitable band positions, was employed to extend the absorption range. In this study, CZTS nanoparticles were direct decorated on TiO_2 NRAs by solvothermal method with metal chloride and thiourea as precursor and ethanol as solvent to improve the conversion efficiency for hydrogen generation.

(436g) Development of New Generation of Sub-Nanometer Catalysts for Sustainable Energy Applications

Wu, Q., Stony Brook University/BNL
Shen, P., Stony Brook University / BNL
Zhao, S., Stony Brook University
Ramakrishnan, G.,
Su, D., Brookhaven National Laboratory
Li, Y.,
Orlov, A., Stony Brook University / BNL
Originally presented on: 11/6/2013 10:30:00 - 10:48:00

Developing novel heterostructured nanomaterials for energy applications as water splitting under visible light has enormous potential for energy production. Currently the existing materials have very low quantum efficiency, which indicate that new approaches are required in catalysts design. In this work we have developed and characterized sub nanometer Au and Pt nanoparticles, which greatly enhanced rate of hydrogen production of CdS. These heterostructured materials were characterized by TEM, STEM, XRD, UV-Vis spectroscopy and DRIFTS. We have also conducted comprehensive studies of these catalysts in photocatalytic reactors, where we observed a substantial hydrogen production under visible light.

(436h) ZrO2 Incorporated ZnO/TiO2 Mesoporous Photocatalyst for H2 Generation

Hussein, A., South Dakota Shool of Mines and Technology
Shende, R., South Dakota School of Mines and Technology
Originally presented on: 11/6/2013 10:48:00 - 11:06:00

The present research deals with the synthesis of ZrO_2 incorporated coupled ZnO/TiO_2 photocatalyst and its effectiveness on H_2 generation from aqueous methanol solution. Ordered mesoporous ZrO_2 supported coupled TiO_2/ZnO materials were synthesized using surfactant as structure directing agent, titanium isopropoxide, zinc acetate dihydrate and zirconium n-propoxide precursors. The ZrO_2 incorporated coupled ZnO/TiO_2composites were characterized by XRD, UV-Vis spectra, SEM/TEM, XRF, Raman and N_2 adsorption-desorption isotherm. Photocatalytic hydrogen evolution was carried out in a 7 mL quartz reactor at ambient temperature. The focused intensity and area on the quartz cuvette were 0.60 W and 2 cm^2, respectively. In a typical experiment, 2.0 mg of $ZrO_2/ZnO\text{-}TiO_2$ photocatalyst was dispersed in 3.5 mL of H_2O containing 0.5 mL methanol or 10 mg biomass as a sacrificial agent. A 300 W xenon lamp (Oriel) equipped with a UV-filter (240 nm - 400 nm) was employed as the light source. Prior to irradiation, the suspension of the catalyst was degassed with ultrapure nitrogen for 30 min to completely remove the dissolved oxygen and to ensure the reactor in an anaerobic condition. From 3 mL reactor overhead space, 0.1 mL of gaseous samples were withdrawn periodically and analyzed using the gas chromatograph (GC), from which photocatalytic effectiveness towards H_2 generation was evaluated and compared with coupled TiO_2/ZnO material.

Thomas Baron Award Lecture: Shear Thickening & Gelation in Colloidal Dispersions - Nonequilibrium States and Their Applications in Personal Protective Equipment

Wagner, N., University of Delaware

Originally presented on: 11/6/2013 12:30:00 - 13:30:00

Colloidal gels and shear thickening fluids are challenging to formulate and process as they are often out of equilibrium states, but as will be shown, provide opportunities for engineering protective materials with a novel mechanical response. Developing fundamental structure property relationships for concentrated particle dispersions that exhibit shear thickening and/or gelation that derive from the properties of the particles that comprise the dispersion, their interparticle interactions, and their interactions with the suspending medium is scientific challenge with technological benefits. In this presentation I will discuss advances in measurement techniques that enable developing such relationships and demonstrate how this understanding can facilitate the development of novel personal protective equipment. Measurements of the microstructure commensurate with the viscosity and normal stress differences in shearing colloidal suspensions provide an understanding of how to control the viscosity, yield stress, shear thinning, and shear thickening rheological behavior. This is achieved through a combination of model system synthesis, rheological, and small angle neutron scattering (SANS) measurements under flow, as well as simulation and theory. Although many technological applications of concentrated suspensions are hindered by shear thickening behavior, novel field-responsive nanocomposites have been developed around shear thickening fluids (STFs). Ballistic, stab and impact resistant flexible composites are synthesized from colloidal & nanoparticle shear thickening fluids and gels for application as PPE. These rheological and microstructural investigations and micromechanical modeling serve as a framework for the rational design of STF-based materials to meet performance requirements.

Lectureship Award Lecture: Wetting the Bed: Liquid Feed Injection and Dispersion in Fluidized Beds

Berruti, F., Western University

Originally presented on: 11/6/2013 13:30:00 - 14:30:00

The presentation will review the experimental and analytical techniques developed at the Institute for Chemicals and Fuels from Alternative Resources (ICFAR) for the characterization of the interaction between liquid feed jets and fluidized solid particles, especially in relation to the Fluid Coking™ technology for the upgrading of heavy oils extracted from the oil sands. The research and development work described will include the experimental methodologies developed using small scale simulated systems operated at room temperature as well as small scale fluid beds operating at reaction temperatures and large scale room temperature simulated systems employing full scale commercial nozzles.

(499a) SiC Photoanodes for Solar Water Splitting

Bohn, C., National Institute of Standards and Technology
Sharifi, F.,
Kang, M. G.,
Szalai, V., National Institute of Standards and Technology
Originally presented on: 11/6/2013 12:30:00 - 12:49:00

AIChE 2013 Session: Nanomaterials for Hydrogen Production and Fuel Cells

Title: SiC Photoanodes for Solar Water Splitting

Authors: C.D. Bohn, M.G. Kang, F. Sharifi, V.A. Szalai

SiC ceramics offer a rich field for exploration for solar water splitting.Â Polytypes include 4H-SiC with a bandgap of 3.23 eV, 6H-SiC with a bandgap of 3.05 eV and 3C-SiC with a bandgap of 2.36 eV.Â Single crystal substrates of some of these materials were obtained from commercial vendors.Â The effect of nanoscale morphology and the addition of Pt nanocrystals at the interface between the SiC and electrolyte were systematically investigated in a pH=7 potassium phosphate buffer under AM 1.5 illumination.Â Gas chromatography provided insight into the chemical products formed by the n-type SiC photoanodes.Â Additionally, linear sweep voltammetry, impedance spectroscopy and scanning electron microscopy were used to monitor the structural evolution of the semiconductor surface.Â Results are discussed in the broader context of photoelectrochemical hydrogen production and recommendations for improving the efficiency of water splitting with SiC are proposed.

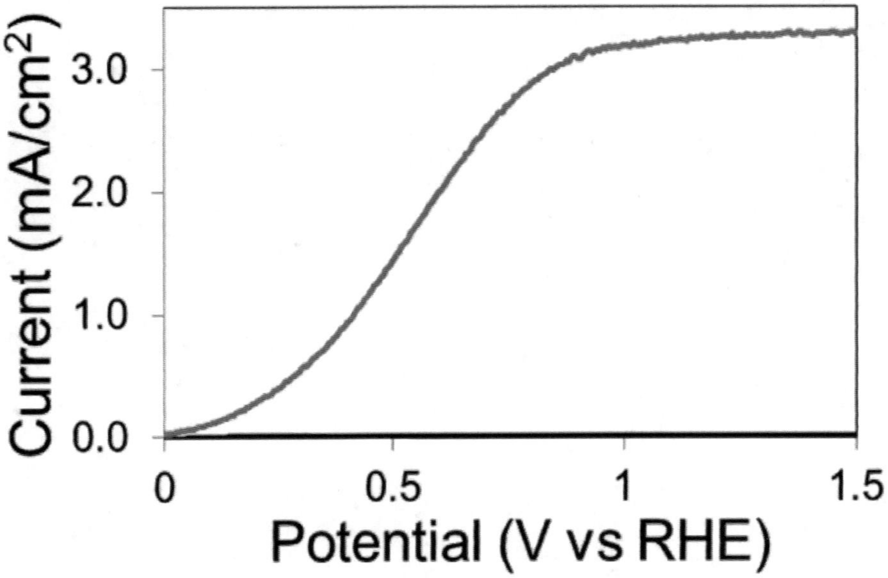

Keywords: solar water splitting, SiC, photoanode

(499b) Advanced Electrocatalysts From Organic Solution Synthesis

Wang, C., Johns Hopkins University

Originally presented on: 11/6/2013 12:49:00 - 13:08:00

Alloy nanoparticles (NPs) have attracted increasing interest due to their superior performance in various catalytic applications. For example, Pt alloys with transition metals (PtM with M = Fe, Co, Ni, etc.) have been found to be highly active for the oxygen reduction reaction in fuel cells. A lot of efforts have been dedicated to the synthesis of alloy catalysts, which are usually in the form of alloy NPs dispersed in high surface area carbon matrix. The approaches mostly include co-precipitation of metal salts in aqueous solution, impregnation, and electrodeposition. In spite of the successful preparation of various types of alloy catalysts, it yet remains challenging to synthesize monodisperse and homogeneous alloy catalysts with robust control over size and composition. Here we introduce our organic solvothermal approach towards high-quality alloy nanomaterials for electrocatalytic applications. Control over particle size, alloy composition and alloy homogeneity has been achieved by matching the different reduction and growth rates of the alloying elements, which has enabled systematic studies of the effects of these parameters on electrocatalysis. Our studies thus provide clues for the design and optimization of alloy nanomaterials for catalytic applications.

(499c) Nanostructured Ti –Fe2O3/Cu2O Bilayered Thin Films for Photoelectrochemical Water Splitting

Sharma, D., Dayalbagh Educational Institute,Dayalbagh

Upadhyay, S., Dayalbagh Educational Institute

Choudhary, S., Dayalbagh Educational Institute

Satsangi, V. R., Dayalbagh Educational Institute

Shrivastav, R., Dayalbagh Educational Institute

Dass, S., Dayalbagh Educational Institute

Originally presented on: 11/6/2013 13:08:00 - 13:27:00

Nanostructured Ti-Fe_2O_3/Cu_2O Bilayered Thin Films for Photoelectrochemical Splitting of Water

Dipika Sharma, Sumant Upadhyay, Surbhi Choudhary, Vibha R. Satsangi[?],

Rohit Shrivastav and Sahab Dass,

Department of Chemistry, [?]Department of Physics & Computer Sciences

Faculty of Science, Dayalbagh Educational Institute (Deemed University),

Agra-282110 (India)

Nanostructured thin films of pure Fe_2O_3, Ti ?Fe_2O_3, Cu_2O, and bilayered Fe_2O_3/Cu_2O, Ti-Fe_2O_3/Cu_2O were deposited on ITO glass substrate using spray pyrolysis method. Among various metal oxide semiconductors, cuprous oxide Cu_2O [Eg=2.4eV] and Iron oxide Fe_2O_3 [Eg=2.1eV] were
chosen for the present study on account of being abundant, non-toxic, environmentally safe, having low band gap and possess staggered type band edges which is suitable for efficient charge separation in water splitting process.

All samples were characterized using XRD, SEM and UV?Vis spectrometry. Photo electrochemical properties were also investigated in a three-electrode cell system. UV-Vis absorption spectrum for pure Fe_2O_3, Ti ?Fe_2O_3, Cu_2O, and Fe_2O_3/Cu_2O, Ti-Fe_2O_3/Cu_2O bilayered thin films shows absorption in visible light region. Nanostructured thin films for all the samples were used as photoelectrode in the Photoelectrochemical cell for water splitting reaction. Maximum photocurrent density of 2.22 mA/cm^2 at 0.95 V/SCE was exhibited by 454 nm thick Ti-Fe_2O_3/Cu_2O bilayered photoelectrode (Fig.1). Increased photocurrent density and enhanced photoconversion efficiency, exhibited by the bilayered thin films may be attributed to improved conductivity and efficient separation of the photogenerated charge carriers at the Ti-Fe_2O_3/Cu_2O interface.

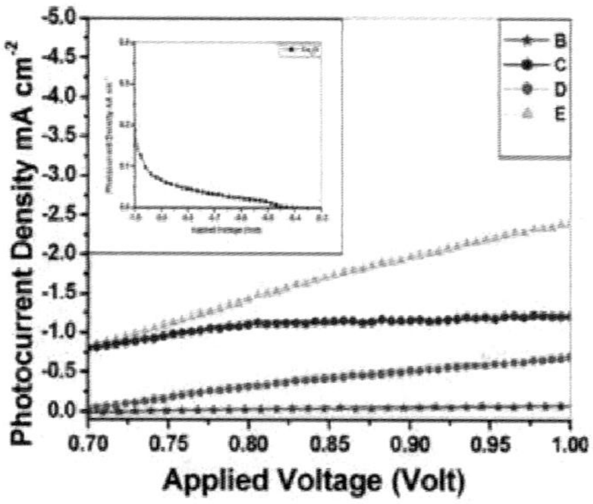

Fig. 1 Photocurrent density vs. Applied potential curve for pure Cu_2O(A), Fe_2O_3(B), Ti ?Fe_2O_3(C), and Fe_2O_3/Cu_2O(D), Ti-Fe_2O_3/Cu_2O(E) bilayered thin films respectively ,under visible light illumination in 0.1M NaOH electrolytic solution using 150W visible light source of irradiation 150 mW cm^{-2} at the position of sample.

Keywords: *Photoelectrochemical, Water splitting, Bilayered, Cu_2O, Fe_2O_3, Ti- Fe_2O_3, Spray-Pyrolysis*

(499d) Synthesis and Characterization of Mn2O3-ZnO Nanocomposites for Hydrogen Generation Via Photoelectrochemical and Photocatalytic Splitting of Water

Singh, N., Dayalbagh Educational Institute
Choudhary, S., Dayalbagh Educational Institute
Satsangi, V. R., Dayalbagh Educational Institute
Dass, S., Dayalbagh Educational Institute
Shrivastav, R., Dayalbagh Educational Institute
Originally presented on: 11/6/2013 13:27:00 - 13:46:00

Synthesis and Characterization of Mn_2O_3-ZnO Nanocomposites for Hydrogen Generation *via* Photoelectrochemical and Photocatalytic Splitting of Water

Nirupama Singh, Surbhi Choudhary, Vibha R. Satsangi[#], Sahab Dass and Rohit Shrivastav*

Department of Chemistry, [#]Department of Physics & Computer Science,

Dayalbagh Educational Institute, Dayalbagh, Agra-282 005, India

*E-mail: rohitshrivastav_dei@yahoo.co.in

The growing consumption of energy, which is now mainly produced from fossil fuels leads to a further depletion of the limited fossil reserves, increased dependency on fossil fuel imports from often politically unstable counties and furthermore gives rise to an increased concentration of green house gases mainly CO_2 in the atmosphere. The report deals with the synthesis of Mn_2O_3-ZnO nanocomposites, their characterization, and possible use in photoelectrochemical and photocatalytic splitting of water for hydrogen generation. Powdered samples of nanocomposites were prepared by co-precipitation at varying Mn/Zn molar ratios. XRD analysis revealed the evolution of wurtzite ZnO and Mn_2O_3 phases in samples, while optical absorption threshold was found shifted towards higher wavelength with increasing Mn amount. The Zn/Fe molar ratio in nanocomposites was optimized, yielding significantly enhanced photoelectrochemical current and photocatalytic hydrogen yield. Plausible explanations for observed change in photoresponse of samples have been offered.

(499e) High Performance Fluorine Doped Oxygen Evolution Reaction (OER) Electro-Catalysts for PEM Based Water Electrolysis

Kadakia, K., University of Pittsburgh
Datta, M. K., University of Pittsburgh
Velikokhatnyi, O.,
Kumta, P. N., University of Pittsburgh
Originally presented on: 11/6/2013 13:46:00 - 14:05:00

Identification of electro-catalysts containing non noble metal or significantly reduced amounts of expensive noble metals (e. g. RuO_2) is highly desirable. Development of such a catalyst with comparable electrochemical performance to the standard noble metal oxide counterpart for the proton exchange membrane (PEM) based water electrolysis would constitute a major advancement in generation of hydrogen by water electrolysis. We have accordingly developed a two-pronged theoretical first principles and experimental approach to generate a nano-structured solid solution of SnO_2:10wt.% F containing only 20at. % RuO_2 [*e.g.* $(Sn_{0.80}Ru_{0.20})O_2$:10F] that displays a remarkably similar electrochemical activity and moreover, comparable or even much improved electrochemical stability and durability compared to pure the noble metal counterpart, RuO_2. Results of these studies will be presented and discussed.

(499f) Multiwalled Carbon Nanotubes Drive the Activity of Metal@Oxide Core-Shell Catalysts in Modular Nanocomposites

Cargnello, M., University of Pennsylvania
Originally presented on: 11/6/2013 14:05:00 - 14:24:00

Multiwalled Carbon Nanotubes Drive the Activity of Metal@oxide Core-Shell Catalysts in Modular Nanocomposites

M. Cargnello,[1,2] M. Grzelczak,[1] B. Rodriguez-Gonzalez,[3] Z. Syrgiannis,[1] K. Bakhmutsky,[4]

V. La Parola,[5] L. M. Liz-Marzán,[3] R. J. Gorte,[4] M. Prato,[1] and P. Fornasiero[1]

[1] *Department of Chemical and Pharmaceutical Sciences, ICCOM-CNR, Consortium INSTM, University of Trieste, via L. Giorgieri 1, 34127 Trieste, Italy*

Â [2] *Department of Chemistry, University of Pennsylvania, 231 S. 34th Street, Philadelphia, PA 19104, USA*

[3] *Departamento de QuÃmica FÃsica, Universidade de Vigo, 36310 Vigo, Spain*

[4] *Department of Chemical and Biomolecular Engineering, University of Pennsylvania, 311A Towne Building, 220 S. 33rdÂ Street, Philadelphia, PA 19104, USA*

[5] *Istituto per lo Studio dei Materiali Nanostrutturati (ISMN-CNR), Via Ugo La Malfa 153, Palermo I-90146, Italy*

The ability to build hierarchical structures by arranging different building blocks with nanometer-scale precision is one of the most useful aspects of nanotechnology. This concept can produce novel materials with properties that are different from those expected from the simple sum of the individual blocks. Indeed, the interactions between the constituent parts in nanometer-scale ensembles can lead to novel electronic, optical, or catalytic properties not available in the initial building blocks. In the field of heterogeneous catalysis in particular, the interactions between the active phase, supports, and promoters is critical for obtaining high performance materials. These interactions can be both electronic and geometric. Furthermore, the demands in terms of geometry and binding energy of the active sites for different catalytic reactions can be very different.

In this contribution, rational nanostructure manipulation has been used to prepare nanocomposites in which multiwalled carbon nanotubes (MWCNTs) were embedded inside mesoporous layers of oxides (TiO2, ZrO2, or CeO2), which in turn contained dispersed metal nanoparticles (Pd or Pt). We show that the MWCNTs induce the crystallization of the oxide layer at room temperature and that the mesoporous oxide shell allows the particles to be accessible for catalytic reactions. In contrast to samples prepared in the absence of MWCNTs, both the activity and the stability of core-shell catalysts is largely enhanced, resulting in nanocomposites with remarkable performance for the water-gas-shift reaction, photocatalytic reforming of methanol, and Suzuki coupling. The modular approach shown here demonstrates that high-performance catalytic materials can be obtained through the precise organization of nanoscale building blocks.

(499g) Separation of Sulphur Dioxide and Oxygen in Thermochemical Hydrogen Production

Elder, R. H., University of Sheffield
He, G., University of Sheffield
Cumming, D. J.,
Allen, R. W. K., University of Sheffield
Originally presented on: 11/6/2013 14:24:00 - 14:43:00

Dense Yttria-stabilised-zirconia (YSZ) membranes have been shown to have great potential for separating SO_2 and O_2 in the sulphur family of thermochemical cycles for hydrogen production. Use of a catalytic YSZ membrane reactor in the sulphuric acid decomposition stage of these cycles could drive the reaction forward and significantly increase the yield of SO_2 and O_2. Whilst SO_2/O_2 separation has been shown to be feasible at temperatures between 750°C and 950°C, significant performance degradation is seen over time. This performance degradation is shown to be due to sulphur poisoning (from both sulphur adsorption and sulphide formation) of the triple phase boundaries. A similar effect is seen in solid oxide fuel cells, where ppm concentrations of sulphur severely diminish cell performance.

Here we investigate the effect of voltage pulsing on membrane performance recovery. Significant performance recovery is seen after the application of a short voltage pulse. The optimum pulse duration and magnitude were found to be 0.08 s and 10 V respectively. The lower the magnitude of voltage pulse, the longer the duration required to achieve maximum recovery. Impedance spectroscopy and SEM results are presented, showing the contribution of different resistances and the effect of the voltage pulsing on the membrane cell.

Consideration of oxygen reduction mechanisms at the triple phase boundary shows that as voltage across the membrane increases, the rate controlling step changes from oxygen adsorption and surface diffusion, to oxygen availability in the feed. The higher voltage pulses are beneficial to performance recovery as polarisation resistance and charge transfer resistance decrease due to the removal of sulphur from the surface. Pulses that have too great a magnitude, or are too long, dramatically promote oxygen reduction and lead to concentration polarisation due to lack of oxygen in the feed.

Longer term testing (20 hours) showed that repeated, regular voltage pulsing can significantly improve membrane performance compared to those that receive no voltage pulsing.

(499h) Chemisorption, Physisorption, and Hysteresis of Hydrogen On Carbon Nanotubes

Barghi, S., University of Southern California

Tsotsis, T. T., University of Southern California

Sahimi, M., University of Southern California

Originally presented on: 11/6/2013 14:43:00 - 15:02:00

We used a magnetic suspension balance, coupled with a residual gas analyzer, to investigate hydrogen adsorption/desorption on carbon nanotubes (CNTs) for pressures up to 100 bars at 25 °C. Such a combination of gravimetric and mass analysis systems, used for the first time, is capable of overcoming the shortcomings of the common hydrogen adsorption measurement techniques. The most crucial drawbacks of the conventional measurement methods are as follows. (i) The inability to distinguish between hydrogen adsorption and chemisorptions; (ii) the relatively large errors caused by the leakage of hydrogen out of the system, and (iii) the difficulty of measuring hydrogen uptake capacity under operating conditions that differ from those of hydrogen storage systems used in practice.

In the experiments multi-walled carbon nanotubes (MWCNTs) were exposed to hydrogen, and were subsequently kept under vacuum for 2 consecutive cycles, in order to study hydrogen adsorption/desorption in the nanostructured materials. The difference between the first and the second hydrogenation cycles was in the initial preparation of the MWCNTs. For the first cycle, the MWCNTs were first degasified at an elevated temperatures (120 °C). For the second cycle the MWCNTs were not degasified at the elevated temperatures and, therefore, hydrogen residue from the first adsorption/desorption cycle was still adsorbed on the MWCNTs at the beginning of the second cycle.

The first hydrogenation cycle was started after pretreatment of the CNT at the elevated temperature. Hydrogen adsorption kinetics for the cycle indicates that the adsorption equilibrium occurs within 4 - 5 hours. Careful examination of the recorded adsorption kinetics revealed the fascinating fact that a considerable part of hydrogen adsorption in the MWCNTs takes place *instantaneously*. The measured equilibrium data revealed some hysteresis for hydrogen desorption in this cycle, which does not vanish, even when the system's pressure approaches zero. This indicates that hydrogen adsorption in the first hydrogenation cycle is a combination of physisorption and chemisorption.

After the first cycle was finished, the CNTs were exposed to hydrogen for the second time without any pretreatments at the elevated temperature. Measurements indicated no hysteresis for during the second cycle. Therefore, only physisorption took place during the second cycle. Comparison between the first and the second hydrogen isotherms, together with the measured hydrogen adsorption kinetics, made it possible to quantitatively distinguish between the physisorption and chemisorption portions of hydrogen adsorption in the MWCNTs. According to our data, for pressures less than 20 bars physisorption and chemisorption are both important, with the former becoming the dominant mechanism as the pressure increases. To the best of our knowledge, this is the first time that hydrogen physisorption and chemisorption on carbon nanotubes have been measured separately.

(503a) Selection and Crystallization Development of a Practical Salt Form for a Pharmaceutical Compound

Yang, B. S., Boehringer Ingelheim Pharmaceuticals, Inc.
Li, Z., Boehringer Ingelheim Pharmaceuticals, Inc.
Jiang, M., Massachusetts Institute of Technology
Kim, S., Boehringer Ingelheim Pharmaceuticals, Inc.
Wang, X.,
Yee, N., Boehringer Ingelheim Pharmaceuticals, Inc.
Senanayake, C., Boehringer Ingelheim Pharmaceuticals, Inc.
Originally presented on: 11/6/2013 12:30:00 - 12:50:00

This paper presents a fully integrated solid form screen and crystallization development workflow for the development of an active pharmaceutical ingredient (API) at BI. Salt I of an in-licensed compound was initially used in the Phase I and Phase IIa studies. However, Salt I was later characterized as an ethanolate/hydrate and had several development issues, including the high and variable moisture content and unstable physical/chemical properties under various conditions (heat, humidity, mechanical and light). As a result, a new salt form that can provide better developability was required that triggered extensive salt/polymorph screen and selection activities, in order to select a new salt form to be used for Phase IIb and beyond.

After a thorough solid form screen and selection process, Salt II, with its superior physicochemical properties to the other crystalline salts identified, was chosen as the new development form. In the initial development, a workable process (Process I) was developed in EtOH/Heptane that was able to produce multiple kilo scale of Salt II to supply for the formulation development. The particle size was controlled by dry milling initially. Later, a more comprehensive crystallization development was conducted to address several development challenges, including the purification of the major impurity from the synthesis, hydrolysis of the compound in the crystallization system, color removal and powder property control, etc. A new modified process (Process II) was successfully developed, enabled by the selection of appropriate solvent system and design of crystallization condition that can inhibit the hydrolysis of the compound during crystallization. Process II used milled seeds and controlled crystallization to directly produce the desired particle size distribution without milling. Process II was successfully scaled up to multiple 30 kg batches in the Pilot Plant.

(503b) In-Situ Monitoring of Resolution of Chiral Compounds Using Process Analytical Technology

Chen, Z., National Engineering Research Center for Industrial Crystallization Technology
Gao, Y., National Engineering Research Center for Industrial Crystallization Technology, School of Chemical Engineering and Technology, Tianjin University
Wang, Y., National Engineering Research Center for Industrial Crystallization Technology, School of Chemical Engineering and Technology, Tianjin University
Yin, Q., National Engineering Research Center for Industrial Crystallization Technology, School of Chemical Engineering and Technology, Tianjin University
Zhang, M., Tianjin University
Hou, B., Tianjin University
Hao, H., National Engineering Research Center for Industrial Crystallization Technology, School of Chemical Engineering and Technology, Tianjin University
Originally presented on: 11/6/2013 12:50:00 - 13:10:00

In-situ Monitoring of Resolution of Chiral Compounds using Process Analytical Technology

Zhen Chen, Yuan Gao, Yongli Wang, Qiuxiang Yin, Meijing Zhang, Baohong Hou, Hongxun Hao*

National Engineering Research Center for Industrial Crystallization Technology, School of Chemical Engineering and Technology, Tianjin University, Tianjin 300072, P.R. China

* Corresponding author: Tel:+86-22-27405754 Fax: + 86-22-27405754

Email: hongxunhao@tju.edu.cn

Abstract

Since chirality was discovered and defined hundreds of years ago, the resolution and purification of chiral compounds has become one main challenge that scientist have to face. Most physicochemical properties of the enantiomers are nearly the same or pretty close. So it is quite difficult to find an appropriate method to resolve the chiral compounds, especially in industry scale. As one main resolution method of chiral compounds, preferential crystallization has been widely used not only in the labs but also in industries. To develop and design an optimum crystallization processes for chiral compounds, it is crucial to understand the thermodynamic and kinetic mechanism of the processes. Process analytical technologies (PAT) can be used to monitor the preferential crystallization processes in line and are very helpful to further understand the mechanism behind the phenomena. In this paper, several PAT tools were successfully used to in-situ monitor the resolution of L-lactide. Raman Spectroscopy and attenuated total reflectance Fourier transformation infrared (ATR-FTIR) were used to monitor the solid phase and liquid phase concentration respectively. Focused beam reflectance measurement (FBRM) and particle vision measurement (PVM) were used to monitor the particle size and morphology respectively. Some other offline analytical tools, such as X-ray diffraction (XRD), differential scanning calorimetry (DSC), Automatic polarimeter were also used to confirm the online analysis results. Solubility data of L-lactide were also measured experimentally. The effect of operating parameters, such as seeding, cooling profile and the initial concentration, on the preferential crystallization processes was also investigated. It was found that all of these operating parameters can affect the resolution results. The reasons and the mechanism of the influence were discussed in detail. Using these knowledge obtained in this paper, an reliable optimum resolation method with high output capacity and good repeatability for L-lactide were finally developed.

Keywords: Resolution, Chiral compounds, Process analytical technology, Preferential crystallization, L-Lactide

(503c) A Model for Crystal Growth for Solids Crystallizing From Solutions Containing Multiple Conformers and Evidence of Growth Inhibition By the "Wrong Conformer"

Derdour, L., Bristol-Myers Squibb co.
Sivakumar, C., Carnegie Mellon University
Skliar, D., Bristol-Myers Squibb Co.
Pack, S., Bristol-Myers Squibb Co.
Lai, C., Bristol-Myers Squibb Company
Vernille, J., Bristol Myers Squibb
Kiang, S., Bristol-Myers Squibb Company
Originally presented on: 11/6/2013 13:10:00 - 13:30:00

The validity of a model derived for crystal growth rate for solids crystallized from solutions containing multiple conformers is evaluated. A simple and straightforward method for determining crystal growth rates using FBRM and FTIR/ATR is utilized to maximize growth rate data collected from a single experiment. Crystal growth rate data obtained experimentally were in good agreement with model predictions. In particular, the data indicated that when the approach of the right conformer (Derdour et al. Chem Eng Sci., 66(1), pp. 88-102, 2011) applies, crystal growth rate is inversely proportional to crystal size. Further, the results indicate that growth rates determined from supersaturation measurements can be over-estimated because of surface nucleation and/or agglomeration that can result from supersaturation spikes during anti-solvent addition. Lastly, crystal growth data obtained at different temperatures indicated that the slow growth rate observed at low temperatures may be caused in part by surface poisoning by the wrong conformer.

(503d) API Particle Engineering for Drug Substance: Crystallization & Wet Milling to Control API Powder Properties

Schild, R., Bristol-Myers Squibb
Engstrom, J. D., Bristol-Myers Squibb Company
Hubert, M., Bristol-Myers Squibb Company
Jones, J.,
Qian, X.,
Wu, Y., Bristol-Myers Squibb Co.
Leane, M.,
Gamble, J., Bristol-Myers Squibb
Wang, S. S. Y., Bristol-Myers Squibb Company
Originally presented on: 11/6/2013 13:30:00 - 13:50:00

In the pharmaceutical industry, target drug substance powder properties, such as particle size and flow are defined for a given formulation to ensure robust drug product processing. Batch crystallizations are primarily designed to maximize the active pharmaceutical ingredient (API) yield and product purity and frequently produce API's with non-ideal size and morphology. In this investigation, the effect of seeding, supersaturation, and post crystallization wet milling were used to produce API batches with a range of particle sizes and flow properties to define the initial design space of acceptable content uniformity and powder flow. For laboratory and pilot plant scale-up experiments, the particle size was measured real-time during crystallization and wet milling using focused beam reflectance measurement (FBRM) and correlated to spot slurry samples analyzed by laser light scattering as well as image based particle characterization. These results were compared with empirical correlations to account for initial particle size, mill generator configuration, rotor or tip speed, milling residence time, and rotor/stator gap or tooth dimensions to aid successful crystallization and wet milling scale-up.

(503e) Hysteresis in Solid-Liquid Suspensions in Optimax®: Using CFD Simulations and Experiments With FBRM

Vedantam, S., National Chemical Laboratory

Ranade, V. V., National Chemical Laboratory

Originally presented on: 11/6/2013 13:50:00 - 14:10:00

Title: **Hysteresis in solid-liquid suspensions in OPTIMAX®: Using CFD simulations and experiments with FBRM**

*Authors: Vedantam, S., Ranade, V.V.**

Affiliation: Industrial Flow Modeling Group (iFMg), CEPD Division, National Chemical Laboratory (NCL), Pune, INDIA

**corresponding author: vv.ranade@ncl.res.in*

ABSTRACT

Introduction:

Crystallization is the most widely used separation and purification process employed throughout the chemical and pharmaceutical industry for chemical products that are solids at room temperature and pressure. Today, crystallization modeling tools are limited by the availability of methods to reliably predict solubility and super saturation for chemical systems of industrial importance. There is also limited capability for describing nucleation and crystal growth, and mixing and design scale-up for industrial crystallizers. Tools are needed to model and optimize crystallizer process performance with respect to product quality, throughput and efficiency. It is essential to understand and relate key thermodynamic, kinetic and hydrodynamic aspects to crystallizer performance not just in terms of yield but also in terms of product quality (characterized by particle size distribution, morphology, polymorphism and the amount of strain as well as the uptake of solvent or impurities in the crystal lattice). Process Analytical Technology (PAT) tools are also on the rise in these applications and published in the literature (Daneau and Steele., 2005; Samad *et al.*, 2013).

Hydrodynamic modeling, one of the key aspects, is the focus of the work presented in this abstract. In literature, Joshi et al (2011), presented a comprehensive review on CFD modeling studies in stirred tanks, stating that it is worth noting that dense suspensions require a correct modeling of both solid?liquid and particle? particle interactions. However, as an initial step, an automated reactor called OPTIMAX® (launched by Mettler-Toledo Ltd.,) is being used for the study of various aspects of crystallization and the quantitative effect of all key aspects on the process. OPTIMAX® (OM), a synthesis workstation provides a platform for product and process development and is coming into use in bio-pharmaceutical synthesis applications. OM is typically an automated reactor, offering heating electrically and cooling on the basis of Peltier technology. Crystallization is one of the wide hopes of applications to be carried out in this particular workstation. In this particular set-up, OM is designed to accommodate internals such as Focused Beam Reflectance Measurement (FBRM) probe (used for estimating the Chord Length Distributions (CLD) and thus, Crystal Size Distributions (CSD)); Temperature probe (Tr probe), Fourier-Transform Infra Red (FTIR) spectroscopy probe; turbidity probe; pH probe; impeller and baffles as well. However, as for this study, the internals used are FBRM probe; Tr probe and 4-bladed down-pumping PBT as impeller are used. It is understandable that the reactor geometry, operating conditions and the flow characteristics govern the hydrodynamic aspects of the crystallization process thus subsequently also affecting the kinetic aspects. Solid-liquid suspension studies on solids distribution and the effect of operating conditions has been simulated and experimentally verified.

Methodology, flow characterization and mixing time:

Commercial software Ansys Fluent (14.0) has been used for all CFD simulations. Multiple Reference Frame (MRF) approach is used for modeling the impeller. Multiple phases have been simulated using Eulerian approach. 1000 mL reactor volume has been used as the domain for simulations and experiments, with an inner diameter of 0.101 m, and a height of 0.137 m. The impeller used is a 45^0, four-bladed, down pumping PBT. A mesh size of 4, 60,000 cells has been used for the entire domain. Simulations are carried out for impeller speed ranging from 100 to 1000 rpm, and varying solids hold-up of the particles used. The geometry indicating the internals, tank and inner and outer regions of the domain, clearly demarked for MRF approach are shown in the Figure 1. The commonly used validation check points such as Power number (N_P, from turbulent energy dissipation rate and moments), Pumping number (N_Q) are estimated from the CFD predictions, and satisfactory validation has been obtained, which gives a go ahead for further studies on crystallization in this reactor.

Experiments are carried out using FBRM to estimate the distribution of particles, and verify the hysteresis likely to occur with the variation in impeller speed. Neutrally buoyant particles of 350 micron are used as solid particles. Hold-up has been varied from 1 % to 10 % in the reactor. Impeller speeds are varied from 100 rpm to 1000 rpm and the particle size distribution has been tracked using FBRM. Similar conditions have been simulated using CFD.

Conclusions:

Solid-liquid suspension studies in OPTIMAX® are carried out using simulations as well as experiments. The CFD study is validated with the established parameters such as N_P (from estimated turbulent energy dissipation rate and the torque), N_Q from the CFD predictions. Possibility of hysteresis has been verified using FBRM and CFD simulations.

References:

Samad, N.A.F.B.A., Sin, G., Gernaey, K.V., Gani., R. 2013, ?A systematic framework for design of process monitoring and control (PAT) systems for crystallization processes?, Computers & Chemical Engineering, 54, 8-23. http://dx.doi.org /10.1016/j.compchemeng.2013.03.003

Buwa, V. V. and Ranade, V. V., 2003a Characterization of dynamics of gas-liquid flows in rectangular bubble columns, AIChEJ, 50, 2394.

Deneau, E., Steele, G, 2005, An in-line study of oiling out and crystallization, Org. Proc. Res. Dev., 9, 943-950.

Joshi, J.B., Nere, N.K., Rane, C.V., Murthy, B.N., Mathpati, C.S., Patwardhan, A.W., Ranade, V.V., 2011, CFD Simulation of Stirred Tanks: Comparison of turbulence models (Part II: Axial Flow Impellers, Multiple Impellers and Multiphase Dispersions), The Canad. J. of Chem. Eng., 89, 754-816.

Ranade, V. V., Bourne, J. R. and Joshi, J. B., 1991. Fluid mechanics and blending in agitated tanks. Chem. Eng. Sci., 46, 1883.

Figures:

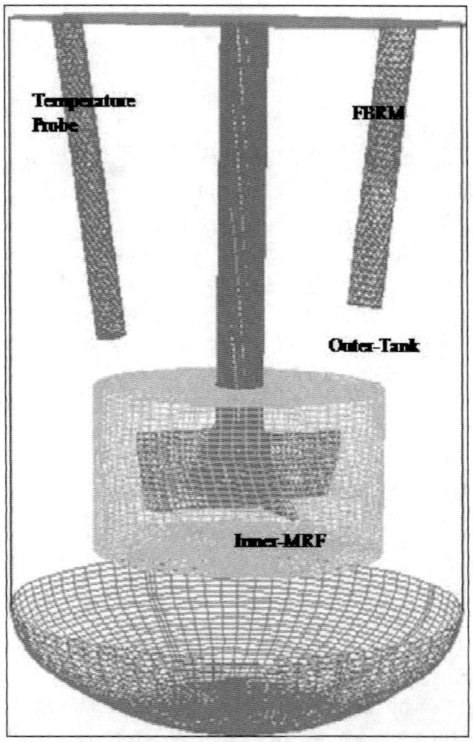

Figure 1. Computational domain showing inner and outer regions of MRF approach

(503f) Development of a Model for Anisotropic Crystal Growth From Solutions

Derdour, L., Bristol-Myers Squibb co.
Chan, E. J.,
Originally presented on: 11/6/2013 14:10:00 - 14:30:00

A model for anisotropic growth was developed. It is based on an isotropic growth model recently developed for crystal growth from solutions containing multiple conformers. The model presented herein implies that growth rates are constrained in one spatial dimension, thus only two linear growth dimensions, namely length and width, are considered. Crystal growth is assumed to occur via step propagation. The model takes into consideration step advance velocity dependence upon supersaturation and solvent/crystal interaction. An approach which uses two solute integration coefficients is introduced that describes crystal growth kinetics in each growth dimension. These coefficients are functions of the linear growth of the faces and are inclusive of solvent/crystal facet interactions. A parameter estimation algorithm to determine these coefficients is described. The algorithm requires experimental data of concentration variation during crystallization and final aspect ratio. The model's predictions of temporal variation of supersaturation and final crystal elongation factor were in agreement with experimental data for a Bristol-Myers Squibb proprietary compound. Finally, model simulations allow for the estimation of integration coefficients for any given aspect ratio.

(503g) A Straightforward Technique to Control the Oxidation of Amorphous Cobalt Nanoparticles During Synthesis Using DMSO

Duggan, J. N., Auburn University
Roberts, C. B., Auburn University
Chaturvedi, A.,
Suzuki, T.,
Originally presented on: 11/6/2013 14:30:00 - 14:50:00

Magnetic nanoparticles have attracted a great deal of interest lately because of their unique magnetic properties and have a wide range of potential applications. For example, nano-scale magnetic particles have enhanced magnetic susceptibility and saturation compared to their bulk-scale counterparts. Cobalt nanoparticles can exhibit a superparamagnetic behavior, which makes these particles particularly useful as contrast agents in Magnetic Resonance Imaging (MRI). However, many of the techniques commonly used to produce well-defined cobalt nanoparticles involve the use of harsh reagents and synthesis conditions, and may require days to produce a single sample of nanoparticles. Furthermore, it is often difficult to produce cobalt nanoparticles because they have a tendency to become overly oxidized, and can subsequently agglomerate and precipitate from the solution.

We have developed a single-step method to synthesize and stabilize cobalt nanoparticles using an FDA-approved, functional solvent, dimethyl sulfoxide (DMSO). Moreover, this research demonstrates that DMSO can be a multi-functional component in nanoparticle synthesis by serving concomitantly as a solvent and as a stabilizing agent that effectively suppresses nanoparticle growth and prevents excessive oxidation of the nanoparticle surface. The size and morphology of the particles were determined using TEM imaging analysis. In addition, we have performed a variety of different characterization techniques to investigate the interaction of the cobalt nanoparticles with the DMSO stabilizing molecules, including FT-IR, XPS, and XRD. Using these techniques, we found that both the sulfur and oxygen functional groups on the DMSO molecules interact with the nanoparticle surfaces via a resonance hybrid structure that stabilizes the nanoparticles. We have also found that these cobalt nanoparticles are amorphous when synthesized at room temperature. Ongoing studies in our lab are focused on subjecting the cobalt nanoparticles to a variety of different temperatures post-synthesis in order to determine the minimum temperature needed to induce oxidation and crystallinity within the particles. In addition, the magnetic behavior of these amorphous cobalt nanoparticles compared to the magnetic behavior of crystalline cobalt and cobalt oxide nanoparticles will be discussed.

(526a) Quiescent and Agitated Redispersion of Griseofulvin Nanoparticles From Nanocomposite Microparticles in Water: Impact of a Wet-Milled Swellable Dispersant

Bilgili, E., New Jersey Institute of Technology
Bhakay, A., New Jersey Institute of Technology
Azad, M., New Jersey Institute of Technology
Dave, R. N., New Jersey Institute of Technology
Originally presented on: 11/6/2013 15:15:00 - 15:35:00

Reduction of drug particle size down to nanometer or sub-micron scale has been an effective way of increasing the dissolution rate of poorly water-soluble drugs[1]. Production of nanoparticles leads to an increase in the surface area and thereby increases the dissolution rate as per the Noyes–Whitney equation[2]. Wet stirred media milling has been commonly used as a top down approach to produce drug nanoparticle suspensions, also referred to as nanosuspensions[3-6]. Nanoparticles are prone to aggregation in suspensions. To ensure chemical and physical stability of the suspensions during a potential long-term storage and meet patient needs/demands, drug nanosuspensions are typically dried into nanocomposite microparticles (NCMPs), in the form of powders, which are then incorporated into various solid dosage forms such as tablets, capsules, sachets, etc. using standard pharmaceutical unit operations. Unfortunately, during the drying processes, drug nanoparticles may form hard aggregates or drug–polymer agglomerates within the structure of NCMPs[6,7]. Another problem is that even if the drug nanoparticles do not form hard aggregates, they may not be recovered from the NCMPs completely and quickly upon redispersion in fluids during *in vivo* or *in vitro* dissolution[6]. In either case, the latent large surface area of the nanoparticles will not be exposed. As a result, the dissolution rate and bioavailability of the drug can be significantly reduced even though the drug has been wet-milled to nanoparticle scale. Therefore, an understanding of nanoparticle recovery during the redispersion of NCMPs in water or other biorelevant media may be of fundamental importance in explaining and/or predicting the dissolution and bioavailability of poorly water-soluble drugs[6].

A great majority of prior work in pharmaceutical literature on drug NCMPs did not focus on the redispersion phenomenon at all. In fact, some researchers inferred the redispersion behavior inversely from the standard dissolution testing on drug NCMPs[8,9]. However, such studies completely disregarded the fundamental mechanisms associated with the redispersion. In redispersion tests, NCMPs are suspended in an aqueous medium (usually water) and subjected to relatively low shear/agitation while nanoparticles are released from the NCMPs. Drug solubility in the aqueous medium is usually very low; hence, little to no drug is dissolved. The aqueous redispersion can be performed either in a cylindrical vessel with a stirring element or sonication probe followed by particle size analysis of several time samples taken during the redispersion[6,10] or in the sample cell of a particle size measuring equipment with size measurements taken over a period of time[11]. Comparison of the particle size distribution of the wet media milled suspension and that of the aqueous redispersion as a function of time reveals the extent–dynamics of the drug nanoparticle recovery from the NCMPs. It is clear that the majority of the redispersion testing in pharmaceutical literature relies on particle sizing.

In this study, the redispersion behavior of nanocomposite microparticles (NCMPs) incorporating griseofulvin (GF, model drug) nanoparticles and wet-milled swellable dispersant particles[12,13] along with other dispersants was investigated using two agitation methods, i.e., no external agitation (near-quiescent condition) and paddle-stirring, with several *independent* characterization methods. *In the first method*, NCMPs were added to initially-quiescent water inside a test tube, and the redispersion dynamics were studied by measuring the turbidity and analyzing the images of the particle population as recorded by a camera. Similarly, NCMPs were added to initially-quiescent water inside the cuvette of a dynamic light scattering equipment, and the particle size of the redispersion was measured. To get a better understanding of the redispersion under near-quiescent conditions, the response of a single NCMP exposed to a drop of water was observed under an optical microscope. *In the second method*, NCMPs were added to water agitated by a paddle stirrer, and both the particle size distribution and the GF assay of the centrifuged redispersion were determined. NCMPs used in this redispersion study were prepared with several dispersants that modulate the redispersion behavior. Superdisintegrants can be used as inexpensive, model, swellable dispersant particles with or without wet-milling along with drugs. Suspensions of GF along with various dispersants produced by wet-milling were coated onto Pharmatose®to prepare core-shell type NCMPs in a fluidized bed process. Hydroxypropyl cellulose (HPC, neutral polymer) alone and with sodium dodecyl sulfate (SDS, anionic surfactant) was used as base-line stabilizer/dispersant during milling. Croscarmellose sodium (CCS, anionic superdisintegrant) and Mannitol (sugar alcohol) were used as additional dispersants to prepare surfactant-free NCMPs. Unmilled (sieved CCS) and wet-milled CCS, co-milled with GF for different times, were used to investigate the impact of CCS particle size on the redispersion response.

The images taken during the redispersion of NCMPs in a test tube and optical microscopy images of a single redispersing NCMP illustrated several stages of the redispersion phenomenon: particle wetting and sinking, swelling, erosion, and dispersion of the nanoparticles. Furthermore, the turbidity after the near-quiescent redispersion in the test tube correlated well with the extent of GF nanoparticles recovered during the redispersion test with a paddle stirrer. Both agitation methods yielded a similar rank-ordering of the redispersibility of the different NCMP formulations. A comprehensive analysis and interpretation of the redispersion test results also enabled us to develop a fundamental understanding of the roles of various dispersants used in the NCMP formulations. In the absence of CCS or SDS, NCMPs containing no dispersants, HPC alone, or HPC/Mannitol were not redispersible because of the poor wettability of these NCMPs and slow dissolution of the HPC embedding the relatively hydrophobic GF nanoparticles. Incorporation of SDS enhanced the wettability and HPC dissolution, which in turn allowed weakening/breakage of the NCMP shell and resulting clusters from this breakage fast. Due to the extensive swelling capacity of CCS, incorporation of wet-milled CCS into the NCMPs caused effective breakage of the NCMP structure and bursting of nanoparticle clusters, leading to fast release of the GF nanoparticles. It was also found that a mixture of colloidal and unmilled CCS particles with a broad multimodal size distribution promoted the GF nanoparticle recovery better than either the unimodal, colloidal CCS particles or the unimodal, unmilled, micron-sized CCS particles.

With the ultimate goal of improving the bioavailability of poorly water-soluble drugs, the results demonstrate that a precursor suspension of wet co-milled drug–superdisintegrant particles and unmilled superdisintegrant particles in the presence of soluble polymers may be used to produce fast redispersible, surfactant-free drug NCMPs. The GF nanoparticle recovery during the redispersion tests also correlated significantly with the GF dissolution using water and paddle-stirring in both experiments, thus confirming the predictive capability of the redispersion tests. The near-quiescent and agitated redispersion tests presented here are expected to be widely used by formulators/engineers in designing robust NCMP formulations.

References

1. Muller R., Peters K., 1998. Nanosuspensions for the formulation of poorly soluble drugs: I. Preparation by a size reduction technique. Int. J. Pharm. 160, 229–237.
2. Noyes A.A., Whitney W.R., 1897. The rate of solution of solid substances in their own solutions. J. Am. Chem. Soc. 19, 930–934.
3. Bilgili E, Afolabi A., 2012. A combined microhydrodynamics–polymer adsorption analysis for elucidation of the roles of stabilizers in wet stirred media milling. Int J. Pharm. 439, 193–206.
4. Bhakay A., Merwade M., Bilgili E., Dave R.N., 2011. Novel aspects of wet milling for the production of microsuspensions and nanosuspensions of poorly water soluble drugs. Drug Dev. Ind. Pharm. 37, 963–976.
5. Knieke C., Azad M.A., Dave R.N., Bilgili E., 2013. A study of the physical stability of wet media-milled fenofibrate suspensions using dynamic equilibrium curves. Chem. Eng. Res. Des., doi: 10.1016/j.cherd.2013.02.008
6. Bhakay A., Dave R., Bilgili E., 2013. Recovery of BCS Class II drugs during aqueous redispersion of core-shell type nanocomposite particles produced via fluidized bed coating. Powder Technol. 236, 221–234.
7. Lee J., 2003. Drug nano- and microparticles processed into solid dosage forms: physical properties. J. Pharm. Sci. 92, 2057–2068.
8. Van Eerdenbrugh, B., Froyen, L., Van Humbeeck, J., Martens, J.A., Augustijns, P., Van den Mooter, G. 2008. Drying of crystalline drug nanosuspensions –The importance of surface hydrophobicity on dissolution behavior upon redispersion. Eur. J. Pharm. Sci. 35, 127–135.
9. Van Eerdenbrugh B., Froyen L., Humbeeck J., Martens J., Augustijns P., Mooter G. 2008. Alternative matrix formers for nanosuspension solidification: Dissolution performance and X-ray microanalysis as an evaluation tool for powder dispersion. Eur. J. Pharm. Sci. 35, 344–353.
10. Schwarz C., Mehnert W., 1997. Freeze-drying of drug-free and drug-loaded solid lipid nanoparticles (SLN), Int. J. Pharm. 157, 171–179.
11. Cheow W., Ng M., Kho K., Hadinoto K., 2011. Spray-freeze-drying production of thermally sensitive polymeric nanoparticle aggregates for inhaled drug delivery: Effect of freeze-drying adjuvants. Int. J. Pharm. 404, 289–300.
12. Bhakay A., Azad M., Vizzotti E., Dave R., Bilgili E., 2013. Enhanced recovery and dissolution of griseofulvin nanoparticles from surfactant-free nanocomposite microparticles incorporating wet-milled swellable dispersants. Drug Dev. Ind. Pharm. under review.
13. Bilgili E., Bhakay A., Azad M., Dave, R., 2013. System and Method for Superdisintegrant-Based Composite Particles for Dispersion and Dissolution of Active Pharmaceutical Agents, U.S. Patent Application, Application No. 13/829,948, March 13.

(526b) Immobilization of Enzymes By Spray-Drying Method in Core-Shell Structured Chitosan Carriers Cross-Linked By Novel "in-Situ" Method

Kaspar, O., Institute of Chemical Technology, Prague
Stepanek, F., Institute of Chemical Technology, Prague
Originally presented on: 11/6/2013 15:35:00 - 15:55:00

This work deals with the encapsulation of model enzymes by a 3-fluid kinetic nozzle. This type of nozzle allows us to produce particles with a core-shell structure. In a case of 2-fluid nozzle chitosan-enzyme solution is cross-linked before spray-drying process ("ex-situ") and thus enzymatic activity could be negatively affected by the cross-linker presence. In the case of 3-fluid nozzle the enzyme is in intimate contact with the cross-linker for a very short time during spray-drying formation and the range of possible chitosan and cross-linker concentration is higher that for "ex-situ" cross-linking. This could be very significant to prevent enzyme deactivation during long storage and reduce undesirable cross-linker content in the final product. Enzymatic activity, encapsulation efficiency and stability of the carriers in aqueous medium are compared with particles produced by 2-fluid nozzle. The next step is the production of structured particles by 3-fluid nozzle where a core of substrate is surrounded by enzyme in a cross-linked chitosan shell. This approach allows us to produce particles with substrate and immobilized enzyme in one carrier.

(526c) Formation and Mechanism Study of Micro-Sized Core-Shell Particles Via Cyclic Gas-Solid Reactions

Sun, Z., The Ohio State University
Zhou, Q., The Ohio State University
Wang, Z., the Ohio State University
Mao, X., the Ohio State University
Fan, L. S., The Ohio State University
Originally presented on: 11/6/2013 15:55:00 - 16:15:00

This work reports and discusses the formation and mechanism of core-shell structured micro-particles via cyclic gas-solid reactions. Unlike most conventional core-shell formation strategies involving coating and coating-like processes, this formation strategy uses the irreversible solid-phase ionic diffusion that occurs during a gas-solid reaction cycle (*e.g.* reduction and oxidation of Fe) to extract the shell species from a core and shell powders mixture to the surface of micro-particle. Without the need for solvent as opposed to many conventional processes, this novel process only involves gas-solid reactions, thereby posing minimum environmental impact. To prove this proposed strategy, a micro-particle (micron-level) made of Fe_2O_3 and Al_2O_3 powder mixture is first reduced by H_2 and oxidized by O_2 for 50 cycles at 900 °C, where the reactions proceed mainly through the diffusion of Fe cations. SEM and EDX analyses show clearly the formation of a desired Al_2O_3 core - Fe_2O_3 shell structure. To support the proposed formation mechanism, a micro-particle made of Fe_2O_3 and TiO_2 powders mixture go through the identical 50 cycles, where the reactions proceeds mainly through diffusion of O anions. The second sample shows no formation of core-shell structure, which highlights the dominating role of solid-phase ionic diffusion. In addition, a 2-D continuum diffusion model is applied to simulate the inter-Fe-particle bridging and directional product layer growth during an oxidation reaction, which is vital to this core-shell formation strategy.

(526d) Direct Small Molecule Sequestration By Macroporous Polymer Resin Synthesis
Drake, I. J., University of California, Berkeley
Originally presented on: 11/6/2013 16:15:00 - 16:35:00

The goal of this work was to sequester a small molecule in a single synthetic step using approaches adopted from macroporous polymer resin synthesis. Unlike macroporous resin synthesis that traditionally produces particles greater than 100 um; a method has been introduced that produces particles between 400 and 3000 nm. A wide range of synthetic parameters were studied to both direct the interfacial and interior structure of the nanoparticles. These parameters included monomer type and composition ratios, particle size, surfactant type, stabilizing agent, initiator type, and small molecule loading. The ideal structure has been imagined to be a macroporous polymer that has a self-assembled cross-linked shell capable of controlling the diffusion of the small molecule in application requiring controlled release. Conditions to reproduce or deterministically direct shell synthesis will be discussed.

A series of synthetic experiments were performed to prepare a stable sequestered small molecule in a macroporous polymer. It was determined that the highest level of intrinsic pore filling by the small molecule of interest was achieved for systems approaching 100% divinyl benzene (DVB) cross-linker. However, imaging by SEM suggested a lack of surface shell or "skin" layer which correlated with an inability to control diffusion. Shell layers have been shown to self-assemble in prior work to produce conventional macroporous polymer resins. Therefore, experiments were rationally designed to examine if these conditions could be reproduced for resins in the size range of 400-3000 nm using the small molecule of interest as the porogen. Using SEM, it was shown that by increasing the hydrophilicity of the monomer, the surface of the particle can be smoothed resulting in a thin shell. In addition, if a second monomer emulsion is introduced, the monomer will diffuse and react at the pore entrance to form a capping layer.

(526e) A Comparison of the Effectiveness of Solid and Solubilized Dioctyl Sodium Sulfosuccinate On Oil Dispersion Using a Baffled Flask Test
Nyankson, E., Auburn University
Originally presented on: 11/6/2013 16:35:00 - 16:55:00

2013 American Institute of Chemical Engineers Annual Meeting,

San Francisco, CA

November 3-8, 2013

A Comparison of the Effectiveness of Solid and Solubilized Dioctyl Sodium Sulfosuccinate on Oil Dispersion Using a Baffled Flask Test

Emmanuel Nyankson, Courtney A. Ober, Matthew J. DeCuir, and Ram B. Gupta

Department of Chemical Engineering, Auburn University, Auburn, AL 36849

The application of chemical dispersants is one of the few feasible response measures for minimizing the impact of a large oil spill at sea. Effective use of dispersants can accelerate oil biodegradation by naturally occurring microorganisms and prevent the oil spill from coming ashore and damaging sensitive coastal ecosystems. However, indiscriminate application of dispersant to subsea and surface oil spills results in a significant amount of dispersant not coming into contact with the oil. The aqueous solubility and miscibility of oil spill dispersants and inevitable sea currents cause dispersant to be washed away and wasted. To compensate for inefficiency, large amounts of dispersants are used resulting in high costs and a large environmental load. For example, an estimated 2.2 million gallons of dispersant were applied to the Gulf of Mexico during the Deepwater Horizon oil spill of 2010. To increase dispersant efficiency, the use of solid water-insoluble polymer particles which contain the active ingredient in chemical dispersants, a surfactant, are being studied for their potential in oil dispersion. In such a formulation the surfactant is only released when in contact with oil, at which time the polymer matrix dissolves in the oil. Furthermore, the ability of particulates to orient at the oil-water interface may allow for continuous release of surfactant directly where it is needed. In this study, the effectiveness of oil dispersion using the surfactant dioctyl sodium sulfosuccinate (DOSS) was compared by delivering the surfactant from polymer particles and liquid solvents using the EPA's baffled flask procedure. The polymer particles were prepared by ultrasonically spray drying polystyrene, polydimethylsulfoxide, and DOSS from a dichloromethane solution varying the polymer to surfactant ratios. The amount of encapsulated DOSS was determined by suspending the particles in water and measuring the amount of DOSS released using a methylene blue complexation procedure. Liquid delivery of DOSS was accomplished by dissolving the surfactant in either dichloromethane, water, or propylene glycol. The results of this study show the potential of surfactant-polymer particles to replace liquid dispersants in oil spill remediation.

Particle Technology Forum: 03A04 Applications of Engineered Structured Particulates

(526f) Design of Polylactide Microparticle Morphology By Internal Phase Separation

Watanabe, T.,
Kimura, Y.,
Ono, T., Okayama University
Originally presented on: 11/6/2013 16:55:00 - 17:15:00

We report a facile approach to prepare monodisperse polylactide (PLA) microparticles and design their surface and internal morphology via solvent diffusion coupled with internal phase separation. With the aid of microfluidics, monodisperse oil-in-water emulsion droplets were produced, in which PLA and perfluorooctyl bromide (PFOB), non-solvent for PLA were dissolved in the dispersed oil phase. The emulsion droplets travelled towards exit of the microfluidics and then mixed with massive amount of pure water that extracted ethyl acetate (EA) from the droplets due to high solubility of EA in water, which gave rise to shrinkage of the droplets and eventually induced internal phase separation between PLA-rich phase and PFOB-rich phase within each droplet. Further removal of EA from the droplets led to precipitation of PLA on the interface and formation of monodisperse microcapsules with a liquid PFOB core and thick PLA shell. We investigated the effect of the flow rate, content of the PFOB in the dispersed phase and molecular weight of PLA on the final shape of the particles. The diameter of the microcapsules was easily controlled by the flow rates while keeping the shell thickness to volume (T/R) ratio constant. The T/R ratio was modulated by changing the PFOB concentration in the dispersed phase independent of the particle size. We found that using higher molecular weight PLA for the matrix gave rise to structural complexity of the microcapsules during the formation, resulting in the microcapsules covered with small PFOB droplets, which was similar to colloidsome. In addition, removal of the PFOB by freeze-drying provided the microcapsules with dimpled surface like golf-ball. Confocal laser scanning microscopy observation revealed that the microcapsules also had core-shell morphology. These results show that this droplet to particle technology has great advantage to produce not only compact particles but also microcapsules with complex architecture. The presentation will give an overview of the fabrication process, mechanism of the morphology transition, and further recent update.

(526g) The Effect of Particle Structure and Formulation On the Protection of Photosensitive Molecules

Steeley, K. G., Rutgers University
Morabito, K., Brown University
Li, D., University of Massachusetts Dartmouth
Calvert, P., University of Massachusetts Dartmouth
Mello, C. M., U.S. Army Natick Soldier Research Development and Engineering Center
Tripathi, A., Brown University
Shapley, N. C., Rutgers, the State University of New Jersey
Originally presented on: 11/6/2013 17:15:00 - 17:35:00

Photosensitivity to ultraviolet (UV) radiation impacts many industrial products, including cosmetics, pharmaceuticals, automotive coatings, clothing, paints, and dyes. Novel methods of protecting photosensitive materials continue to be in constant demand as product photodegradation continues to be a problem. The addition of UV protective species to a formulation is a common method for UV protection. Encapsulation in a polymer matrix has also been shown to protect photosensitive molecules. A combination of the two methods is investigated as a means to provide substantial protection to a molecule sensitive to UV radiation.

Two types of core-shell polymeric microparticles were created, one with a solid core and the other whose core is a suspension of uniform polymer particles. The photosensitive molecule beta-carotene was added to the core of all particles, and UV absorbers and an antioxidant were added to various compartments. The shell thickness of the particles was also varied. Particles were exposed to UV radiation for predetermined amounts of time before the amount of undegraded beta-carotene was determined via Ultraviolet / Visible Spectrophotometry. Extent of degradation was analyzed using degradation curves and rate constants fit from the curves.

Encapsulation in uniform particles offers protection to beta carotene, and the extent of protection increases with the addition of UV absorbers or antioxidants. Both core-shell particle geometries offer additional protection without the addition of UV protectants, although the core consisting of many particles protects better than the solid core. In most cases UV protection increased as protectants were added to the formulation and as concentration of protectant was increased. Proximity of the protectant molecule to beta-carotene was shown to be a significant factor by comparing degradation of the uniform particles and the core-shell particles.

(526h) Adsorption and Release of Active Species Into and From Multifunctional Ionic Microgel Particles

Chen, H.,
Dai, L., Arizona State University
Originally presented on: 11/6/2013 17:35:00 - 17:55:00

In this report, we synthesize monodisperse ionic microgel particles which undergo large volume phase transitions in response to environmental stimuli such as pH and temperature. In addition, the study elucidates the effective encapsulation and release of rheology modifiers from these microgel particles to alter the bulk viscosity of a surrounding fluid. Moreover, these ionic microgel particles demonstrate abilities to adsorb and repel iron oxide nanoparticles (Fe_3O_4-NPs) upon pH variation. The extent of the loading of Fe_3O_4-NPs within the colloidal particles and morphology can be manipulated by tunable interactions between the Fe_3O_4-NPs and ionic microgel particles.

(543a) Homogeneous Fluidization of Geldart D Particles in a Gas-Solid Agitated Fluidized Bed

Wang, J. J., Zhejiang University
Han, Y., Zhejiang University
Feng, L. F., Zhejiang University
Originally presented on: 11/6/2013 15:15:00 - 15:37:00

The gas-phase polymerization process has been succeeded to manufacturing a variety of olefin polymers. The process is featured with a fluidized bed reactor, in which the monomer in the gas phase is polymerized with a co-monomer and hydrogen under a specific catalyst in the solid phase. Good fluidization quality is necessary for ensuring gas–solid contact and mass transfer. However, it is difficult to maintain smooth fluidization due to the agglomeration of polymer particles, especially for high-impact polypropylene ones whose surfaces may be sticky.

Mechanical agitation was introduced to fluidized beds for the purpose of improving fluidization performance. Various types of blades were used in gas-solid fluidized beds to meet the needs of different experimental and industrial processes, such as control of granule diameter in granulation, increase in drying rate, speed-up of the crystallization rate, disruption of bakers' yeast cells and separation of flotsam and jetsam particles. Although mechanical agitation can improve fluidization, it is uncertain to how and what extent it affects hydrodynamics and bubble behaviors.

We conducted experimental and numerical studies on the effect of agitation of a frame type impeller on the fluidization quality in a stirred fluidized bed. A 3D unsteady CFD simulation is implemented based on the Eulerian-Eulerian frame coupled with the kinetic theory of granular flow (KTGF) for the solid phase. The multiple reference frame (MRF) approach is incorporated to deal with the rotation of the impeller.

1. Bed pressure drop

The agitation speed of the frame impeller does not have a significant effect on the bed pressure drop. For the frame impeller, the effect of agitation on the axial component of particle velocity was small and its impact on the bed pressure drop was negligible.

2. Pressure fluctuation

For a given agitation speed, the standard deviation of pressure fluctuation increases linearly with the superficial gas velocity, indicating a more heterogeneous fluidization because of an increase in the bubble size. For a given gas velocity, an increase in the agitation speed improves the fluidization performance with a lower standard deviation of pressure fluctuation. The minimum fluidizing velocity remains almost unchanged whatever the agitation speed of the frame impeller. However, the minimum fluidizing velocity increases with the agitation speed. It is important to point out that the transition from aggregative fluidization to homogeneous fluidization for Geldart D particles is achieved by the agitation of the frame impeller. An increase in the agitation speed enlarges the homogeneous fluidization regime to an extent that is common to, liquid-solid fluidized beds, gas-solid fluidized beds with Geldart A particles or high operating pressure.

3. Solid volume fraction distribution

When the agitation speed of the frame impeller is 10 rpm, the fluidization performance in the stirred fluidized bed is very close to that in a classical one (0 rpm). Bubbles split and coalesce continuously as the fluidization proceeds. The particles exhibit aggregative fluidization for the whole range of the superficial gas velocity, and the bubble size increases with increasing superficial gas velocity. However, the agitation of the frame impeller seems to have little effect on the fluidization quality and its inefficiency may be attributed to the fact that its speed is low.

When the agitation speed exceeds 20 rpm, the fluidization performance is much improved. At low superficial gas velocity, bubbles do not exist in the bed and the Geldart D particles exhibits homogeneous fluidization. As the superficial gas velocity increases, the fluidized bed goes from bubbleless fluidization to bubbling fluidization. At high superficial gas velocity, an increase in the agitation speed brings about a decrease in the bubble size because the agitation of the frame impeller forces particles to enter the bubbles. When bubbles shrink or vanish completely, the fluidization undergoes a transition from aggregative fluidization to homogeneous fluidization. In addition, an increase in the agitation speed increases the upper limit of the gas velocity for homogeneous fluidization.

4. Particle velocity

In terms of the particle velocity range of 0 to 0.5 m/s, a higher agitation speed yields an increase in the portion of magnitude of particle velocity. It may be attributed to the fact that the agitation has given a component of tangential velocity to the particles. However, when considering the range with the particle velocity magnitude larger than 0.5 m/s, the portion of particle velocity is reduced with the increasing agitation speed of frame impeller and the particle velocity in the fluidized bed tends to be a normal distribution. The agitation of frame impeller improves the homogeneity of particle velocity distribution and decreases the regions with small or zero particle velocities. As a result, the solid mixing is improved.

We performed experimental work and numerical simulations in order to investigate the fluidization performance in a stirred gas-solid fluidized bed filled with Geldart D particles. The 3D unsteady CFD model which incorporates the two-fluid model, kinetic theory of granular flow and multiple reference frame model, is validated with experimental data with respect to the bed pressure drop. The transition from aggregative fluidization to homogeneous fluidization is achieved by the agitation of frame impeller, and this phenomenon is predicted and illustrated by the CFD simulation.

(543b) Effect of Electrostatics On Bubbles in Gas-Solid Fluidized Beds

Jalalinejad, F., UBC
Bi, X. T., University of British Columbia
Grace, J. R., University of British Columbia
Originally presented on: 11/6/2013 15:37:00 - 15:59:00

The generation of electrical charges, reported in gas-solid fluidized beds for over sixty years, can cause serious problems like wall sheeting in polyolefin reactors, leading to costly shutdown, electrical shock hazards and even explosions. Understanding the associated phenomena plays an important role in learning how to avoid these problems. In this study an attempt has been made to broaden the understanding of electrostatics in gas-solid fluidized beds by adopting computational fluid dynamics (CFD), using the Two-Fluid-Model in MFIX (open-source Multiphase Flow with Interphase eXchange software originated by the U.S. Department of Energy). The Maxwell equations were incorporated in the MFIX code in this work. The resulting model is then used to investigate how electrostatic charges modify bubble shape, size, velocity and interaction for three different cases: (a) single bubbles, (b) bubble pairs in vertical and horizontal alignment, and (c) freely bubbling bed. In each of these cases, a two-dimensional column, partially filled with mono-sized particles, is simulated for uncharged and charged particles.

In case (a), it is predicted that single bubbles elongate and rise more quickly in charged particles than in uncharged ones. For Case (b) electrostatics cause asymmetry of coalescence for a pair of vertically-aligned bubbles, while it causes the migration of a side bubble towards the axis of the column and changes the leading-trailing role for a pair of horizontally-aligned bubbles. Finally in case (c), the simulation predicts that electrostatic charges decrease bubble size and frequency in the free bubbling regime, accompanied by a change in the spatial distribution of bubbles, causing them to rise more towards the axis of the column.

An attempt was also made to test experimentally the single bubble simulation results. To reach this goal, a two-dimensional fluidization column was built with a central jet to inject single bubbles. The setup is also equipped with a novel Faraday cup device to measure the charge density of particles accurately. The experimental results show a decrease in bubble size and an increase in bubble height-to-width ratio with increasing charge density, accompanied by an increase in particles raining from the bubble roof.

(543c) Extended Conditional Quadrature-Based Moment Method for Polydisperse Gas-Particle Flows With Size-Conditioned Velocity

Kong, B., Iowa State University
Passalacqua, A., Iowa State University
Fox, R. O., Iowa State University
Originally presented on: 11/6/2013 15:59:00 - 16:21:00

Extended conditional quadrature-based moment method for polydisperse gas-particle flows with size-conditioned velocity

Bo Kong, Alberto Passalacqua and Rodney O. Fox

Department of Biological and Chemical Engineering & Ames Laboratory

Iowa State University, Ames, IA, USA

Keywords: Extended conditional quadrature-based moment method, polydisperse gas-particle flows, size-conditioned velocity, generalized population balance equation

Abstract:

Polydisperse gas-particle flows are common in many fields of engineering, such as in fluidized beds and risers, which are widely used in a variety of chemical processes. Due to the various physical and chemical interactions between particles and gas, such as particle collisions and breakage, gasification and chemical reactions, the individual particle size and the overall size distribution of particles changes as the flow develops. Since different size particles have different velocity distributions, and different particle velocity distribution again lead to different size changes, the joint number density function of particle size and velocity has to be modeled in a gas-particle flow. In this work, the extended conditional quadrature method of moments (ECQMOM) is developed to treat the size-velocity coupling in numerical solutions to the generalized population balance equation (GPBE) for particles. ECQMOM is a multivariate moment-inversion algorithm that combines the conditional quadrature method of moments (CQMOM) (Yuan and Fox, 2011) and the extended quadrature method of moments (Yuan et al., 2012). CQMOM is based on the concept of conditional moments, which can be used to solve the distribution of a conditional variable by using univariate inversion. EQMOM is a univariate inversion method, which replaces the delta functions with smooth kernel density functions by adding one additional moment. In this work, a beta kernel function was chosen for particle size distribution and a Gaussian kernel function was used for the particle velocity distribution. By reconstructing the size-conditioned velocity distribution function, the spatial fluxes in the moment equations are treated using a kinetic-based finite-volume solver. The particle-phase volume-fraction and momentum equations were then coupled with the Eulerian solver for the fluid phase. The computational algorithm is implemented in an open-source CFD package and tested in one dimension and then two dimensions with small and large standard deviations in the particle size.

(543d) Experimental Study On Fluidization Behavior of Geldart B and D Particle Systems in Deep Tapered Beds

Molnar, M. J., Hemlock Semiconductor Corporation
Karri, S. B. R., Particulate Solid Research, Inc.
Dehtiar, M. E., Dow Corning Corporation
Braun, E. D.,
Arrington, Y., Particulate Solid Research, Inc.
Cocco, R. A., Particulate Solid Research, Inc.
Gelderbloom, S. J.,
Originally presented on: 11/6/2013 16:21:00 - 16:43:00

For large Geldart B and Geldart D particles, the working height attainable in cylindrical bubbling fluidized beds is typically limited by growth of bubbles. Bubbles coalesce rapidly in these systems, and as a consequence, violent motion associated with bubble breakage at the freeboard can lead to uneven fluctuating forces on the vessel. For Geldart D particles, spouted beds are typically employed to avoid large bubbles, but in a reacting system, use of a spouted bed does not allow uniform solids contact with the reagent gases. While baffles can be used to reduce the size of bubbles in Geldart B systems, for the intended application, erosive wear could not be tolerated for the scale-up opportunity. Likewise, reductions in average particle size or superficial velocity or decreases in bed height to accommodate larger bed cross-section were all undesirable for the intended operating targets. For the latter, the available external surface area for heating from the wall is also compromised when compared with the heat duty needed for scale-up.

An exploratory investigation of bubbling fluidization with tapered geometries was conducted for Geldart D and large Geldart B silicon particles. Two separate, tapered fluidized bed systems, one 8-in. × 18-in. diameter and a second 17.5-in. × 40-in. diameter, were constructed from plexiglass and used for experimental studies of bubble growth, bubble size, bed density, and aggregate fluidization behavior of these particles. Acoustical monitoring of pressure fluctuations and accelerometer-based vibration measurements were employed in addition to optical bubble probes and high speed cameras to characterize the bed hydrodynamics over distinct conditions of gas superficial velocity and bed height. Transition conditions for slugging behavior were identified for two distinct particle size distributions. Modifications to overall fluidization behavior were demonstrated with design changes to gas delivery, thus allowing enhancement of the overall bed inventory while avoiding the slugging regime. Comparisons of the experimental data with calculations of fluidization behavior with commercially available simulation software are also presented.

(543e) Impact of Collisional Vs. Viscous Dissipation On Flow Instabilities in Gas-Solid Homogeneous Cooling Systems

Yin, X., Colorado School of Mines
Zenk, J. R., University of Colorado at Boulder
Mitrano, P. P., University of Colorado at Boulder
Hrenya, C. M., University of Colorado at Boulder
Originally presented on: 11/6/2013 16:43:00 - 17:05:00

Flow instabilities encountered in the homogeneous cooling of a gas-solid system are considered via lattice-Boltzmann simulations. Previous studies using dissipative molecular dynamics considered the effect of normal restitution coefficient and solid fraction on the onset and evolution of the instabilities in homogeneous cooling granular systems[1,2]. In this study, the relative contribution of the two mechanisms leading to instabilities, the viscous dissipation (gas-phase effects) and the collisional dissipation (particle-phase effects), is assessed. The results indicate that the instabilities encountered in the gas-solid system mimic those previously observed in their granular (no fluid) counterparts, namely a velocity vortex instability that precedes in time a clustering instability. We further observe that the onset of the instabilities is quicker in more dissipative systems, regardless of the source of the dissipation. Somewhat surprisingly, however, a crossover of the kinetic energy levels is observed during the evolution of the instability. Specifically, the kinetic energy of a more dissipative system is seen to become greater than that of a less dissipative system after the vortex instability sets in. This crossover of kinetic energy levels between a more dissipative system and a less dissipative system can be explained by the alignment of particle motion found in a vortex. Such alignment leads to a reduction in both collisional and viscous energy dissipation due to the more glancing nature of collisions. More disspative system therefore is able to preserve its kinetic energy due to the early onset of vortex instability.

References

[1]Mitrano, P. P., Dahl, S. R., Cromer, D. J., Pacella, M. S. & Hrenya, C. M. (2011) Instabilities in the homogeneous cooling of a granular gas: A quantitative assessment of kinetic-theory prediction. *Phys. Fluids* 23:093303.

[2]Mitrano, P. P., Garzo, V., Hilger, A. M., Ewasko, C. J. & Hrenya, C. M. (2012) Assessing a hydrodynamic description for instabilities in highly dissipative, freely cooling granular gases. *Phys. Rev. E* 85:141303.

(543f) Modeling of Sub-Grid Heterogeneity of Gas-Solid Suspension in Bubbling Fluidized Beds

Schneiderbauer, S., Johannes Kepler University
Pirker, S., Johannes Kepler University
Originally presented on: 11/6/2013 17:05:00 - 17:27:00

Due to increasing computer power the numerical simulation of fluidized and moving beds has become feasible. However, while kinetic theory based CFD (Computational Fluid Dynamics) has become a valuable design tool for modeling pilot plant scale gas-solid fluidized bed reactors, a fully resolved simulation of industrial scale reactor is still nearly unfeasible. Many sub-grid drag modifications have, therefore, been put forth by academic researchers to account for the effect of small unresolved scales on the resolved meso-scales in this case. However, all these models significantly differ in terms of their dependencies on the void fraction and on the particle slip velocity. In a first step, we, therefore, thoroughly implemented the sub-grid drag models of (i) EMMS[1], (ii) Kuipers[2], (iii) Sundaresan[3] and (iv) Simonin[4] and compared them to (v) our relation[5,6](CD-Lab model) and to (vi) the homogenous drag law of Wen and Yu in case of a three dimensional bubbling fluidized bed. The results are verified by a fine grid reference simulation. It is shown that

1. Applying the homogenous drag law of Wen and Yu, which ignores unresolved sub-grid structures, fails to predict the hydrodynamics of the bubbling fluidized bed using coarse meshes.
2. Applying each of the discussed sub-grid drag modifications reveals the bed expansion adequately.
3. The bubble size is estimated suitably by the investigated sub-grid drag closures.
4. However, the bubble rise velocity is significantly overestimated by these closures, which indicates the requirement of sub-grid stress modifications for the frictional regime.
5. Compared to the fine grid simulation the computational demand is reduced by approximately two orders of magnitude using the coarse grid for equal time step sizes. Coarse meshes, however, allow larger time steps that additionally improves the computational efficiency by approximately one order of magnitude in our study.

Since it appears that the sub-grid stress contribution is strongly connected to the predicted bubble rise velocity, we study the contribution of the sub-grid heterogeneities to the effective particle stresses in a second step. In this study, we derive closures for the unresolved parts of the drag and the solids stresses starting from the filtered two-fluid model equations. In the kinetic-collisional regime these are based on the assumption that the heterogeneity inside fluidized beds is triggered by the formation of local clusters. In contrast, in the frictional regime sub-grid stresses (SGS) arise from strain-rate and solids phase fluctuations of sub-grid shear layers mainly. These sub-grid correlations are then applied again to the three dimensional bubbling fluidized bed of fine glass particles. This study reveals that

1. The contribution from the sub-grid stresses has minor impact on the hydrodynamics of the fluidized

bed. Especially, bed expansion, solids phase distribution, gas flow, particle mass flux and bubble sizes are insensitive to the magnitude of the particle stresses.

2. However, without consideration for the contribution from the sub-grid stresses the effective particle stresses are underestimated significantly by the coarse grid simulations. It is shown that the bubble rise velocity is, therefore, overestimated considerably in this case.
3. Applying the presented sub-grid modifications yields fairly good agreement of the stresses and the solids phase velocity fluctuations with the fully resolved simulation and, therefore, the correct rise velocity of bubbles and slugs. Hence, a sub-grid stress modification for the frictional stresses is indispensable to estimate the bubble rise velocity in bubbling fluidized beds precisely.

To conclude, this study demonstrates that the presented sub-grid drag modifications apply well to the coarse grid simulation of a bubbling fluidized bed of fine particles. Nevertheless, the sub-grid contribution to the frictional stresses has to be taken into account to compute the bubble rise velocity correctly.

Finally, several tasks remain. For example, the models should also be validated at large scale fluidized beds (O(10)m). Thus, the general validity of the models must be further studied. These will be discussed in future publications.

References:

1. Lu, B, Wang, W & Li, J. Searching for a mesh-independent sub-grid model for CFD simulation of gas-solid riser flows. *Chemical Engineering Science* **64**, 3437–3447 (2009).
2. Wang, J, Van der Hoef, MA & Kuipers, JAM. Coarse grid simulation of bed expansion characteristics of industrial-scale gas-solid bubbling fluidized beds. *Chemical Engineering Science* **65**, 2125–2131 (2010).
3. Igci, Y & Sundaresan, S. Constitutive Models for Filtered Two-Fluid Models of Fluidized Gas-Particle Flows. *Industrial & Engineering Chemistry Research* **50**, 13190–13201 (2011).
4. Parmentier, J-F, Simonin, O & Delsart, O. A functional subgrid drift velocity model for filtered drag prediction in dense fluidized bed. *AIChE Journal* **58**, 1084–1098 (2012).
5. Schneiderbauer, S, Schellander, D & Pirker, S. A filtered frictional-kinetic model for gas-solid fluidized and moving beds. *Proceedings of the 9th International Conference on CFD in the Minerals and Process Industries* 7 (2012).
6. Schneiderbauer, S, Puttinger, S & Pirker, S. Comparative analysis of sub-grid drag modifications for dense gas - particle flows in bubbling fluidized beds. *AIChEJ (accepted for publication)* (2013).

(543g) Hydrodynamics of Liquid-Solid Fluidization of Ternary Mixtures in Tapered Beds

Surepalli, R. M. R.,
Talapuru, S., UNIVERSITY COLLEGE OF TECHNOLOGY, OSMANIA UNIVERSITY
Mohd, N. F.,
Originally presented on: 11/6/2013 17:27:00 - 17:49:00

Tapered columns have good application in industrial process like biological treatment of waste water, immobilization of bio-film reactions, crystallization and roasting of ores etc.. In contrast to cylindrical columns tapered columns have improved quality of mixing and better fluid-solid mixing and minimal entrainment of fine particles. Fine particles are present by themselves or formed during fluidization operation. Peng and Fan [1] were the first to investigate the characteristics of fluidization of mono component solids in tapered beds. They reported various flow regimes in the beds through measurements of pressure drop across the bed at various flow rates of fluidizing liquid using different tapering angles of the bed and also through visual observation. It is to be noted that as the cross sectional area of the tapered bed increases continuously from the bottom to the top, the drag force exerted by the fluidizing fluid on the particle decreases along the vertical direction. Koloini and Farkas [2] studied the pressure drop of fixed bed, minimum fluidization velocity and bed expansion in tapered beds of monocomponent particles. Pitt et al [3] measured the expanded bed height and hydraulic pressure drop at several apex angles for mono component particles. There have been some investigations on fluidization behavior of binary mixtures in cylindrical columns (Murthy et al [4], Pruden and Epstein [5], Kennedy and Bretton [6] and Bruno Formisani, et al [7]) .In the present study, liquid solid fluidization experiments have been carried out with mono component and ternary mixture solids. As mentioned above, the bed during operation may contain varied size of particles for which the hydrodynamics have not been investigated thoroughly, so it is necessary for mixtures beyond binary need to be investigated. Hence, in the present work an attempt has been made to study the behavior of ternary mixture beds in different tapered angle columns and also particle sizes.

Experimental setup consisted of calming section, distributor plate and tapered test column. The test column was a diverging tapered vessel of square cross section. All the sides of the column were made of plexi glass and ends of the surfaces of glass plates were held in steel L-plates. The height of the column was 600 mm and the bottom cross section was 50 x 50 mm and it was provided with 5 mm pressure taps to measure the pressure drop, one above the distributor and the other at a distance of 450 mm from the distributor at the top. The bed material was glass beads of different sizes, 0.5-0.8 (A), 1.3-1.5 (B) and 2.5-3.0 (C) mm. Total weight of the bed charged was 1000, 1400 and 1800 grams. Apex angles of the column ranged from 10° to 20°. Combinations of solids A, B and C were as follows, in the same order: 20:20:60, 40:20:40, 60: 20:20, 20:60:20 and 20:40:40. Water was the fluidizing medium.

Behavior of the mono component fluidized bed was observed to be similar to that of other investigators, Peng and Fan [1] and Murthy et al [2]. The description of the bed behavior in the case of ternary mixtures is as follows; To start with the bed was fluidized and defluidized. Later as the flow rate of water was slowly increased there was movement of small particles (A) to the top of the bed followed by intermediate particles (B) and the large particles (C) settling at the bottom, thus bed segregated into three layers. As the flow rate was increased further particles A started fluidizing at the top surface of the bed. On further increase of flow rate, a small cavity appeared at the distributor. The cavity and the fluidization of particles A increased with water flow rate. Later, the particles A were observed to be fluidizing completely along with some of the intermediate sized particles B. This prevailed till the complete bed started expanding indicating the on set of full fluidization.

For ternary mixture fluidization, the critical fluidization velocity, U_{cmix}, the critical pressure drop, P_{cmix} and the velocity for full fluidization, U_{tfmix} increased with the weight of the bed. As the coarse particle composition decreased in the mixture, U_{cmix}, P_{cmix} and U_{tfmix} decreased. This is due to the overall average particle diameter of the mixture decreases with decreased composition of large particles C. The same trend was observed in all the tapered columns. With change in angles of the columns, U_{cmix}, P_{cmix} and U_{tfmix} decreased with increase in the value of the angle. This can be attributed to the decrease in the height of the bed for the same weight of the solids as the tapered angle increase. Following are the correlations developed for mono component and ternary mixture solids.

Mono component particles

$$Re_c = 0.55\,(Ar)^{0.68} * (\tan(a/2))^{0.7} * (H/W_0)^{0.7}$$

$$Re_{tf} = 0.127\,(Ar)^{0.67} * (\tan(a/2))^{0.7} * (H/W_0)^{0.7}$$

Ternary mixture particles

$$U_{cmix} = 0.2829[(U_{c1} * X_1)^{0.8} + (U_{c2} * X_2)^{1.25} + (U_{c3} * X_3)] * [1/\tan(a/2)]^{0.5}$$

$$U_{tfmix} = 0.17[(U_{tf1} * X_1) + (U_{tf2} * X_2) + (U_{tf3} * X_3)] * [\tan(a/2)]^{0.8}$$

Re = Reynolds number

Ar = Archimedes number

H = Static height of the bed

W_0 = Width of the distributor

U_c, U_{tf} = Mono component critical and total fluidization velocity

References

1. Y Peng and L T Fan, "Hydrodynamic characteristics of fluidization in liquid solid tapered beds", Chem. Eng. Sci., 1977, 52, 4, 2211-2290.
2. Koloini, T. and E. J. Farkas ; Can. J. Chem. Engg., **51,** 499(1973).

3 Pitt, Jr., W. W., C. W. Hancher and H. W. Hsu : A.I.Ch.E. Symp. Ser., No. 181,**74**, 119(1988).

4 Sankarshana Talapuru, J.S.N.Murthy, S.Nandana, Phani Kiran S.V.G.R. and V.Ramesh, " Hydrodynamic

characteristics of liquid-solid fluidization of binary mixtures in tapered beds" ,2010 ECI Conference on 13th International conference on fluidization ,article 41,Vol.RP6.

5 Pruden B.B and Epstein N., Chem. Engg.Sci.1964 **14** 696.

6 Kennedy S.C. and Bretton R.H., A.I.Ch.E .J. 1966 **12** 24 30.

7 Bruno Formisani, Rosella Girimonte and Tizinia Longo: "The Fluidization Pattern of density – segregating Two –solid beds" ,ECI Conference on the 12th International Conference on Fluidization – New Horizons in Fluidization Engineering, Vancouver, Canada ,2007.

(553a) A Multiple-Radioactive-Particle Tracking Technique for Mixing Applications

Rasouli, M., Polytechnique Montreal
Bertrand, F., Ecole Polytechnique de Montreal
Chaouki, J., Ecole Polytechnique de Montreal
Originally presented on: 11/6/2013 15:15:00 - 15:35:00

Radioactive Particle Tracking (RPT) is a powerful and nonintrusive measurement technique to track particles in a multiphase system. In particular, the high penetrating power of gamma rays makes RPT an appealing technique that overcomes the limitations of other conventional methods (e.g. LDA and PIV) in the case of opaque particulate systems. RPT has been successfully used to investigate flow dynamics and mixing in a variety of blenders and stirred tanks (cylindrical drum, V-blender, Rushton turbine) as well as fluidized bed reactors.

Although RPT has shown efficient non-invasive tracking abilities in different flow situations, it is limited to one single tracer and relies on the so-called ergodicity assumption (the time average of one tracer is equal to a population average) to investigate mixing characteristics in a system. However, tracking more than one particle simultaneously is of interest because it could then yield direct information on both the relative translational and rotational motions of these particles. In particular, such rotational motion may play a role for instance on the lift force exerted on particles and then have an impact on their distribution. In systems where heat transfer is involved, the rotational movement of solid particles may change the rate of heating or cooling by increasing the interphase heat transfer coefficient. Moreover, tracking more than one particle could be used to assess the accuracy of a simulation model based for instance on the discrete element method (DEM).

This work presents a multiple-radioactive-particle tracking technique (MRPT) for the tracking of two (free or linked) tracers in a system. First, the proposed method is presented in detail. Its accuracy and precision are then investigated on the basis of experiments that have involved stationary or moving tracers in a cylindrical drum. Finally, the application of MRPT to mixing systems and other processes will be discussed.

(553b) Deagglomeration of Cohesive Materials Using Intensifier Bar Blending

Schrad, M., Eli Lilly and Company
Sloan, J. T., Eli Lilly and Company
Sommer, T., Eli Lilly and Company
Pletcher, T., Eli Lilly and Company
Hilden, J., Eli Lilly and Company
Originally presented on: 11/6/2013 15:35:00 - 15:55:00

Pharmaceutical companies reduce drug substance particle size to improve in vivo exposure and to ensure unit dose uniformity of low dose drug products. One consequence of particle size reduction is the formation of cohesive drug substances that tend to agglomerate. These agglomerates can cause significant dose uniformity issues for low dose drug products. Tumble bin blenders are often times ineffective at breaking up and dispersing these agglomerates during blending. Tumble bins with intensifier bars have been used successfully to break up and disperse drug substance agglomerates. Unfortunately, ensuring that all drug substance agglomerates have been removed is a challenge. Development studies have demonstrated the utility of using red iron oxide as a surrogate for agglomerated drug substance. Red iron oxide bulk contains fine particles that readily form large (>200 microns) agglomerates. These agglomerates can be visualized using x-ray imaging techniques. Formulations containing red iron oxide were blended with tumble bins and tumble bins with intensifier bars. The intensifier bar blender was observed to break and disperse the red iron oxide agglomerates through mechanical agitation. The extent of blending was characterized by the fraction of intact red iron oxide agglomerates remaining in the blend over time. Studies have shown that the deagglomeration rate can be characterized by an ideal mixing model. This model was used to determine the blending endpoint of a low dose formulation.

(553c) De-Agglomeration Rate Determination Using Specific Energy Input for Intensifier Bar Blending

Sommer, T., Eli Lilly and Company
Pletcher, T., Eli Lilly and Company
Hilden, J., Eli Lilly and Company
Schrad, M., Eli Lilly and Company
Sloan, J. T., Eli Lilly and Company
Originally presented on: 11/6/2013 15:55:00 - 16:15:00

Tumble bin blending with intensifier bars (I-bars) has gained popularity in the pharmaceutical industry due to the increased shear provided to the blending process by the I-bar. This increased shear promotes de-agglomeration of blend components and uniform dispersion in low drug load formulations. Previous development studies had shown that this de-agglomeration rate could be modeled as a first order ideal mixing process to result in a de-agglomeration rate equation and associated rate constant.

In the work presented here, an alternative first principles model was derived leveraging analogues from comminution theory to characterize the first order de-agglomeration rate constant as a function of the I-bar specific energy input. Studies were then run at varying bin sizes, bin fill levels, and I-bar speeds to test this model using red iron oxide as a surrogate for agglomerated drug substance. Results of the study are analyzed to determine the validity of using measured cumulative specific energy input to determine de-agglomeration endpoint. Additionally, the results of this specific energy model are compared to those of the ideal mixing model.

(553d) Experiment and DEM Based Investigation of Breakage of Pharmaceutical Granules

Naik, S. S., University of Connecticut
Chaudhuri, B., University of Connecticut
Originally presented on: 11/6/2013 16:15:00 - 16:35:00

Particle size reduction of dry granular material by mechanical means, also known as milling or comminution, is undoubtedly a very important unit operation in pharmaceutical, agricultural, food, mineral and paper industries. As comminution is a stochastic and a non-linear process, an attempt was made to understand this complicated process by conducting parametric studies experimentally and computationally using Discrete Element Method (DEM). Studies were performed lactose spheres to understand the effect of hammer speed (rotational), feed rate and hammer-wall tolerance. The size and shape of the resultant progeny of particles were analyzed by sieve analysis and microscope/image analysis techniques respectively. Greater size reduction was observed at higher speeds and low feed rates owing to the greater centrifugal force experienced by the particles and longer mean free path lengths respectively. Particle shape analysis revealed fragmentation to be the dominant mechanism of size reduction at higher speeds. Increase in impeller wall tolerance resulted in rolling mode regime of powder bed which was found to be significant at low impeller speeds. DEM simulations were carried out to study the effect of mill parameters on fragmentation. Below a critical hammer tip speed, a blending action rather than comminution was observed. The feed rate determines the hold up of material in sizing chamber and hence energy required for size reduction. At low hold-up, longer path lengths are achieved by particles resulting in higher impact velocity and hence a finer size distribution. At higher hold up the number of collisions is high, but the kinetic energy per particle is low leading to poor breakage probability. To develop an understanding of how grain breakage evolves in a granular system under impact the components of stress tensor were also determined from the DEM simulations.

(553e) Energy-Efficient Rate-Based Particle Separation

Lievano, D. M., University of Pittsburgh
Bhattacharya, T., University of Pittsburgh
McCarthy, J. J., University of Pittsburgh
Originally presented on: 11/6/2013 16:35:00 - 16:55:00

The effective separation of particles is key to numerous processes and industries handling solid materials. Despite this fact, particle separations techniques remain typically quite "low tech" and often are energy-intensive (e.g., sieving) or environmentally unfriendly (e.g., froth floatation) or both. Rate-based separation processes, on the other hand, represent a unique approach to particle separation that has the potential to be more flexible, more efficient, and more environmentally friendly than existing "low tech" techniques. There are a number of potential applications of this type of technique being examined in our lab, including a "granular chromatograph" where particle-wall cohesion leads to "adsorption" rates that dictate a particle's traversal down a channel; as well as both passive and active "granular ratchets" where variable particle diffusion is geometrically induced within a flow and serves as a means of altering transport rates. In the present paper, we highlight passive granular ratchets, where particles of differing properties flow through a device often called a Galton board. In this type of device, the gravity-driven flow of particles down an inclined plane causes collisions between the particles and the evenly distributed pegs along the board. Dissipative collisions between particles as well as between paticles and pegs results in a diffusion-like motion of particles perpendicular to the flow. The extent of separation (i.e., how far one type of particle is removed from another) depends on the different distances traversed by the two types of particles and, ultimately, on the collision rate and energy dissipation for particle-peg events. A simple theory, based on statistics and single-collision mechanics, will be set forth for comparison with our results.

(553f) Product Design Methodology to Overcome Segregation

Johanson, K. D., Material Flow Solutions Inc.

Originally presented on: 11/6/2013 16:55:00 - 17:15:00

Segregation is a mechanistic driven phenomenon. Materials may segregate during processing and handling as fine particle sift through the matrix of coarse particles. Air currents can carry finer particles to different parts of the process equipment. Differences in particle scale friction can cause a range of velocities as material slides down a pile. Each of these mechanisms provides a means of separating particles of different size, shape, and surface characteristics. Ideally the segregation problem solution is to design a material insensitive to segregation effects. However, a bulk solid mixture is always a discreet system where particle scale properties control the ability of this mixture to remain well blended. Each mechanism reveals a distinct segregation pattern. If one can understand all the separation tendencies that occur with a particular mixture based on particle scale properties, then one can modify those properties to minimize the separation tendency. This paper presents a methodology of relating the segregation intensity of a multi-component mixture to particle scale properties such as particle size distribution, particle surface friction, and particle density. The sample material studied is a typical drug mixture containing five key components. The repose angle, particle scale friction, particle size, and particle density were measured for the key components in the mixture. The segregation profiles and intensities were also measured using a spectral technique. Models were developed to describe sifting segregation, angle of repose segregation, and air entrainment segregation. These models were used with the segregation measurements and particle scale measurements to predict the expected segregation profile of all key components. Finally, a methodology was implemented to minimize segregation of the mixture by modifying key particle scale properties to change the type and magnitude of segregation. The model was used to guide the product design process. In some instances the methodology provided a means of eliminating segregation due to a particular mechanism. In some instances the methodology was used to optimize counteracting segregation mechanisms to create an overall segregation pattern that is more mixed.

(553g) Kinematics and Segregation of Bidisperse Granular Flow in Quasi-2D Bounded Heaps

Fan, Y., Northwestern University
Schlick, C.,
Umbanhowar, P. B., Northwestern University
Ottino, J. M., Northwestern University
Lueptow, R. M., Northwestern University
Originally presented on: 11/6/2013 17:15:00 - 17:35:00

Segregation of size bidisperse granular materials in quasi-two-dimensional (quasi-2D) bounded heap flow generates various final particle distributions which are parameter dependent. Accurate prediction of these distributions, however, requires a detailed knowledge of the flow kinematics. In this research, we first present a computational study of the kinematics of quasi-2D bounded heap flow using the discrete element method. The experimentally validated computational results reveal a universal functional form for the streamwise velocity profile for both mono- and bidisperse systems when velocities and coordinates are scaled by the local surface velocity and the local flowing layer thickness, respectively. The universal form holds true regardless of streamwise location, feed rate, particle size distribution and, most surprisingly, local particle concentration for bidisperse flows. Based on the kinematics, we develop a segregation model free of arbitrary fitting parameters that considers the interplay of advection, segregation, and diffusion. Compared with previous segregation models, excellent quantitative agreement with both experimental and simulation results is achieved. The model yields two dimensionless control parameters which depend only on the physical parameters of feed rate, particle size ratio, and system size. One, the Peclet number, captures the interplay of diffusion and advection, while the second characterizes the competition between segregation and advection. A parametric study demonstrates that final particle configurations can be controlled by properly tuning both dimensionless parameters, and the model can be readily adapted to more complicated flow geometries.

(598a) Yield Loci of Lactose Powders At Low Consolidation Stresses: Measurement and Modeling

Davies, C. E., Massey University
Saw, H. Y., Massey University
Jones, J. R., Massey University
Brisson, G., Riddet Institute, Massey University
Paterson, A. H. J., Massey University
Originally presented on: 11/7/2013 8:30:00 - 8:49:00

We have previously demonstrated a correlation for model lactose powders in which cohesion is related to surface-volume mean diameter, *in situ* bulk density at the preconsolidation stress, and the preconsolidation stress normalized with the minimum value used; mean diameters ranged from 30 microns to 223 microns, and the preconsolidation stresses spanned the range, 0.31 kPa to 4.85 kPa. This work is now extended using the observation that (for the lactose test powders) the yield loci are essentially linear, and their slope does not exhibit much dependence on powder size or preconsolidation stress; for bulk density we have found that the value under compaction is well related to loose-poured density, by a logarithmic model. It is thus possible to reconstruct (linear) yield loci using estimates of cohesion, mean particle diameter, bulk density and preconsolidation stress, using size distribution information and loose poured density. The utility of reconstructed yield loci has been assessed by comparing estimates of minimum hopper exit opening, with values determined from primary measurements; in both cases, the angle of wall friction was for 304 stainless steel with a 2B mill surface finish. Exit dimension estimates from primary yield loci and reconstructed yield loci were generally in good accord.

(598b) Effect of Moisture Content On the Bulk and Flow Properties of Fine and Coarse Glass Beads

Scicolone, J. V., New Jersey Institute of Technology
Metzger, M. J., Rutgers University
Anderson, K. V., Rutgers University
Koynov, S., Rutgers, The State University of New Jersey
Glasser, B. J., Rutgers, The State University of New Jersey
Muzzio, F. J., Rutgers, The State University of New Jersey
Originally presented on: 11/7/2013 8:49:00 - 9:08:00

The way particles pack and flow is an essential area of study in particle technology. Poor packing and flow can lead to segregation, clogging, and large product variability, which adversely affect the powder handling, processing, and production stages. In this work, the effect of water on the packing and flow properties of fine (cohesive) and coarse (non-cohesive) particles was experimentally investigated. Four different particle sizes of glass beads, ranging from <10mm to 275mm, were studied as dry particles and with various weight percent of water. The dry coarse particles flowed smoothly while the fine particles required agitation to initiate flow. The poor flow associated with fine particles is due to the formation of agglomerates. Agglomerates form when the interparticle forces (primarily van der Waals forces) are greater than the mass of the particle. Soft agglomerates can form with coarse particles when small amounts of moisture are added to coarse particles, creating liquid bridges between the particles. When enough moisture is added to the samples, the voids in the agglomerates fill up with water, creating a coarse dispersion or slurry.

Since sample characterization is dependent on many different conditions and properties, a multi-faceted approach was undertaken. With the use of a Freeman FT4 Powder Rheometer, measured changes in bulk density, porosity, compressibility, flow function coefficient, cohesion, and flow energy were collected as a function of water content and particle size. The results show that both water content and particle size distribution play significant rolls on the packing and flow of the powders. With a greater understanding of the many bulk properties associated with wetted material, changes can be implemented to improve how these materials are handled and processed, effectively lowering manufacturing costs.

(598c) Study of the Flow Performance and Physical Properties of a New Direct-Compression (DC) Grade of HPMC and How Its Performance and Attributes Compare to Those of the Existing Methocel(TM) CR Grades

Rogers, T., The Dow Chemical Company
Hoelzer, B., The Dow Chemical Company
Balwinski, K., The Dow Chemical Company
Rokosz, A.,
Kent, M.,
Hewlett, K., The Dow Chemical Company
Theuerkauf, J., The Dow Chemical Co.
Originally presented on: 11/7/2013 9:08:00 - 9:27:00

The purpose of this study was to investigate the flow performance and physical properties of a new direct-compression (DC) grade of HPMC designed to streamline manufacture of matrix tablets and how its performance and attributes compare to those of the existing METHOCEL™CR grade.

Multiple lots of the improved DC material and METHOCEL CR were characterized via bulk powder flow, particle size and shape analysis, bulk/tapped density, sieve separation, and powder avalanche testing. In addition, a 2-kg sample from each lot of HPMC powder was tumbled in a V-blender for 10 minutes, and powder aliquots were subsequently collected at three depths (surface, 7-10 cm, and 18 cm below surface). Each of the aliquots was then characterized via QICPIC analysis (Dynamic Image Analysis DIA system QICPIC™). Finally, characterization of each powder was attempted using an internally developed powder flow test, found to be discriminatory from one morphology to another and indicative of powder flow through an industrial hopper.

Where the improved DC material was successfully characterized within the powder flow apparatus, METHOCEL CR rat-holed and plugged the apparatus, which had to be tapped from the outside to clear the CR material. Although K100M grades generally exhibited higher Carr indices than the corresponding K4M grades, the DC materials consistently exhibited lower Carr Indices than METHOCEL CR, regardless of molecular weight grade. Mean times between powder avalanches were lower for the DC materials as well. Upon separation through a 63-µm sieve, the fines content was lower in the DC material (<50% vs. >60% passed through). Both scanning electron microscopy and QICPIC analysis confirmed the presence of granular, more spherical particles in the DC material. There was no sign of segregation of DC particles after tumbling for 10 min.

Because the DC material is more granular and has a more spherical morphology, this material is more flowable and facilitates direct compression. Furthermore, a complementary paper will demonstrate the utility of the DC material in a dry granulation process to manufacture matrix tablets containing an active pharmaceutical ingredient and commonly utilized excipients. Both sets of findings will assist formulators in choosing the morphology grade of HPMC best for their standard tablet manufacturing processes.

(598d) Effect of Physical Properties and Shearing Rate On Powder Flow During Die Filling From a Hopper

Tardos, G. I., The City College of the City University of New York
Kheiripour Langroudi, M., Merck & Co., Inc.
Woldu, A., Merck&Co., Inc.
Rajniak, P., Merck & Co., Inc.
Originally presented on: 11/7/2013 9:27:00 - 9:46:00

Powder flow in the "Intermediate" regime where collisions and particle friction co-exist, is the most prevalent in industrial practice and is the object of the present work. The ability of a powder to dilate or compact during flow at different shearing rates, i.e. to exhibit compressibility, plays a critical role in its rheology. This is important, as most of previous works on powder flows were developed under the main assumption that powders flow at the "critical condition" where the they are considered to be incompressible (with slight dilation). The basic motivation for the present work was to study flows of different pharmaceutical excipients through circular orifices of different size at varying shearing rates to understand the role of compressibility.

Two versions of a lab-scale die filling apparatus were used: one with intermittent (periodic shear) and the other with continuous flow. A capacitance probe, that measures solid fraction based on the dielectric permittivity of the media, is mounted close to the die to enable the measurement of density and density fluctuations so that volumetric flow rates can be obtained directly. This is an essential feature since the bulk density of the powder in the shearing zone is variable and also fluctuates in time.

We performed experiments with free flowing materials to verify the prediction of the Beverloo equation *at no shear.* We also used free-flowing and somewhat cohesive powders to provide modifications to this equation when shear is present. We found that there is a strong correlation between powder flowability and its density while filling the die. This observation stresses the importance of continuous monitoring and careful control of powder density during die-filling as density fluctuations cause tablet weight uniformity issues in the tablet press. To complete the picture we also used cohesive powders and found that in this case, although the powder was not flowing through any orifice sizes under no shear conditions, its flowability improved significantly under shear. We demonstrated that for this kind of powder, flow instabilities set in as low shear is applied, which in turn causes erratic flow. The flow becomes more stable either when the powder is subject to shear for a longer period of time (we obtained this by substituting intermittent feed by a continuously sheared hopper) or as shear is increased beyond a critical value.

(598e) Measurement of the Axial Dispersion Coefficient in a Rotating Cylinder: Dependence On Material Properties

Koynov, S., Rutgers, The State University of New Jersey
Glasser, B. J., Rutgers, The State University of New Jersey
Muzzio, F. J., Rutgers, The State University of New Jersey
Originally presented on: 11/7/2013 9:46:00 - 10:05:00

Rotating drums are encountered in numerous industrial applications, e.g. calciners, impregnators, coaters, cement mixers, etc. In all of these devices, the rotation of drum is used to engender mixing of the granular material in the radial direction. Axial mixing, despite its significantly lower rate, can also have an impact on the process performance, especially when residence time control is essential. Typically, the particle dynamics in rotating drums are quantified as a function of process conditions, such as rotation speed, fill level, and cylinder size, as well as particle properties – usually limited to the characterization of particle size.

In this work, the quantification of the axial particle dynamics has been expanded to include bulk flow properties. Fick's law is used to describe the axial dispersion behavior of the particles. Therefore changes in behavior can be characterized using the axial dispersion coefficient. A method has been developed to measure the axial dispersion coefficient and correlate it to bulk particle flow properties, such as cohesion.

(598f) Application of a Newly Developed Blendability Model in Determining Stages of Mixing

Shi, W., Bristol-Myers Squibb
Sprockel, O. L., Bristol-Myers Squibb

Originally presented on: 11/7/2013 10:05:00 - 10:24:00

The current study presents a new technique of determining blendability and differentiating various stages of mixing using a previously developed powder mixing kinetic model [Powder Tech. 235 (2013), pp 400-404]. The model defines the resistance to mixing as the cohesive energy of a key ingredient to be mixed and the driving force of mixing as the shear energy of the mixing environment. Using iron powder as the key ingredient and a magnetic field to manipulate the cohesive energy, it was found that the rate constant follows an Arrhenius relationship between the cohesive energy and the shear energy. After the shear energy at a given operational condition is quantified by using iron powder, iron powder is substituted with active pharmaceutical ingredients (APIs). The cohesive energy of the API is then derived with the assumption that the shear energy is maintained in the mixture when the key ingredient is at a low concentration. With the application of process analytical tools (PAT), the cohesive energy of API is further used to characterize the shear energy at various operational conditions, including fill fraction and size of the mixing equipment. Additionally, the model shows that the mixing of powder does not always follow a single-stage process, but rather a multi-staged process. The multi-staged process is likely related to the physical change in aggregation status of powder during mixing, i.e., from the aggregated status to the non-aggregated status. Such differentiation of these stages is dependent on the strength of aggregates and is not captured using other mixing indices. Further fine-tuning of the model is required to understand the transitional regions between these stages and to reveal other possible stages during mixing, such as demixing. In summary, the study demonstrates that the model is capable of differentiating blendability of APIs and projecting mixing performance upon scale-up.

(598g) Intensifier Bar Blending Analysis and Scale-Up Using X-Ray Tracer Particles

Hilden, J., Eli Lilly and Company
Schrad, M., Eli Lilly and Company
Sloan, J. T., Eli Lilly and Company
Sommer, T., Eli Lilly and Company
Pletcher, T., Eli Lilly and Company
Originally presented on: 11/7/2013 10:24:00 - 10:43:00

Of the various unit operations used to manufacture solid oral dosage forms such as tablets, bin blending is a crucial and not well characterized unit operation. Bin blending is frequently employed and multiple blending steps are often required to produce a single batch of tablets. Bin blending is also particularly sensitive to scale-up in comparison with other unit operations (e.g. tablet compression and roller compaction) because it requires a larger bin and cannot just be run for longer periods of time. The effects of gravity and associated shear stresses in larger bins are difficult to model/simulate.

The incorporation of intensifier bars (I-bars) into bin blenders has gained in popularity because I-bars provide aggressive agitation of powders. I-bars typically spin at 2000-3000 RPM, with tip speeds of 15-20 m/s and are effective in forcefully mixing powders and dispersing statically-held agglomerates.

In this work, Fe_2O_3 agglomerates were added to I-bar equipped blenders. The I-bar caused breakage and dispersion of the agglomerates and the dispersion rates were measured using a novel 3D X-ray imaging technique. A mathematical model was derived from first principles and used to predict dispersion rates at different scales and under different blending conditions. In total, 59 experiments were conducted, generating nearly 300 samples. Results were analyzed and are presented in order to assess the validity of the scale-up model.

(598h) Enhanced Centrifuge-Based Approach to Powder Characterization

Thomas, M. C.,
Beaudoin, S. P., Purdue University
Originally presented on: 11/7/2013 10:43:00 - 11:02:00

Improvement in solids processing performance depends on a fundamental understanding of powder behavior, which is driven by the adhesion characteristics of the powder's particulate ensemble. Key physical properties of the particulate system, chiefly size, geometry, and surface roughness, affect particle adhesion. These particulate properties can be extracted through particle adhesion measurements to effectively describe the powder. Ultimately, adequate descriptions of the powder can in turn be used to exert control over powder behavior by optimizing process conditions.

A centrifuge-based technique is a practical method that measures particle adhesion. The technique is simple to perform and describes the adhesion behavior of an ensemble of particles. Further enhancement of the centrifuge-based technique can be developed to provide valuable insight at the particle scale by determining the aforementioned particulate properties. This is accomplished through specially designed substrates that allow the extraction of particle characteristics.

Substrates with hemispherical indentations provide a means by which particle size characteristics can be determined. Adhesion forces were determined *in silico*, utilizing an in-house van der Waals force simulator, to describe the effects of the ratio of particle to indentation diameter for indentation diameters between 10 and 50 μm. In this study, the ratio of the particle diameter to the indentation diameter was varied from 0.1 to 2 in 0.1 increments. Idealized spherical geometries and smooth surfaces were used for the particle and indentation. Simulations of a single particle and single indentation, symmetrically oriented along a centralized vertical axis, were the basis of the adhesion calculations. By systematically varying the indentation size, particle size can be extracted due to large discrepancies in the particle-indentation adhesion forces. Once a particle size distribution has been determined, systematic variation in roughness on the indentations affords a means for evaluating the size-specific roughness distribution on the particles. This study establishes a viable strategy to enhance the centrifuge-based approach for powder characterization.

(602a) Gasifier Performance Modeling Including Mass Transfer Effects

Breault, R. W., National Energy Technology Laboratory
Nicoletti, P., URS COrporation

Originally presented on: 11/7/2013 8:30:00 - 8:50:00

In the development of multiphase reacting CFD codes, a number of simplifications were incorporated into the codes and models. One of these simplifications was the use of a simplistic mass transfer correlation for the faster reactions and omission of mass transfer effects completely on the moderate speed and slow speed reactions such as those in a fluidized bed gasifier. Another problem that has propagated is that the mass transfer correlation used in the codes is not universal and is being used far from its developed bubbling fluidized bed regime when applied to riser reactors. These problems are true for the major codes, which at this time does not consider mass transfer effects.

To alleviate this problem, a mechanistic based mass transfer coefficient algorithm has been developed based upon earlier work by Breault et al. [1, 2, 3 and 4]. This fundamental approach uses the local hydrodynamics to predict a local, time varying mass transfer coefficient. The algorithm previously developed has been generalized to work on cells of any volume. As was the case for the non-generalized cell volume, the predicted mass transfer coefficients and the corresponding Sherwood numbers agree well with literature data and are typically about an order of magnitude lower than the correlation noted above. The incorporation of the new mass transfer model into the C3M code gives the expected behavior for all the gasification reactions evaluated in the paper. The new algorithm should be used such that the conversions are not over predicted. Additionally, its behavior with changes in solids flow rate are consistent with the changes in the hydrodynamics.

1. 1. Breault, R.W, "A review of gas–solid dispersion and mass transfer coefficient correlations in circulating fluidized beds", Powder Technology 163 (2006) 9–17
2. 2. Breault, R. W. and Guenther, C. P., "Mass transfer in the core-annular and fast fluidization flow regimes of a CFB", Powder Technology 190 (2009) 385–389
3. 3. Breault, R. W. and Guenther, C. P., "Mass transfer coefficient prediction method for CFD modeling of riser reactors", Powder Technology 203 (2010) 33–39
4. 4. Breault, R. W., Li, Tingwen, and Nicoletti, P., "Mass Transfer Effects in a Gasification Riser", submitted to Powder Technology

(602b) An Acoustic Method for the Measurement of Minimum Fluidization and Bubbling Properties of Group A Solids

Issangya, A. S., Particulate Solid Research, Inc.
Hays, R. C., Particulate Solid Research, Inc.
Cocco, R., Particulate Solid Research, Inc. (PSRI)
Knowlton, T., Particulate Solid Research, Inc. (PSRI)
Karri, S. B. R., Particulate Solid Research, Inc. (PSRI)
Originally presented on: 11/7/2013 8:50:00 - 9:10:00

Minimum fluidization and bubbling densities and velocities are widely used in fluidized bed design and modeling. The methods presently used for measuring these parameters are, however, time consuming and labor intensive. Traditional methods have often relied on visual observations to determine the minimum bubbling point, which is likely to introduce some subjectivity in the data and, also necessitates the use transparent columns. Such test units often will not withstand high temperature and pressure conditions. This presentation discusses results obtained in a new 6-in (0.15-m)-diameter test unit in which the minimum fluidization and bubbling points were detected by the use of acoustic waves. The test materials were FCC catalyst particles of various fines contents (% < 44 mm). Tests were conducted at low-to-moderate system pressures. Gas flowrate was automatically adjusted at a chosen time interval using a mass flow controller. Results showed that fines content had no significant effect on the minimum fluidization velocity. The minimum bubbling velocity, however, increased rather sharply when the fines content exceeded about 5% less than 44 microns. Both the minimum fluidization density and minimum bubbling density of FCC catalyst particles were found to decrease slowly with increasing fines content. For the pressure range used in these tests, there was no significant effect of pressure on both the minimum fluidization velocity and the minimum bubbling velocity. Furthermore, pressure had very small influences on both the minimum fluidization and minimum bubbling densities.

(602c) Fluid Dynamic Evaluation of Circulating Fluidised Beds for Carbon Capture

Zaragoza, F., University of Edinburgh
Brandani, S., University of Edinburgh
Fan, X., University of Edinburgh
Zhang, K.,
Originally presented on: 11/7/2013 9:10:00 - 9:30:00

Carbon capture by adsorption in amine-functionalised solids has the potential to become a less energy-demanding alternative to the conventional amine scrubbing. Such solid sorbents present higher CO_2 capacity and lower regeneration energy than amine solutions, characteristics that can be exploited using the proper gas-solid contactors to achieve rapid thermal cycling. Circulating fluidised beds (CFBs) are an interesting option in this respect, as they feature the high efficiency in mass and heat transfer of the batch-like fluidised beds and the possibility of handling the adsorption/desorption cycle continuously.

The University of Edinburgh (UoE) and North China Electric Power University (NCEPU) run together the FOCUS (Fundamentals of Optimised Capture Using Solids) project, co-funded by the Engineering and Physical Sciences Research Council (EPSRC) and the National Natural Science Foundation of China (NSFC). The FOCUS project aims to develop a novel process based on CFB technology using supported amine sorbents for post-combustion carbon capture in coal-fired power plants. Lab-scale CFBs are being constructed at the UoE in order to study the fluid dynamics of the process. Experiments are being run to determine a) flow regimes in the different sections of the CFBs, b) operational gas and solid flow rates, and c) gas residence time distributions (RTDs) and solid RTDs using Positron Emission Particle Tracking (PEPT). These measurements will be used to validate compartmental models of the adsorption process and 3-D CFD models of the gas solid flows. Accurate models are required given the stringent purity and recovery requirements of the carbon capture process.

(602d) Spout Shape and Stability in a Flat-Base Spouted Bed Using a High Speed X-Ray Tomography

Yang, X., Delft University of Technology

van Ommen, J. R., Delft University of Technology

Wagner, E., Delft University of Technology

Mudde, R. F., Delft University of Technology

Originally presented on: 11/7/2013 9:30:00 - 9:50:00

Time-resolved measurements of the shape and position of a spout were done with a high speed X-ray CT scanner. This system contains 3 X-ray sources and detector arrays, which makes it possible to scan the whole cross section of spouted bed of 24cm diameter simultaneously. We reconstructed the images of the spout in several cross sections. The spout size and position is found from the raw data with a temporal resolution of 2500 fps and spatial resolution of 1 mm.

The time averaged spout size is validated with McNab's prediction. From the spout data we estimated the particle circulation in the bed. It shows good agreement with theoretical predictions. When $U/U_{ms} = 0.96 \sim 1.44$ (with U_{ms} the minimum spout velocity), there is very small fluctuation of the spout diameter, with an amplitude less than 10% of the spout diameter. We analyzed the fluctuation using the Power Spectrum Density. It is shown that the spouts are most stable when $U/U_{ms} = 1.20 \sim 1.27$. We also analyze the movement of the center of the spouts, and find that there is an oscillation of the spout position when it is close to surface of the bed.

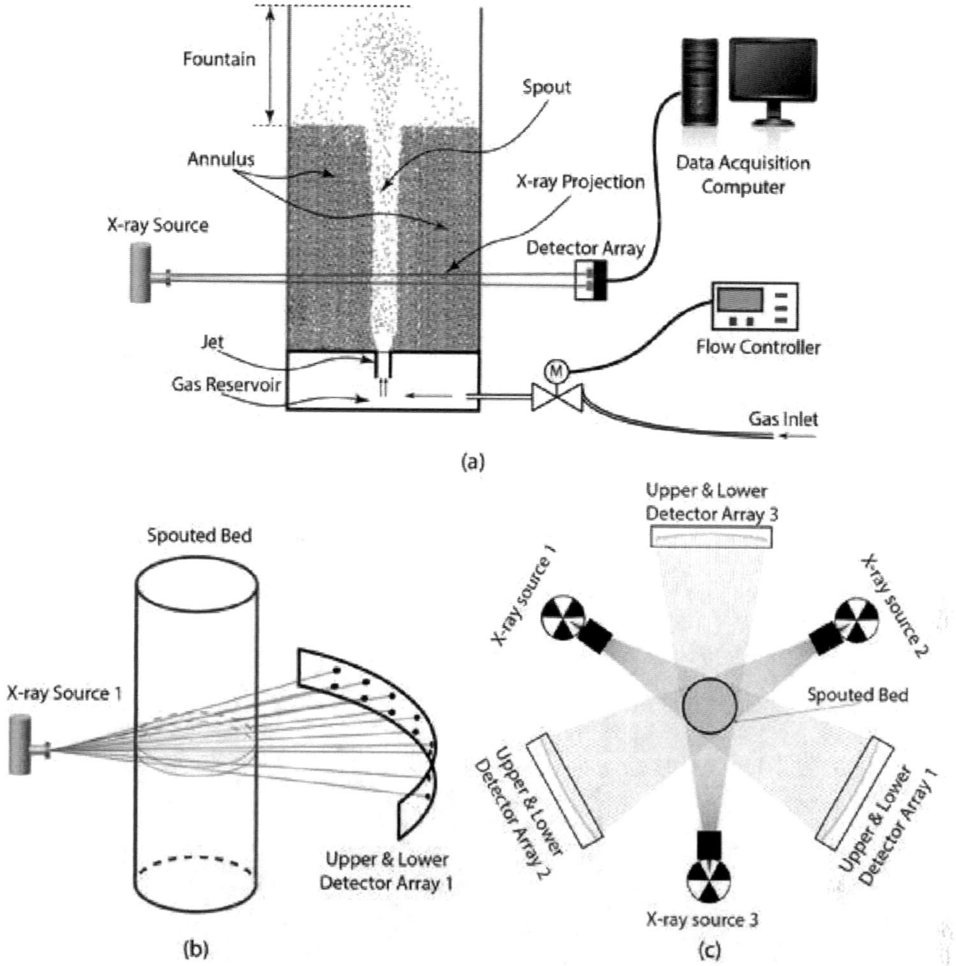

Figure 1. Experiment setup: (a) Overall; (b) 3 D view; (c) Top view.

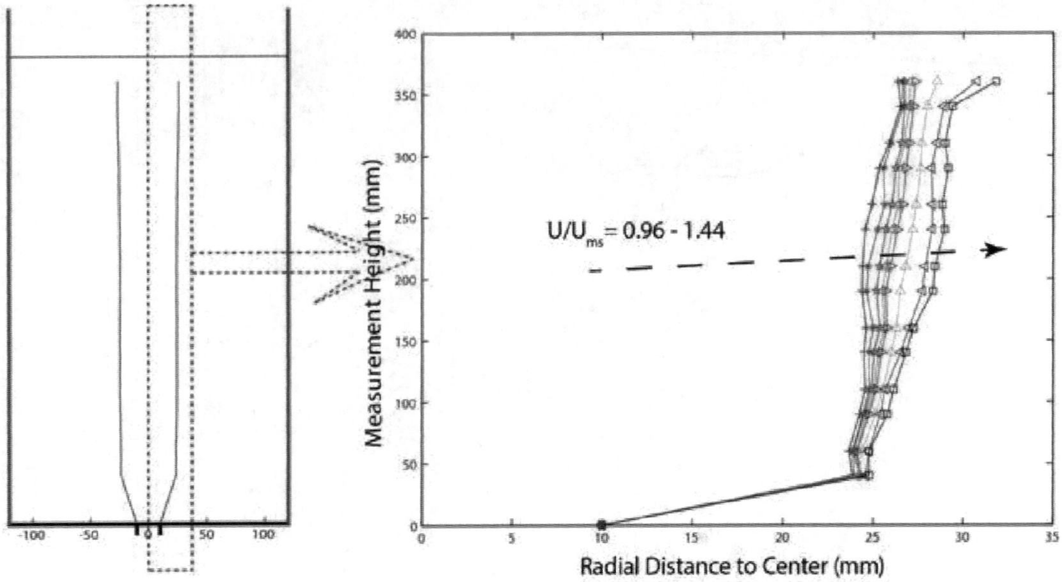

Figure 2. The time-averaged profile of the spouts.

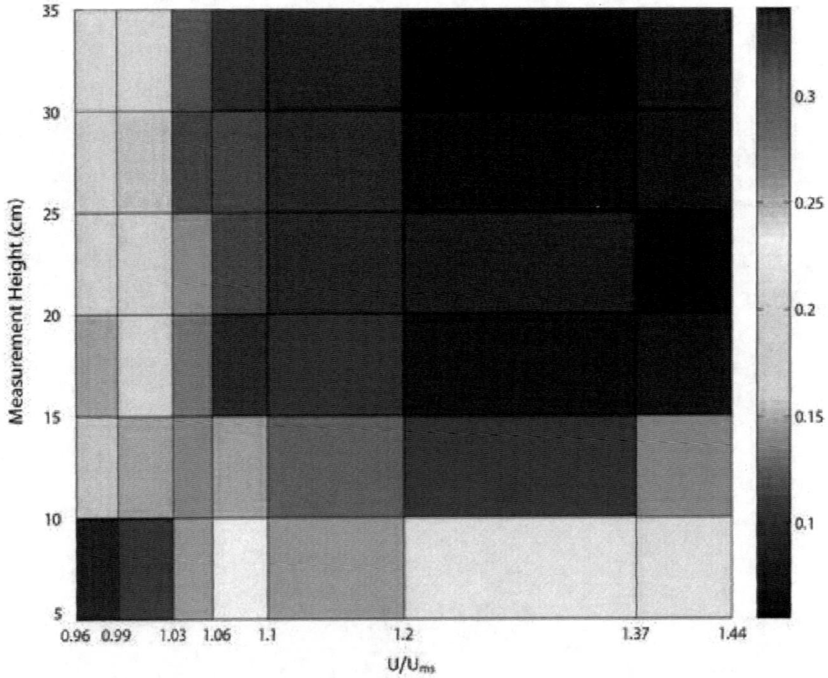

Figure 3. The stability of the spouts' size indicated with average Power Spectrum Density (Red -> Blue: Instable -> Stable).

(602e) Scale-Up and Experimental Investigations of Loop-Seal in Chemical Looping Combustion (CLC) Process: Solid Flowrate Regulation and Gas Tightness

Forret, A., IFPEN
Yazdanpanah, M. M., Total
Gauthier, T., IFP Energies Nouvelles
Originally presented on: 11/7/2013 9:50:00 - 10:10:00

The current work is defined in the frame work of CLC project at IFPEN, a cooperative development project joint by TOTAL.

Chemical looping combustion (CLC) is a novel combustion process with inherent CO_2 capture resulting in a minimum capturing energy penalty. CLC process is an oxy-combustion technology in which the oxygen required for the combustion is supplied by metal oxides in so called "Fuel Reactor". The metal oxides, known as oxygen carriers, are then sent to a separate reactor (Air Reactor) where they are re-oxidized in contact with air. The oxidized particles are sent back to Fuel Reactor, hence forming a loop.

Control of solid flow rate and gas-tightness are two essential parameters in chemical looping combustion process. This paper presents experimental results for use of loop-seals as a pneumatic solid flow rate controlling device which minimizes gas leakage between two regions in a circulating fluidized bed. The pressure drop variations in different elements of the loop-seal are analyzed as function of gas and solid flow rate and pressure balance across the solid circulation loop. Moreover, minimization of gas leakage and formation of the moving solid column in the supply chamber is investigated in terms of gas flow rate, solid flow rate, external aerations, and pressure drop downstream of a small loop-seal. The horizontal solid and gas flow in the loop-seal and the impact of the first and second loop-seal aerations are also investigated.

Loop-seal of different internal diameters (0.021 m and 0.150 m) are used in this work. Ilmenite and two sand particles are used with corresponding density of 4750 kg/m^3 and 2650 kg/m^3 and average diameters ranging from 128 to 368 μm for the small loop seal. In case of large loop-seal, a metallic oxide called BMP and a mixture of BMP and FCC catalyst are used with corresponding density of 3300 kg/m^3 and 3100 kg/m^3 and average diameters ranging from 150 to 200 μm. The mock-up used in this study is a circulated fluidized bed, constituted of a reactor (dense bed), a L-valve that fix the solid flowrate, a riser and a loop-seal.

Loop-seals are commonly used in circulating fluidized bed systems to convey particles from a low pressure region to a high pressure region [1]. One of the principal functions of a loop-seal is to avoid the undesirable inverse gas flow and to provide gas tightness. Formation of a moving solid bed in the supply chamber of the loop-seal provides the primary functions of the loop-seal in terms of pressure drop balance and gas leakage minimization. Pressure drop of the moving bed in the supply chamber of the loop-seal is developed as results of the relative solid and gas flow and is a function of the overall pressure balance.

Three types of pressure drops can be distinguished in this part of the system including: reactor pressure drop (DR) which is a function of solid inventory in the reactor, pressure drop in the horizontal pipe (HP), recycle chamber (RC), recycle pipe (RP) of the loop-seal, and cyclone which are function of solid and gas flow rate, and finally pressure drop in the supply chamber (SC) of the loop-seal which is a dependent variable, that varies to adjust the overall pressure balance in the circulation loop.

Calculation of the pressure drop in the horizontal pipe of the loop-seal is a key parameter which is characteristic of a loop-seal. Various correlations have been developed in the literature on the pressure drop in the loop-seals as listed by the reference [2]. These correlations were compared with the actual experimental results of 21 and 150 mm diameter loop-seals. Arena et al. [3] have correlated the pressure drop in horizontal section of an L-valve in terms of solid flow rate (W_s), solid bulk density (ρ_b), pipe diameter (D_{ls}), and particles diameter (d_p). This correlation resulted in the best prediction between the experimental results for all 4 solids and 2 tested loop-seals.

Impact of the first and the second loop-seal aeration (Q_{ls1} and Q_{ls2}) on the loop-seal operation are studied in a continuous solid circulation test for the 150 mm loop-seal. The input solid flow rate was maintained constant by aid of the L-valve. The solid flow rate in the current tests is calculated based on the pressure drop in the riser. A constant pressure drop in the riser is an indication of the constant solid flow rate. The variation of the second aeration (Q_{ls2}) shows no impact on the solid circulation. However, change of the first aeration (Q_{ls1}) has considerable influence on the stability of the loop-seal. Increase of the Q_{ls1} decreases the height of the solid column in the dipleg (ΔP_{SC}). Accordingly, more particles will be accumulated in the downstream reactor and its pressure drop increases.

The variation of the solid repartition in the loop-seal is related to the variation of the solid flow rate in the horizontal section of the loop-seal as well as the change of the gas flow rate in the dipleg (SC) which results in a unique solid height in the dipleg for each pressure and aeration condition in the loop-seal. The impacts of Q_{ls1} and Q_{ls2} are studied in details for the small loop-seal [4].

Solid flow rate in the current loop-seal is controlled in the horizontal section of the loop-seal similar to a L-valve [4]. The solid flow rate through the horizontal section can be directly linked to the gas velocity in the horizontal pipe U_{HP}. The gas flowrate in the horizontal pipe results of the sum of the first loop-seal aeration flowrate and the flowrate in the supply chamber, that can be calculated from the Ergun equation (knowing the linear pressure drop in the SC) [5]. The solid flux in the loop-seal is measured as a function of fluidization number in the horizontal pipe of the loop-seal (U_{HP}/U_{mf}) for 4 solids and 2 loop-seals. This work shows the linear dependency of the solid flux in the loop-seal to fluidization number as the principal solid flow controlling parameter in the loop-seal.

References :

[1] A. Johansson, F. Johnsson, and B. A. Andersson, "The Performance of a Loop Seal in a CFB Boiler," *Journal of Energy Resources Technology*, vol. 128, no. 2, pp. 135–142, 2006.

[2] M. M. Yazdanpanah, "Investigation of a Chemical Looping Combustion (CLC) Configuration with Gas Feed," Dissertation, IFP Energies nouvelles, Lyon, France, 2011.

[3] U. Arena, C. B. Langeli, and A. Cammarota, "L-valve behaviour with solids of different size and density," *Powder Technology*, vol. 98, no. 3, pp. 231–240, 1998.

[4] M. M. Yazdanpanah, A. Forret, T. Gauthier, and A. Delebarre, "An experimental investigation of loop-seal operation in an interconnected circulating fluidized bed system," *Powder Technology*, vol. 237, no. 0, pp. 266–275, 2013.

[5] P. Basu and J. Butler, "Studies on the operation of loop-seal in circulating fluidized bed boilers," *Applied Energy*, vol. 86, no. 9, pp. 1723–1731, 2009.

(602f) Using Complimentary Imaging Techniques: X-Ray and MRI Studies of a Bubbling Gas-Solid Fluidized Bed

Pore, M., University of Cambridge

Materazzi, M., University College London

Chandrasekera, T. C., University of Cambridge

Holland, D. J., University of Cambridge

Lettieri, P., University College London

Sederman, A. J., University of Cambridge

Dennis, J. S., University of Cambridge

Originally presented on: 11/7/2013 10:10:00 - 10:30:00

Introduction

Gas-solid fluidized systems are difficult to measure non-invasively, due to their optical opacity. Advances in instrumentation and computing power have lead to the rise of tomographic techniques, such as Magnetic Resonance Imaging (MRI) (Holland *et al.*, 2008), Electrical Capacitance Tomography (ECT) (Du *et al.*, 2005) and X-ray Computed Tomography (XCT) (Mudde, 2010), for observing fluidization phenomena in 3D systems. An ideal measurement technique would allow multiscale, non-invasive imaging to a high spatial and temporal resolution, and be suitable for on-line use in industrial applications. Since no single technique has all of these capabilities, it is necessary to use complimentary techniques. In order to use multiple measurement methods, the two techniques must first be cross-validated. The complimentary characteristics of the techniques can be used to gain more information about the system than would have been possible if each technique was used in isolation. This study aims to cross-validate MRI and high-speed X-ray imaging and investigate the use of these complimentary techniques for measurements in a bubbling gas-solid fluidised bed.

Experimental

Fluidization Set-Up: The bed was 50 mm I.D., provided with either a) a single orifice (orifice diameter, d_o, of 2 mm) or b) a 10-hole or 14-hole multi-hole (orifice diameter, d_o, of 1 mm) distributor. The distributors were designed to ensure that there were no jet-wall interactions. The particles were poppy seeds (0.5 mm diameter, 200 mm unfluidised bed height) fluidised with air at a superficial gas velocity of 60 - 300 mm/s, as measured at 293 K and 1 bar. The particles were classified as Group B using the Geldart particle groupings (Geldart, 1973).

MR Imaging: Experiments were conducted in a 7.1 T (300 MHz ^{1}H frequency) magnet equipped with a Bruker AV3 spectrometer. The actively shielded gradient system was capable of producing a maximum gradient strength of 77.7 G/cm. A birdcage radiofrequency (r.f.) coil (66 mm i.d.) was used to excite and detect the ^{1}H nuclei within the poppy seeds. The region above the distributor was imaged using 3D time-averaged and 2D ultra-fast imaging sequences.

X-ray Imaging: Experiments were conducted using the high power pulsed X-ray facility at University College London (capable of producing X-rays at 150kV and 450mA, with a pulse length of 0.2ms, and a frame capture rate of 72 fps). For this work, X-rays were generated with 38 kV for the single jet system and

41kV for the multiorifice system, and a current of 400 mA. The X-ray pulses were 500 μs long and images were acquired at 72 frames per second.

Results and Discussion

The two techniques (fast X-ray imaging and MRI) were cross-validated by measuring the length of a single jet formed at a 2mm orifice. Figure 1 shows the length of a single jet with increasing superficial gas velocity for X-ray and MRI measurements. Good agreement was seen between the jet lengths measured using the two techniques.

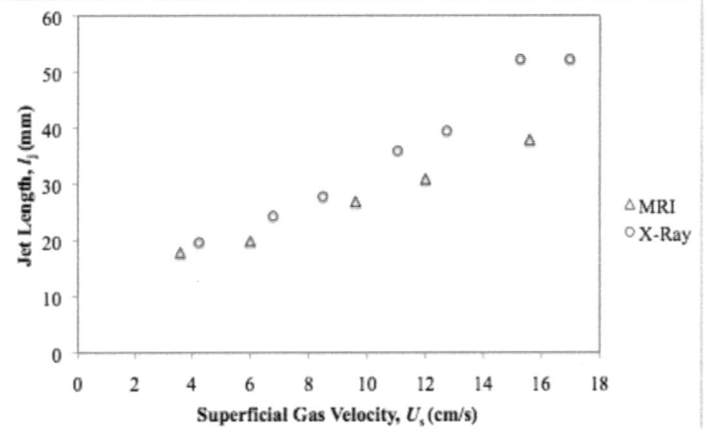

Figure 1 *Jet length as a function of superficial gas velocity for a single jet from a 2 mm orifice in a bed of Group B particles.*

Figure 2 shows consecutive images of the bubbling bed fitted with the 10-hole distributor acquired using X-ray and MRI. The bed was fluidised at a superficial gas flow rate of 0.21m/s. X-ray imaging is advantageous as it can image larger regions and therefore it is possible to image the gas entering as jets and the subsequent formation, coalescence and rising of bubbles in the bed. MRI cannot image as large a region, but it does allow slices of the bed to be imaged non-invasively, thereby providing more detailed information on the gas jets formed at the distributor and the bubble formation at the top of the jets. It is not possible to observe bubbles detaching from the tips of the jets in the X-ray images as the images are a projection through the bed and due to the small size of the bubbles formed, the contrast in the images is low in this region. By combining measurements from both X-ray and MRI measurements it is possible to resolve fluidization phenomena in both the grid region and higher in the bed.

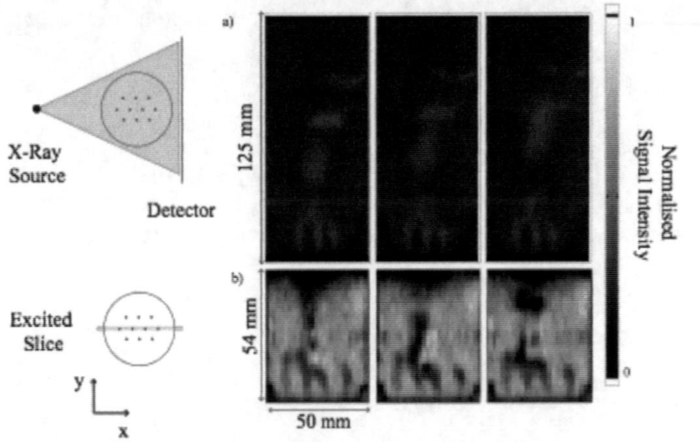

Figure 2 *Fast imaging of a bubbling bed fitted with a 10-hole distributor (superficial gas flow rate, U_S, of 0.21 m/s) using a) X-ray and b) MRI imaging. The separation time between frames was 13.8 and 25 ms respectively for the X-ray and MR images. The X-ray images are of the bed projected along the y-direction. The MR images are of the centre slice of the bed. The MR images show clearly the individual jets and bubble formations from the tips of the jets, which cannot be seen from the projected X-ray images. The gray-scale is inverted between the two sets of images as the particles absorb X-rays, but give an MR signal. Therefore dark regions in the X-ray images and white regions in the MR images indicate a high particle density.*

A time series of projected X-ray data is shown in Figure 3. Images were projected along the *y*-direction to more clearly identify the trajectory of bubbles. High signal intensity indicates the presence of a high voidage region i.e. a bubble or a jet. Bubble rise velocities, bubble frequencies and bubble coalescence (labelled in Figure 3 as a, b and c respectively) can be observed.

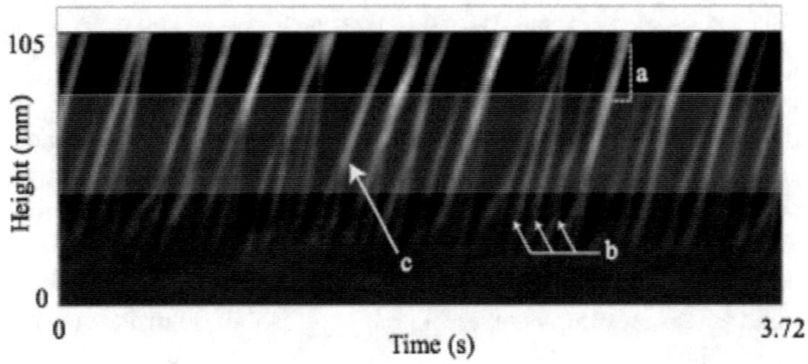

Figure 3: *Time series of X-ray data projected along the x-direction for a bed fitted with a 10-hole distributor and fluidized at a superficial gas flowrate of 0.21 m/s. Several features of a bubbling bed can be extracted: a) bubble rise velocity can be calculated using the gradient of the trajectory of the bubble, b) bubble frequencies can be calculated using Fourier Transform analysis and c) bubble coalescence events can be observed. White regions indicate trajectories of bubbles passing through the bed.*

Bubble frequencies can be calculated from X-ray data using a Fourier transform analysis of the projected time series. Figure 4 shows bubble frequencies as a function of height above the distributor for a 10-hole distributor with a superficial gas flow rate of 0.21 m/s. A bubble frequency of 9.2 Hz occurred at approximately 50 mm above the distributor with a smaller secondary peak at 11.8 Hz. The bubble frequency dropped to 3.5 Hz at 75 mm above the distributor with a smaller secondary peak at 7.0 Hz, indicating that either bubble coalescence or dispersion occurred in the region between 50 and 75 mm above the distributor, both of which can be observed visually in the shaded region in Figure 3. Below a height of 50 mm above the distributor, it is not possible to distinguish bubble frequencies due to the poor contrast of small bubbles in the X-ray images. However, a similar analysis can be conducted on the MR images to calculate bubble frequencies at lower heights and thereby obtain a full description of bubble frequencies in the bed using both techniques in conjunction. This paper will consider quantitative methods for calculating bubble rise velocities, bubble frequencies, bubble coalescence rates and permanent jet heights from MRI and X-ray data. The X-ray analysis will also be extended in a scale-up study of a 140 mm diameter bed.

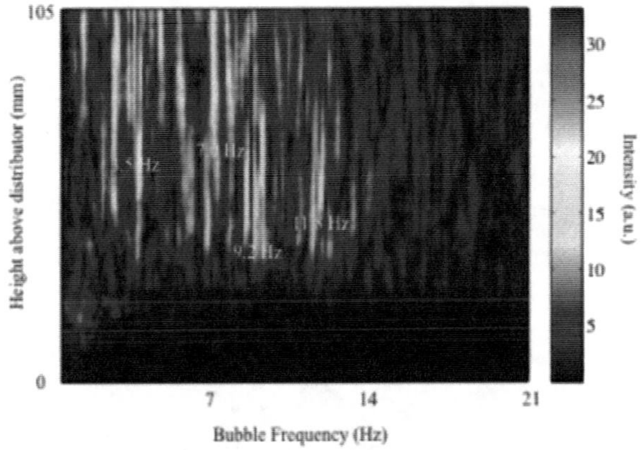

Figure 4 *Bubble frequency as a function of height above the distributor for a bed fitted with a 10-hole distributor and fluidised with a superficial gas flow rate of 0.21 m/s. Peaks are seen at 9.2 Hz at 50 mm (with a secondary peak at 11.8 Hz) and at 3.5 Hz a at 75 mm and higher (with a secondary peak at 7.0 Hz). Due to the poor contrast of small bubbles in X-ray images, it is not possible to resolve bubble frequencies at heights below 50 mm using X-ray data.*

Conclusions

This paper demonstrates the use of two complimentary imaging techniques in measurements of fluidized beds. MRI provides high spatial and temporal resolution data, but is limited in the height of the bed that can be measured. Conversely, fast X-ray imaging can image larger beds to a high-temporal resolution, but is limited in spatial resolution near the distributor. The two techniques can therefore be combined to observe a wider range of length scales than could be observed when using both techniques in isolation.

Acknowledgements

This work is funded by the EPSRC, Grant EP/F041772/1.

References

Du, B., Warsito, W., Fan, L.-S., 2005. ECT Studies of Gas-Solid Fluidized Beds of Different Diameters. Ind. Eng. Chem. Res. 44, 5020–5030.

Geldart, D., 1973. Types of gas fluidization. Powder Technol. 7, 285–292.

Holland, D., M?ller, C., Sederman, A., Mantle, M., Gladden, L., Davidson, J., 2008. Magnetic resonance imaging of fluidized beds: Recent advances. Theor. Found. Chem. Eng. 42, 469–478.

Mudde, R.F., 2010. Time-resolved X-ray tomography of a fluidized bed. Powder Technol. 199, 55–59.

(602g) Interactions Among Operating Conditions, Electrostatics and Particle Entrainment in Gas-Solid Fluidized Beds

Alsmari, T., British Columbia University
Grace, J. R., University of British Columbia
Bi, X., University of British Columbia
Jalalinejad, F., UBC
Originally presented on: 11/7/2013 10:30:00 - 10:50:00

Particle entrainment is a key parameter in gas-solids fluidized beds , but its prediction is extremely difficult, with empirical correlations differing by orders of magnitude, even for standard systems and operating conditions. One likely reason is that electrostatics are not considered in the entrainment experiments and in developing correlations. However, recent studies at UBC have shown that entrained particles carry a net charge which can influence particle motion in the freeboard and bed hydrodynamics, because removal of fine particles of one polarity results in particles in the dense bed being charged oppositely. This paper will present preliminary results of experiments in which we have measured simultaneously entrainment and charges on the entrained particles for binary particle systems and different operating conditions. These experiments are conducted in a three-dimensional pressured vessel of 150 mm inner diameter and 2 m height constructed of stainless steel, equipped with a distributor consisting of two stainless steel perforated plates. The degree of electrification in the bed is measured by four collision ball probes (stainless steel, 5.3 mm O.D.), located at different levels and directly connected to electrometers (Kistler Model 5010B Digital Electrometer). The charge density of entrained particles in the freeboard is measured by a sampling pipe constructed as a Faraday cup. An instrumented stainless pipe (ID=38.1 mm, L= 400 mm) coated with Ni is installed in the reactor exit line to measure the electrical current transferred from entrained fine particles to the metal pipe from collisions.

(624a) Carbon Nanomaterial Dense Layer Formation On Graphite Particles for Li-Ion Battery Anode By Using Colloidal and CVD Combined Process

Kamiya, H., Tokyo University of Agriculture and Technology
Nomura, Y., Tokyo University of Agriculture and Technology
Iijima, M., Tokyo University of Agriculture and Technology
Anoshkin, I. V.,
Nasibul, A. G.,
Kauppinen, E.,
Originally presented on: 11/7/2013 8:30:00 - 8:53:00

Carbon nanomaterials, such as carbon nanotube, CNT and carbon nano fiber, CNF, have various high functional properties to develop Li-ion battery performance. However, hard aggregates of CNT during CVD preparation process were formed in gas phase, and such aggregates hindered to prepare uniform and high performance electrode. Such hard aggregates were not able to be dispersed by conventional physical processes. In this research, new preparation process of carbon nanomaterials uniform and dense layer on the surface of graphite fine particles, which is Li ion battery anode raw material, are developed. Uniform carbon nanomaterials coated layer on graphite will develop the performance of Li battery anode material. Firstly, surface modified CoO nanoparticles, which are precursor of Co nanoparticle catalyst for CNT formation in CVD process, were prepared in liquid phase and adsorbed on graphite particles. By the selection of optimum surface organic molecular structure on CoO nanoparticles, high coverage of CoO nanoparticles on graphite was achieved in toluene. Surface interaction control between CoO nanoparticles and graphite was useful to increase the number density of adsorbed nanoparticles on graphite. After reduction of CoO nanoparticles on graphite in gas phase, carbon nano fiber dense layers were formed on the surface of graphite particles by CVD process. Since the amount of produced carbon nano fiber on graphite was depended on the number density of adsorbed CoO nanoparticles, colloidal surface interaction design between CoO nanoparticles and graphite was important for high performance raw powder materials.

(624b) Insertion and Immobilization of Pre-Synthesized Particles Into Sub-100 Nm Porous Structures By Electrophoresis

Kusdianto, K., Tokyo University of Agriculture and Technology
Sasaki, K.,
Lenggoro, W.,
Originally presented on: 11/7/2013 8:53:00 - 9:16:00

Pulse and direct current (DC) electrophoretic deposition of nanometer-sized particles (average sizes of 10 and 50 nm) into sub-100 nm scaled pore arrays made from anodized aluminum has been investigated. The deposited particles were observed when several parameters such as pH, applied voltage, and particle size were altered. At the applied voltages lower than the decomposition voltage of water (1 V), the number concentration of the particles deposited by conventional (constant) DC was higher than that of pulse DC. The number of deposited particles increased with increasing pH. The deposition efficiency inside the pores was enhanced by applying pulse DC. When the voltage as high as 10 V was applied however, no particles were observed inside the pores even by application of pulsed DC. The adhesion strength (removal behavior) of the deposition was evaluated by application of a particle-detachment system based on ultrasonic energy on the structure. The particles deposited inside the pores in the 1 V samples were not found to be detached in contrast to those on the surface.

(624c) Development Of a Novel Rotating Fluidized Bed For Fine Particle Processing

Watano, S., Osaka Prefecture University
Nakamura, H., Osaka Prefecture University
Originally presented on: 11/7/2013 9:16:00 - 9:39:00

A rotating fluidized bed (RFB) which imparts high centrifugal force to fluidizing particles by horizontally rotating the vessel is a promising technology for fine particles processing. We have already reported that the RFB can be used for uniform fluidization of nano particles, micro granulation/ fine particle coating, synthesis of nano particles and etc. In this study, new devices such as a tapered rotating vessel and an inclined injection of fluidizing air have been developed for the better performance of the RFB. The experimental results confirmed that the tapered rotating vessel improved the particle recovery and the inclined injection greatly improved the particle mixing performance. A numerical analysis using a DEM (Discrete Element Method) and CFD (Computational Fluid Dynamics) was also conducted to analyzed the mechanism how the tapered rotating vessel and the inclined injection improved the particle behaviors and performance. The optimum geometry of the devices was also determined by using the numerical analysis.

(624d) Numerical and Experimental Study of the Correlation Between Bulk Stresses and Interparticle Force in the Biomass Combustion Ash

Kamiya, H., Tokyo University of Agriculture and Technology
Tong, Z.,
Originally presented on: 11/7/2013 9:39:00 - 10:02:00

In recent years there has been high commercial interest in the energy production based on combustion of biomass according to future fuel dependency and in order to keep CO_2-emissions down. However, uncontrolled ash deposit build up on the furnace wall which is necessary to be cleaned at regular intervals lead to long periods without power production and large cleaning costs. The consequent lack of basic understanding of the deposition mechanism is a major hindrance to advance of biomass combustion technology. This paper presents measurements of tensile strength as a function of temperature (25-800°) for ash deposit of palm residues combusted at 750 and 820°. It has been shown that the tensile strength is strongly dependent on the temperature. It is interesting that the ash deposit combusted at 750° slightly increase with temperature while the ash deposit combusted at 820° first sharply increase and then sharply decrease with temperature. In order to fundamentally understand the relationship of macroscopic strength and microscopic adhesion, the Discrete Element Method (DEM) simulations were performed with different primary particle size distribution, bulk solid fraction and bond type (related to temperature). The solid bond and liquid bridge between the particles are modeled using the Bonded Particle Model (BPM) and Capillary Force Model (CFM), respectively. The quantitative information obtained from the numerical simulation can be used to better understand the biomass ash deposition mechanisms.

(624e) Hydrodynamic Behaviors of Sand and Plastic Particles in a Large-Scale Triple Bed Combined Circulating Fluidized Bed

Fushimi, C., Tokyo University of Agriculture and Technology
Ishizuka, M., The University of Tokyo
Guan, G., Hirosaki University
Suzuki, Y., National Institute of Advanced Industrial Science and Technology
Norinaga, K., Kyushu University
Hayashi, J. I., Kyushu University
Tsutsumi, A., The University of Tokyo
Originally presented on: 11/7/2013 10:02:00 - 10:25:00

Hydrodynamic behaviors of binary mixture of silica sand and nylonshot (coal char substitute) were investigated in a triple-bed circulating fluidized bed (TBCFB). The TBCFB consisted of a downer (Φ 0.1 m × 6.5 m), a bubbling fluidized bed (0.75 m × 0.27 m × 1.9 m), a riser (Φ 0.1 m × 16.6 m) and a gas-sealing bed (GSB, Φ 0.158 m × 5 m). The experiments were carried out at ambient temperature and atmospheric pressure by using air. The initial fraction of the nylonshot in the solid mixture ($X_{nylon,i}$) was 15.4% and 30% on mass and volume bases, respectively, or otherwise, 30.7% and 50%. The maximum obtained solids mass flux (G_s) at $X_{nylon,i}$ of 15.4 and 30.7wt% were 394 and 349 kg/m^2 s, respectively, when the gas velocity in the riser (U_{gr}) was 10 m/s. Apparent solids holdups of silica sand and nylonshot were calculated separately from the static pressure gradient across the riser and the downer. The results showed that it is possible to circulate char in the gasifier, which plays a significant role in decomposition of tar from pyrolysis as the primary step of gasification. A novel developed pressure balance model successfully predicted Gs of the binary mixtures in TBCFB well.

(624f) Microscopic Numerical Investigation of Dense Gas-Solid Flows Near a Solid Wall (Effects of wall on the behavior of particles)

Fujihara, S., Osaka University
Tsuji, T., Osaka University
Tanaka, T., Osaka University
Originally presented on: 11/7/2013 10:25:00 - 10:48:00

Flow including dense solid particles often appears in industrial equipments such as fluidized beds and spouted beds used for combustion, catalytic reaction, granulation and coating processes. To enable advanced designing and optimization of operating conditions, it is important to understand the flow behavior in the equipment in detail. However, its behavior is complicated especially in the near-wall region. A solid wall affects the flow directly by wall friction effect and the particle arrangement through the contacts between wall-particle and particle-particle. Due to the difficulty of experimental observations, the effect of wall is not known well at present. In the present study, the influence of solid wall on a flow field including dense moving particle is investigated in detail by means of microscopic direct numerical simulation. Immersed boundary method is combined with discrete element method. In order to investigate the influence of solid wall, sedimentation of dense particles near a vertical wall is simulated. To perform the computations with reduced computational cost, a new mirror boundary condition is proposed, too. It is confirmed that in near-wall region, the existence of the solid wall has significant effects both on particle and fluid flow structures. Moreover, sedimentation velocity of particles near solid wall is decreased and fluid resistance is increased due to the wall.

(624g) Drag Force Relation By Direct Numerical Simulation for Gas-Solid Flows

Zaidi, A., OSAKA UNIVERSITY
Tsuji, T., Osaka University
Tanaka, T., Osaka University
Originally presented on: 11/7/2013 10:48:00 - 11:11:00

Direct Numerical simulations are used to propose the drag correlation for random mono-disperse spheres which are applicable to high stokes number particles which can be used for the estimation of gas-solid flows. The simulated solid volume fraction varies from 0.05 to 0.5 and Reynolds number varies from 0.01 to 1000. To make sure that the current results are reliable, care is taken in the selection of parameters of simulation (e.g. grid resolution, domain size etc.). Average drag force acting on particles in both regular and random arrangements is compared with literature data and good agreement is observed. One particular exception was the simulation results of Beetstra et al. (2007) who used Lattice Boltzmann method which showed large deviation with the current simulations. We think that the grid used by Beetstra is not resolved enough for the range of parameters he studied. At the end we curve fitted our simulation results and proposed a new drag relation. We expect that this new relations can help to improve the numerical modeling of gas-solid flows in mesoscopic simulations.

(637b) Preparation of Monodispersed Chitosan-Coated Poly(lactic-co-glycolic acid) Particles Suitable for Targeted Drug Delivery Applications

Davoodi, P., National University of Singapore
Wang, C. H., National University of Singapore
Srinivasan, M. P., National University of Singapore
Originally presented on: 11/7/2013 8:54:00 - 9:14:00

In recent years, the number of people suffering from cancers of all types is increasing rapidly. While many studies have been conducted for the improvement of conventional therapeutic approaches, one of the major concerns with regard to conventional chemotherapy continues to be the adverse effects of chemotherapeutic drugs on normal cells. Surface functionalization of carriers with ligands with specificity towards particular cells is a promising route to reduce collateral damage. However, the challenge is to bestow reactive properties on surfaces of carriers such as PLGA under conditions that are not adverse to the material or the function and further, to produce surface modified carriers with a narrow size distribution as required for controlled drug delivery. To address this issue, we have investigated the use of chitosan as a surface modifying agent to facilitate binding of target molecules and ligands.

The objective of the present study was to prepare chitosan-coated PLGA microparticles with a narrow particle size distribution. Current techniques have limitations in terms of fabricating uniform size particles in a continuous manner. Moreover, batch-to-batch differences in the size of the particles produced make it difficult to scale up the process. In this work, we have employed electrohydrodynamic atomization together with emulsion-evaporation for the fabrication of PLGA particles concomitantly coated by chitosan to produce monodispersed particles via a semi-continuous process. The chitosan-coated particles were spherical and show smooth surfaces, as observed by scanning electron microscopy (SEM). The core-shell structure was confirmed using laser scanning confocal microscopy (LSCM). Chitosan solution concentration was optimized to get desirable surface charge determined by Zeta potential measurement. Surface protein adsorption and stability of the coated particles were measured for the optimization of surface charge density. Finally, folic acid was chosen as the target molecules and successfully conjugated to the surface of the coated particles.

Simplicity and flexibility of the developed technique will open a new road for the fabrication of monodispersed multi-layered microparticles for different goals.

(637c) Modified Glucan Particles for Drug Delivery and Magnetic Resonance Imaging

Salon, I., ICT Prague
Sarvasova, N., Institute of Chemical Technology
Stepanek, F., Institute of Chemical Technology, Prague
Originally presented on: 11/7/2013 9:14:00 - 9:34:00

Glucan particles are purified *Saccharomyces cerevisiae* (or baker's yeast) cell walls. They typically consist of 85 % 1,3-D-glucan polymers, 2 % chitin, 1 % protein and lipids, the rest being inorganic matter and moisture. They are hollow, porous 2–5 μm microspheres. The 1,3-β-glucan outer shell provides receptor-mediated uptake by phagocytic cells expressing β-glucan receptors. They can be used as a drug carrier for the macrophage drug delivery.

In the present work, glucan particles have been isolated from *Saccharomyces cerevisiae* and purified. Subsequently, the release kinetics of vitamin B12 as a model drug substance from lyophilized vitamin loaded glucan particles was measured. The experiments were made at different pH values (pH = 1; 3; 5; 7,4; 9; 12) and at different temperatures (20 °C, 37 °C, 60 °C) with UV-Vis spectrometry in quartz cuvettes with magnetic stirring. The pH- and temperature-dependent diffusion coefficients of vitamin B12 across the glucan wall were calculated from the experimental results.

The composite material (Glucan-Super Paramagnetic Iron Oxide Nanoparticles, G-SPIONs) for combined drug delivery and Magnetic Resonance Imaging (MRI) was then made. The composite was prepared from pure glucan particles and superparamagnetic iron oxides nanoparticles (SPIONs). Adsorption of SPIONs on the glucan particles was based on their opposite charge (measured by the zeta-potential) achieved by suitable surface modification. The MRI imaging of whole G-SPIONs was made under dynamic flow physiological conditions in a 3D capillary network medium positioned in the MRI device.

(637d) Inhalable Magnetic Nanocomposite Microparticles for Application in Lung Cancer Therapy - In Vitro Safety and Efficacy Studies

Stocke, N., University of Kentucky
Meenach, S. A., University of Rhode Island
Anderson, K. W., University of Kentucky
Arnold, S., University of Kentucky
Mansour, H. M., University of Kentucky
Hilt, D. J. Z., University of Kentucky
Originally presented on: 11/7/2013 9:34:00 - 9:54:00

Chemotherapy results in many adverse side effects when administered systemically through oral or intravenous delivery, and the localization of this therapy can result in improved therapeutic outcomes and enhanced patient response. When treating lung cancer, targeted pulmonary administration of chemotherapy has shown enhanced therapeutic response and reduced side effects [1, 2]. Additionally, dual administration of chemotherapy and hyperthermia can result in a synergistic effect, and the combination of hyperthermia with targeted pulmonary administration of chemotherapy poses a potential improvement on current cancer treatments [3]. In this study, the *in vitro* safety and efficacy of inhalable dry powders was examined on a variety of representative cell lines. These composite powders – so-called magnetic nanocomposite microparticles (MnMs) – were formulated via dilute organic suspension advanced spray drying in closed-mode and consisted of common excipients and anti-cancer drugs intelligently combined with magnetic nanoparticles. These powders were characterized with a variety of physicochemical techniques, and their aerodynamic performance was determined with the Next Generation Impactor (NGI). The resulting MnMs were tested *in vitro* with human alveolar adenocarcinoma (A549) and human bronchioalveolar carcinoma (H358) cells. These inhalable powders were dispersed into cell culture media and exposed to the cells for a given period of time and subsequent cytotoxicity studies and transepithelial electrical resistance (TEER) measurements were performed. Toxicity and TEER studies showed the *in vitro* safety of these materials up to relevant doses. The efficacy of two common lung cancer chemotherapeutic drugs representing two different anti-cancer drug classes, cisplatin and erlotinib, was examined before and after incorporation into these materials and, the results indicated these anti-cancer drugs retained their activity. Additionally, the activity of these drugs in combination with hyperthermia was examined, and results suggest the potential of these MnMs as a novel treatment modality for lung cancer.

References:

[1] G.A. Otterson, M.A. Villalona-Calero, S. Sharma, M.G. Kris, A. Imondi, M. Gerber, D.A. White, M.J. Ratain, J.H. Schiller, A. Sandler, M. Kraut, S. Mani, J.R. Murren, Phase I study of inhaled doxorubicin for patients with metastatic tumors to the lungs, Clinical Cancer Research, 13 (2007) 1246-1252.

[2] G.A. Otterson, M.A. Villalona-Calero, W. Hicks, X.L. Pan, J.A. Ellerton, S.N. Gettinger, J.R. Murren, Phase I/II Study of Inhaled Doxorubicin Combined with Platinum-Based Therapy for Advanced Non-Small Cell Lung Cancer, Clinical Cancer Research, 16 (2010) 2466-2473.

[3] R.D. Issels, Hyperthermia adds to chemotherapy, European Journal of Cancer, 44 (2008) 2546-2554.

(637e) Calcium Stearate Matrix Pellets: Impact of Drying On the Pellet Microstructure and the Ibuprofen Distribution Profiles

Schrank, S., Graz University of Technology
Kann, B., Department of Biopharmaceutics and Pharmaceutical Technology, Saarland University
Windbergs, M., Department of Biopharmaceutics and Pharmaceutical Technology, Saarland University
Glasser, B. J., Rutgers, The State University of New Jersey
Zimmer, A., University of Graz
Khinast, J. G., Graz University of Technology
Roblegg, E., University of Graz
Originally presented on: 11/7/2013 9:54:00 - 10:14:00

Although drying is one of the standard unit operations during manufacturing of solid pharmaceuticals, its potential to modify the final drug and/or dosage form properties is often underrated. Pellets, prepared by wet extrusion/spheronization, contain a certain amount of liquid and need to be dried in a final step. The impact of the drying process may be two-fold: First, the drying process may modify the pellet microstructure by controlling the extent and mechanism of shrinkage. Second, the drying process may alter the active pharmaceutical ingredient´s (API´s) physicochemical properties. APIs that dissolve in the granulation liquid may be subjected to intra-granular (i.e., within a pellet) migration as a function of the drying rate. Once a saturated solution is formed, the API re-crystallizes and may accumulate at certain regions within the pellet and/or undergo changes in its crystal structure.

The current study describes the impact of the drying process on calcium stearate (CaSt) matrix pellets containing ibuprofen as model API. These pellets were previously shown to be spherical in shape and mechanically stable [1]. In a first step the impact of drying on the microstructure was investigated. Therefore, low ibuprofen loadings were used to minimize any effects of dissolved ibuprofen particles on the pellet shrinking behavior. After spheronization pellets were either dried by i) tray drying at 20 °C, ii) fluid-bed drying (FB) at an inlet air temperature of 20 or 50 °C and iii) lyophilization.

We previously showed that CaSt swells in the granulation fluid, i.e., 50% ethanol and is prone to shrinkage during drying [2]. As such, the drying technique determined the pellet porosity, the pore size distribution and the specific surface area [3]. Furthermore, the inlet-air temperature during FB drying modified the pore size distribution and the specific surface area, while keeping the porosity constant. For lower temperatures the specific surface was increased and consequently, the median pore diameter was decreased. FB drying at 50 °C favored the collapse of smaller pores, whereas at 20 °C smaller pores shrunk but still remained intact after drying. Pore collapse was possibly attributed to softening of the material. As soon as the pellets contained dry regions where the temperature exceeded 50 °C, material softening occurred due to formation of a eutectic, which was proven by differential scanning calorimetry (DSC). The ibuprofen in-vitro dissolution rate was a non-linear function of the specific surface area, as the CaSt matrix remained intact and ibuprofen dissolution was diffusion controlled.

In a second step we evaluated the impact of drying on the physicochemical properties of the API. Therefore, pellets were tray dried at 20 and 50 °C. Ibuprofen was applied as dry powder, which was mixed with CaSt and subsequently wetted with ethanol. In a different approach ibuprofen was dissolved in the granulation liquid to ensure that all ibuprofen particles were dissolved and thus, prone to API migration and re-crystallization upon drying. Raman spectroscopy revealed that the final ibuprofen distribution throughout the pellets was inhomogeneous independent on the preparation procedure. Ibuprofen accumulated in regions close to/at the pellet surface. When ibuprofen was applied as solution and drying was performed at 20 °C, the pellet´s interior was entirely depleted from ibuprofen due to a comparatively long constant rate period that promoted transport of dissolved ibuprofen particles towards the surface. When ibuprofen was applied as powder and/or dried at 50 °C, certain amounts of ibuprofen were still found in the pellets′ interior. First, this implies that ibuprofen was not completely dissolved during wetting when applied as powder. Second, increasing the drying temperature to 50 °C elevated the rate of evaporation. Consequently, the constant rate period was shortened and ibuprofen transport towards the surface lowered. In addition, DSC measurements and infrared (IR) spectroscopy highlighted changes in the ibuprofen crystal structure after drying at 50 °C.

Overall, our results indicate that both, the drying technique and the drying conditions may cause severe changes in the final pellet properties and play thus, a crucial role during manufacturing of solid pharmaceutical dosage forms.

[1] E. Roblegg, S. Ulbing, S. Zeissmann, A. Zimmer, Eur. J. Pharm. Biopharm. 75 (2010), 56-62.

[2] S. Schrank, A. Hodzic, A. Zimmer, B. Glasser, J. Khinast, E. Roblegg, AAPS PharmSciTech, 13 (2012), 686-698.

[3] S. Schrank, B. Kann, M. Windbergs, A. Zimmer, J. Khinast, E. Roblegg, *submitted.*

(637f) Growth Phenomena During the Fine Grinding of Active Pharmaceutical Ingredients

Bitterlich, A.,
Komoss, C., TU Braunschweig
Dengler, M.,
Busmann, E.,
Bunjes, H., TU Braunschweig
Kwade, A., Institute for Particle Technology
Originally presented on: 11/7/2013 10:14:00 - 10:34:00

A large percentage of active pharmaceutical ingredients (APIs) being identified through combinatorial screening programs are poorly water soluble [1]. Therefore, improvement of dissolution has been a major development focus during the recent years. The reduction of the API´s particle size down to the nanometer scale using media mills or high pressure homogenizers is one approach to overcome the challenge of insufficient bioavailability [2]. By reducing the particle size, the dissolution rate of the API is increased significantly, according to the Noyes-Whitney/Nernst-Brunner equation which leads to an improved biopharmaceutical profile.

In this study the nanogrinding process of three active pharmaceutical ingredients, naproxen, fenofibrate and cinnarizine, was investigated using two different types of media mills: On the one hand a planetary ball mill (PM 400, Retsch GmbH, Germany) was modified according to Juhnke [3] to enable small scale screening of 12 samples of 1 ml volume, simultaneously. On the other hand, a stirred media mill (SL-12 C nano, VMA Getzmann GmbH, Germany) was applied for larger batch sizes with 600 ml volume.

As the fine grinding of particles increases surface free energy, which would lead to an agglomeration of the particles, stabilizers were required to obtain separated primary particles. Naproxen was stabilized using Kollidon® 30 (BASF), cinnarizine and fenofibrate using a combination of the cellulose Pharmacoat® 603 (Shin-Etsu) and dioctyl sulfosuccinate sodium salt (DOSS, Sigma-Aldrich). Process parameters were varied and the influence on product quality, particularly particle size and shape, was investigated.

Some interesting effects were identified for naproxen in the small scale planetary ball mill experiments. Depending on the chosen grinding media size and material, apparent growth of the API particle size was determined via dynamic light scattering (ZetaSizer nano, Malvern Instruments GmbH, Germany) after a certain grinding time. TEM micrographs were taken to identify the changes which occurred to the samples. When using 100 µm yttrium-stabilized zirconia beads, the product particle shape started to change from spherical to needle-like after about 120 minutes grinding. By increasing grinding time, the needles started to grow into the length direction. By further increasing the grinding time to 16 hours, the particle morphology further changed to platelet-like. However, crystallinity was preserved throughout the comminution process, which was determined via WAXD.

These effects were reproducible but occurred only with a specific process parameter set, regarding bead size, material and number of revolutions of the mill. Obviously (re-)crystallization occurred during the comminution process, depending on the energy transferred to the system. The increase in particle size using 300 μm alumina beads was revealed to have a different origin. On the TEM micrographs really small primary particles (< 100 nm) were observable, which agglomerated strongly and formed a gel network. In contrast to the (re-)crystallization, this effect occurred also in the larger scale inside the stirred media mill. Similar effects (agglomeration and (re-)crystallization, respectively) were observed also for cinnarizine and fenofibrate in the stirred media mill, which will be subject to the presentation.

Thus, during the grinding process three different effects could occur simultaneously: the real grinding of the API, agglomeration and (re-)crystallization.

1. Merisko-Liversidge, E., Liversidge, G.G., Adv. Drug Deliv. Rev. 63, 427-440 (2011)
2. Rabinow, B.E., Nature Reviews, 3, 785–796 (2004)
3. Juhnke, M., Berghausen, J., Timpe, C.: Chem. Eng. Technol. 2010, 33, 1412-1418.

(637g) Decoupling Solvent Penetration Mechanisms for Dissolution Mechanism Elucidation of Pharmaceutical Tablets

Callegari, G., Rutgers University
Gulak, Y., Rutgers, The State University of New Jersey
Cuitino, A., Rutgers University
Drazer, G., Rutgers University
Originally presented on: 11/7/2013 10:34:00 - 10:54:00

In recent years there has been a growing effort to understand the underlying dissolution mechanisms. Although most of the studies were on controlled released (CR) tablets, quality by design of immediate release (IR) tablets is gaining interest in the pharmaceutical industry and there is a need for a reliable methodology to characterize drug release.

The performance of IR tablets, is usually controlled by the dynamics of solvent penetration into the tablet through two competing mechanisms: capillarity and diffusion, inducing matrix swelling. We present dissolution experiments in which we monitor solvent infiltration dynamics for tablets produced at different compaction forces simultaneously with tablet deformation by measuring the mass of liquid uptake and through visualization techniques. We also present finite time dissolution experiments in which tablets are extracted after a certain time in contact with the solvent and thoroughly characterized. We finally present experiments in which, we were able to decouple capillary infiltration and swelling dynamics by using a low-viscosity, non-swelling solvent and also explore different solvent conditions to increase the characteristic swelling time.

(637h) Effect of Surface Modification On Flow and Packing Properties of Fine Powder Under Humid Environment

Karde, V., Indian Institute of Technology Gandhinagar
Panda, S.,
Ghoroi, C., Indian Institute of Technology Gandhinagar
Originally presented on: 11/7/2013 10:54:00 - 11:14:00

Moisture content of any of the formulation component can pose a considerable threat to the performance and quality of pharmaceutical product. Thus protection of pharmaceutical powders from humid or moist conditions has always been an area of prime importance in pharmaceutical industry. In the present study, surface modification through dry coating has been successfully exploited for the improvement of flow and packing properties of ~18 micron corn starch powders under different humid conditions (relative humidity 30% to 90%). Both uncoated and surface modified powders were characterized for flow and packing properties using tap density apparatus and FT4 powder rheometer. While flow and packing properties of uncoated starch powder deteriorated with increasing the relative humidity (RH), surface modified starch powder showed improvement in packing and flow properties as indicated by shear test, compressibility test and bulk density determination. It was found that moisture content of both uncoated and coated starch samples increases uniformly as a function of relative humidity. However, surface modified starch powder showed consistently improved property throughout the experimental range of humidity. The study demonstrates that surface modification can be useful to avoid detrimental effects of dominant interparticle forces like capillary forces at changing humidity conditions.

(646a) Combustion Dynamics of Mechanically Alloyed Al-Mg Powders in Turbulent Flames

Corcoran, A., New Jersey Institute of Technology
Aly, Y., New Jersey Institute of Technology
Dreizin, E. L., New Jersey Institute of Technology
Originally presented on: 11/7/2013 8:30:00 - 8:55:00

Metal powders are widely used as high-energy density additives to energetic formulations. Recent efforts focused on development of new metal-based materials, that are expected to ignite more readily and burn faster than pure metal powders, such as Al, used in many solid propellants, explosives, and pyrotechnics. Full scale tests of energetic systems with multiple new components are expensive and thus predictive models for particle ignition and combustion rates are required in order to establish whether the novel materials are compatible with the present system configurations. Most current models and descriptions are based on laboratory experiments performed in stationary or laminar combustion configurations and with relatively coarse powders of pure metals. Experimental data on various new materials, especially including fine powders burning in practically interesting turbulently mixing environments are lacking. In our recent work, an approach to characterize combustion of fine aluminum powders in turbulently mixing gas flows was discussed. In this paper, the approach is extended to characterize combustion of fine, mechanically alloyed Al·Mg powders. The powders with particle sizes roughly from 1 to 20 μm are prepared by ball-milling elemental aluminum and magnesium in a planetary mill. The Al/Mg ratios are varied to explore powders with different compositions. The experiments are aimed to measure burn times for the alloyed particles burning in environments with different levels of turbulence. For each material composition, the powders are injected using a nitrogen flow into an air-acetylene flame. Experiments are performed with both laminar flame and flames turbulized by auxiliary tangential jets of air. The air flows with adjustable flow rates are used to achieve different controlled levels of turbulence intensity. Optical emission of the burning particles was recorded using filtered photomultiplier tubes. Measured durations of individual particle emission pulses were assumed to represent their burn times; these data were classified into logarithmically spaced time bins and are correlated with the particle size distribution to obtain the burn time as a function of the particle size. It is observed that the effect of particle size on burn rate is stronger for the mechanically alloyed Al·Mg powders compared to that for pure aluminum. Experiments will be presented for powders of different compositions and burning in environments with different levels of turbulence. It is expected that these results will serve to develop a mechanistic model for burn rates of Al·Mg alloys in environments with varied turbulence intensities. The experimental methodology is expected to be useful for a broad range of reactive powders.

(646b) Formation of Reactive Composite Films

Doorenbos, Z., Innovative Materials and Processes, LLC
Leaming, S., South Dakota School of Mines & Technology
Haines, C., US Army RDECOM-ARDEC
Kapoor, D., US Army RDECOM-ARDEC
Puszynski, J., South Dakota School of Mines and Technology
Originally presented on: 11/7/2013 8:55:00 - 9:20:00

The use of metal/metal, metal/metal-oxide reactive or pyrophoric systems for defense, aerospace, commercial and electronic industries have been widely used for over fifty years. As new requirements were established more sophisticated methods for the formation of these reactive materials have been developed. To meet new requirements, specifically faster energy release, many of these technologies require the use of nanoscale materials. In the past decade, the use of nanoscale materials to tune the combustion characteristics of energetic materials has been investigated. The special attention was paid to pyrophoric and nanothermite systems.

This contribution will focus on the formation of thin pyrophoric metals and metal/metal oxide reactive substrates reinforced with carbon nanofibers, MWNTs and/or carbon veils. These reactive substrates are formed using tape casting or pressure filtration techniques. The combustion characteristics of such reactive substrates are determined using pyrometers along with high speed and thermal cameras. The effect of reactive substrate composition (i.e. metal, metal oxide, fiber and binder(s)) along with substrates thickness on the combustion characteristics will be presented. Impact, friction and ESD sensitivity thresholds will also be presented for the metal/metal oxide reactive substrates.

(646c) Mechanically-Alloyed Al-Mg Powders With Customized Particle Sizes and Compositions

Aly, Y., New Jersey Institute of Technology
Schoenitz, M., New Jersey Institute of Technology
Dreizin, E. L., New Jersey Institute of Technology
Originally presented on: 11/7/2013 9:20:00 - 9:45:00

Adding aluminum to propellants, pyrotechnics, and explosives is a common way to boost their energy density. A number of approaches have been investigated that shorten aluminum ignition delay, increase burn rate, and decrease the tendency of aluminum droplets to agglomerate. Previous work showed that particles of mechanically alloyed Al·Mg powders burn faster than similarly sized particles of pure aluminum. However, prepared mechanically alloyed powders were coarser than fine aluminum commonly used in energetic formulations. This work is focused on preparation of mechanically alloyed, composite Al·Mg powders in which both internal structures and particle size distributions are adjusted. Powders with 50 - 90 at. % Al were prepared and characterized. Milling protocol is optimized to prepare equiaxial, micron-scale particles suitable for laboratory evaluations of their oxidation, ignition, and combustion characteristics. Particle size distributions are measured using low-angle laser light scattering. Electron microscopy and x-ray diffraction are used to examine particle morphology and phase makeup, respectively. Combustion of aerosolized powder clouds is studied using a constant volume explosion apparatus. Combustion of individual particles is studied using a laser ignition setup. Oxidation and decomposition are studied using thermo-analytical measurements. For all materials, ignition and combustion characteristics are compared to one another and to those of pure Al. Additionally, ignition and oxidation characteristics are compared to determine which events lead to particle ignition. Compositions with improved performance (i.e., shorter ignition delays, higher burn rates, and faster pressurization rates) compared to pure Al are identified.

(646d) Selection of Pyrotechnic Formulations Based On Data Generated By Thermochemical and Thermodynamic Software Packages (INVITED)

Redner, P., US Army, RDECOM-ARDEC
Oyler, K., US Army RDECOM-ARDEC
Csernica, C., US Army, RDECOM-ARDEC
Mehta, N., US Army RDECOM-ARDEC
Puszynski, J., South Dakota School of Mines and Technology
Originally presented on: 11/7/2013 9:45:00 - 10:10:00

Thermodynamic and thermochemical analyses are a very important aspect of the selection phase for the development of new energetic formulations as they can greatly enhance trial and error explorations in the lab. There are several commercially available software packages, including: Cheetah, PEP Code, NASA_Lewis code, FCT, and HSC Chemistry, but each of them has a number of limitations regarding adequate predictability of the fundamental properties of complex reactive mixtures. This presentation will focus on a comparative study of several binary, ternary and quaternary pyrotechnic mixtures with and without gas generation.

The specific attention will be on condensed phase systems consisting of Al, B, and Si as fuels and oxides, iodates and nitrates as oxidizers, in addition to some other materials of interest. The presentation will summarize the pros and cons of each software package for its potential use for the prediction of thermodynamic properties of condensed phase reactive systems in an array of pyrotechnic applications.

(646f) Extension of the Transferable Potentials for Phase Equilibria to Nitroaromatics

Bhatnagar, N., Wayne State University
Kamath, G., University of Missouri-Columbia
Potoff, J. J., Wayne State University
Originally presented on: 11/7/2013 10:35:00 - 11:00:00

United-atom force fields, based on the Transferable Potentials for Phase Equilibria (TraPPE), are developed for twelve nitro-aromatic compounds, which include 2,4-dinitrotoluene (2,4-DNT), 2,6-dinitrotoluene (2,6-DNT), 3-nitrotoluene (3-NT), 4-nitrotoluene (4-NT), 1,3-dinitrobenzene (1,3-DNB), 1,4-dinitrobenzene (1,4-DNB), 2,4-dinitroanisole (DNAN), 1,3,5-trinitrobenzene (TNB), 2,4,6-trinitrotoluene (TNT), 2-nitroanisole (2-NAN), 4-nitroanisole (4-NAN) and n-methyl-p-nitroaniline (MNA). 1-Octanol-water and air-water partition coefficients are predicted for the optimized TraPPE-UA force field with adaptive biasing force molecular dynamics simulations, and compared to available experimental data. Log Kow values are predicted with an average absolute deviation of 0.2 log units, while Henry's law constants are predicted to with an average absolute deviation of 0.5 log units. Two additional models are presented for energetic materials with five membered rings for which no experimental data are available in the open literature: 3,5-dinitropyrazole (DNP) and 3-nitro-1,2,4-triazole-5-one (NTO). Investigation of the local microstructure around each solute reveals that 1-octanol is able to form hydrogen bonded chains around the solute, while little organized microstructure was observed around the solutes in water.

(661a) Video Analysis Techniques to Visualize Particle-Particle Mixing

Kingston, T. A., Iowa State University
Heindel, T. J., Iowa State University
Originally presented on: 11/7/2013 12:30:00 - 12:50:00

Particle-particle mixing has a significant influence on the yield and/or quality of the desired products in numerous industrial processes including energy generation, food processing, and pharmaceutical production. A fundamental problem commonly encountered during the mixing process is the tendency for mixtures to segregate due to differences in particle size, shape, and/or density. Many attempts to characterize the mixing effectiveness of granular applications rely solely on sampling the material after exiting the mixer and it alone does not provide insight into the mixing dynamics during the mixing process. In this study, an optically transparent double screw mixer was used to perform particle-particle mixing visualization studies consisting of a binary mixture of low density and high density granular materials. Four temporally and spatially synced video cameras, as well as optical access around the entire periphery of the double screw mixer, allowed for qualitative observations of the entire mixing process. The effect of four factors on the mixing dynamics was investigated: (i) screw rotation speed, (ii) dimensionless screw pitch, (iii) screw rotation orientation, and (iv) material injection configuration. Dynamic mixing videos and qualitative observations will be presented for a number of different operating conditions, emphasizing the effect these four factors have on the mixing process. The techniques developed in this study provide a thorough method to visualize particle-particle mixing processes and has implications in both batch and continuous systems including fluidized beds, rotating cylinders, screw pyrolyzers, and horizontal impeller mixers.

(661b) Multivariate Statistical Algorithm to Monitor and Control a Continuous Powder Mixing Process Using Online NIR Spectroscopy

Quinones, L., University of Puerto Rico - Mayaguez Campus

Originally presented on: 11/7/2013 12:50:00 - 13:10:00

NIR spectroscopic was used to get on-line spectra and obtain immediate results of API concentration at the output of a continuous powder mixer in order to assess the efficiency of the mixing. The results were compared with values obtained with ultraviolet spectrophotometer. Calibration curves were developed prior to continuous mixing experimentation for both UV and NIR methods. Considering that the mixing experimentation was performed between 0 and 100%w/w of Naproxen, it was created eight different Partial least squares (PLS) calibration models for NIR ranging from 0 to 100% instead of using only one calibration model so that a better prediction were obtained. The NIR precision and exactitude achieved were about 2-5% and 1-3% w/w respectively depending on the value of concentration studied. The calibration models were generated taking into consideration the variability of the process working at different RPMs of the mixer and at different feed rates of the feeders and considering as well the powders properties mainly the powder cohesion variation in the mentioned range of API concentration.During the process of mixing three different step changes were applied to the feeders to obtain three different API concentrations in the range 0-30% to characterize the dynamic response of the continuous mixer. The results obtained by UV and NIR showed stable mixing at the different API concentrations and achieved a reasonable mixing with values of RSD lower than 6%. The error obtained with the NIR and UV increases when increasing the API concentration and it was attained to the increase in powder cohesion which made more difficult the process of mixing. The parameters of the dynamic response were similar to those obtained theoretically.

(661c) Analysis of Powder Phenomena Inside a Fette 3090 Feed Frame Using NIR Spectroscopy
Mateo, D., University of Puerto Rico Mayaguez Campus
Originally presented on: 11/7/2013 13:10:00 - 13:30:00

Pharmaceutical Tablets are the most common solid dosage form due to its many advantages such as simplicity of manufacture, ease of dosing and stability. These tablets are manufactured using rotary tablet presses, where powder is filled into the dies with a feed frame system and then compressed with punches, decompressed and finally, the tablet is ejected. The die filling process is a complex and dynamic process and is also an essential part of controlling the final properties of tablets. Tablet characteristics are very important parameters in pharmaceutical industry. Fette 3090 tablet presses are commonly used in the pharmaceutical industry nowadays. A near-infrared (NIR) probe was successfully installed into a Fette 3090 feed frame to evaluate the die filling process in-line. The Fette 3090 mass exits contribution from the top stage to the bottom stage was measured at different paddle wheel speeds using pure lactose. It was found that the resultant amount of mass from the top stage of the feed frame vary significantly from the bottom left side to the bottom right side of the feed frame by changing the paddle wheel speed. Therefore, a second NIR probe was installed at each side of the feed frame to measure the effect of the shear applied and residence time distribution at each bottom side of the feed frame. A powder wave behavior inside the feed frame due to the paddle motion was also observed. It was found that at higher paddle wheel speed the frequency of the wave behavior increased. Differences in spectra base line are mainly affected by mass hold-up inside the feed frame and paddle wheel speed. These findings are important for the model construction of NIR inside a feed frame. Studies using blends were performed to develop a calibration model based on the feed frame system dynamic to determine acetaminophen potency variability during the die filling process. Experiments to evaluate content uniformity at the entrance and exit of the feed frame system were done and the output concentration of the feed frame versus the concentration at each side of the feed frame was evaluated.

(661d) Development of a Novel Microwave Sensor for Improved Process and Quality Control

Austin, J. S. III, Purdue University
Gupta, A., Purdue University
Reklaitis, G. V., Purdue University
Harris, M. T., Purdue University
Originally presented on: 11/7/2013 13:30:00 - 13:50:00

Process analytical technologies play a key role in many solids manufacturing industries. With the advent of the FDA's PAT initiative, analytical methods are becoming increasingly important in the pharmaceutical industry as well. Commonly, solids engineers turn to near infrared spectroscopic techniques to meet their PAT needs. Although NIR spectroscopy is a powerful sensing platform, it suffers from several significant drawbacks. Most notably, an NIR sensor can only monitor the material properties of the first few millimeters of a sample. This can be a substantial drawback in wet granulation and other unit operations where caking over the sensor can occur. In addition, NIR sensor calibrations do no transfer between formulations and are sensitive to many physical as well as chemical properties. Furthermore, NIR sensors require complex chemometric software and trained technicians to operate them. Microwave sensing promises a viable alternative that does not suffer from these significant drawbacks.

The introduction of microwave sensors to solids processing has mostly been driven by a desire to monitor either the moisture content or the density of a material. While on-line microwave sensors excel when used to monitor these properties, there has been little effort to expand their capabilities. In this study, a novel microwave sensor was designed to investigate the feasibility of monitoring other material properties, such as material composition. Furthermore, the design of this sensor allowed material properties to be monitored spatially across a particulate material.

(661e) UV-Vis Online Analysis and Modeling of Chlorination Process of PVC Particles Using a Gas-Solid Method With UV Intensification

Lu, W., Tsinghua University
Bai, L., Tsinghua University
Yang, Q.,
Yan, B., Tsinghua University
Cheng, Y., Tsinghua University
Originally presented on: 11/7/2013 13:50:00 - 14:10:00

Chlorinated poly (vinyl) chloride (CPVC) is a high performance thermoplastic, produced by further chlorination of PVC resin. As the chlorine content rises from 56.8 wt% to 63-71 wt%, CPVC products have better physicochemical properties of heat stability, flame retardant and so on in comparison with PVC. The aqueous suspension method has been successfully applied to manufacture CPVC using a batch operation. However, inevitable discharge of waste water and gases is encountered. From the viewpoint of environmental concern, gas-solid method is acknowledged as a much cleaner one, because dry chlorination process does not need water and is easy to handle CPVC products and effluent gases.

Dry chlorination of PVC particles is known as a gas-solid, non-catalytic reaction, which can be intensified by heat, UV light[1] and cold plasma[2]. A chlorination platform was established in our lab, including a vibrated-bed photoreactor and a UV-Vis online analysis system to quantitatively record instantaneous reaction rate from the very beginning to the end of chlorination process. Influences of operating parameters such as temperature, chlorine concentration and UV intensity can be illustrated vividly from the measured dynamic characteristics. Firstly, temperature and chlorine concentration are the two key factors determining the reaction rate and the chlorination degree. Secondly, UV radiation has significant intensification effect on the process, where both 253.7 and 350 nm UV lights work well to enhance the chlorination reaction. It can be deduced from the curves of chlorination rate that the chlorination process obeys a reaction-limited regime, but the intrinsic kinetics derived by radical quasi-steady state hypothesis of Cl· and -CH· in the former work[3] cannot be well-fitted. Theoretical analysis and characterization results by Pyrolysis GC-MS further demonstrated in different ways that the activation energy (E) in the rate constant (k) may not be constant as a result of steric hindrance in dense solid-state polymer phase. We therefore proposed a Steric-hindrance Active Energy ($E\sigma$) related to chlorination degree to correct Ein order to characterize its change during the chlorination. A theoretical model considering the reaction, diffusion and solution of chlorination in the polymer phase was accordingly established, showing good accordance between the model predictions and experimental results.

References:

[1] Weben JP, Cloud S, Assenat M, Fons S, Vrillon C, Superchlorinated polyvinyl chlorides and methods of preparation. 1966. US3532612.

[2] Lu W, Yang Q, Yan B, Cheng Y, Plasma-assisted synthesis of chlorinated polyvinyl chloride (CPVC) characterized by online UV–Vis analysis. *Chem. Eng. J.* 2012, 207-208, 923-930.

[3] Barriere B, Glotin M, Leibler L, An analysis of the reaction-diffusion mechanism governing the chlorination process of poly(vinyl chloride). *J. Polym. Sci., Part B: Polym. Phys.* 2000, 38, 3201-3209.

(661g) Preparation of Silica Particles By Using of Slug Flow

Cech, J., Institute of Chemical Technology, Prague

Originally presented on: 11/7/2013 14:30:00 - 14:50:00

In recent years there has been an increasing development in the field of microtechnology. The main reasons are (i) to provide well-defined reaction conditions, (ii) to save costly material and chemicals, (iii) to eliminate transport resistance and increase reaction rate, (iv) integration of several unit operation in a single microchip. One type of microfluidic devices are microsystems employing formation of regular segments of immiscible phases. Droplets of two immiscible phases form segments one and the other phase in microchannel. Droplets created in microfluidic devices are usually of uniform size and shape and can be used to form gel emulsions, colloidal particles, and complex geometric structures. Properties of produced capsules can be varied by the addition of micro-or nanoparticles in a dispersed phase.

Monodisperse silicon particles are prepared by EISA method (evaporation-induced self-assembly method). Size of the prepared particles is in the order of tens of micrometers (20-50 μm). Particle morphology is adjusted by mixing of reaction precursor with silicon particles with the diameter of about of 1 micron. The pore diameter of these microspheres is about 6 nm. In this way, we created a hybrid particles with a dual structure with characteristic dimensions 1 μm and 50 μm. With the use of slug flow, which produces droplets of the same size. We can influence the concentration of 1 μm microspheres in the resulting hybrid particles by changing their concentration in the precursor.

(672a) Effect of Particle Size Distributions in Modeling Calcium Fluoride Flash Drying

McDowell, K. W., E. I. du Pont de Nemours and Company, Inc.
Krakowski, S. F., E. I. du Pont de Nemours and Company, Inc.
Nopper, R. W. Jr., E. I. du Pont de Nemours and Company, Inc.
Originally presented on: 11/7/2013 12:30:00 - 13:00:00

Calcium fluoride is a widely used inorganic compound that occurs naturally and is produced synthetically as a precipitated by-product of various processes. The moisture content and drying behavior of calcium fluoride are important for its utility in the downstream production of other chemicals like hydrogen fluoride and in the various manufacturing operations in which it is used. For many processes, it is important that the moisture content and particle size distribution of the raw material be within an acceptable range; so, a drying unit operation is often used to condition the calcium fluoride feed material prior to its use. This work applies flowsheeting software to model the drying behavior of calcium fluoride wetcake in a cocurrent, pneumatic conveying "flash" dryer. The particle size distribution of the calcium fluoride feed is the key input parameter varied in the drying simulations while drying air velocity and relative humidity are held constant. The simulation results are also compared with thin-layer drying kinetics literature models fitted to experimental data generated at different drying temperatures (50°C, 80°C, 110°C) for calcium fluoride wetcake with an initial moisture content of 30–35%w.b. This work can be used to study the effect of feed material particle size distribution on size reductions that may occur as a result of drying in a pneumatic conveying operation.

(672b) Scale-Up From Lab to Production Spray Dryer Using Levitated Single Droplet Analysis (DKA) and Computation Fluid Dynamics (CFD)

Walter, M., GEA Process Engineering Inc.

Originally presented on: 11/7/2013 13:00:00 - 13:30:00

Scale-Up from Lab to Production Spray Dryer Using Levitated Single Droplet Analysis (DKA) and Computation Fluid Dynamics (CFD)

Since its recent introduction, the single droplet analyzer (DKA - Digital Kinetics Analyzer) and Computational Fluid Dynamics (CFD) has allowed optimization of spray dryer design.Â Prior to utilizing DKA combined with CFD, one would have to theorize how a powder would behave during drying under various conditions and design the spray dryer accordingly. Â Â New Computational Fluid Dynamics modeling power combined with a novel means to examine in detail how a single drop of feed dries has allowed spray dryers to be more efficient, safer, smaller, of higher capacity, and with fewer operational headaches than in the past.

Â

In addition to improvements in spray dryer design, DKA allows us to examine how individual particles dry and how different drying parameters affect the dried product.Â This allows the spray dryer to produce a ? custom particle? through adjustment of spray drying parameters determined by DKA.Â Â For example, some might want a strong particle not easily damaged during transport; others might require a lightweight powder that dissolves easily. Â Use of DKA can determine if a given feed will dry into hollow spheres, shriveled particles, ?doughnut? shapes, etc. while providing insight into design of drying equipment and process parameters for attaining a certain type of particle.

(672c) Use of Feed Conditioning to Address Operational Issues in Industrial Dryers

Jacob, K., The Dow Chemical Company

Originally presented on: 11/7/2013 13:30:00 - 14:00:00

Industrial dryers often suffer from plugging, rinding and doughy phases which prevent them from operating at maximum capacity and achieving specification moisture. The use of feed conditioning and dry product recycle can be a powerful technique to overcome all of these problems. This paper will focus on how to determine optimal levels of feed conditioning using shear testing (both internal cohesion and wall friction) in conjunction with Kofler plate techniques. Several full scale industrial case studies will be examined where these techniques are applied to solve operational issues.

(672d) Using a PBM to Describe the Drying Behaviour of Wet Pharmaceutical Granules: Scenario Analysis and GSA

Mortier, S. T. F. C., Ghent University
Nopens, I., Ghent University
De Beer, T., Ghent University
Gernaey, K. V., Technical University of Denmark (DTU)
Originally presented on: 11/7/2013 14:00:00 - 14:30:00

During the production of pharmaceutical tablets, a drying step is necessary when using wet granulation. Fluidized bed drying is a commonly used technique for this purpose. However, the drying technique and drying process settings may have a significant influence on the resulting properties of the granules and tablets [1]. In view of the trend towards continuous production, process knowledge is required and, as such, the necessity to develop mechanistic models is emerging. These models can assist in gaining process knowledge and, at a later stage be helpful for introducing process control based on on-line measurements and real-time adaptations of sensitive process variables. Indeed, mechanistic models include the impact of input variables on the process variables of interest, i.e. the output variables.

The case under study is a fluidized bed drying system as part of a full continuous from-powder-to-tablet manufacturing line (ConsiGma^{TM}, Collette^{TM}, GEA Pharma Systems, Wommelgem, Belgium). The system consists of three parts: a continuous twin screw granulator (high shear), a six-segmented fluidized bed dryer system and a discharge system. The drying system consists of six segments, which are fed consecutively with continuously produced wet granules, hence allowing continuous drying.

A model describing the drying behaviour of single pharmaceutical granules was developed earlier [2]. This model consists of two sub-models, one for the fast drying phase and one for the second and slow drying phase. The model was calibrated and validated using experimental data, after which a model reduction was performed resulting in an empirical reduced model [3]. This step was necessary in order to incorporate it in a Population Balance Model (PBM). The PBM predicts the evolution of the moisture content distribution as function of gas temperature and velocity. Degrees of freedom of the PBM are the particle radius, initial moisture content and filling time.

In this work we investigate the influence of the gas temperature and gas velocity on the distribution of the moisture content after drying. This distribution is of particular interest as it directly influences the quality of the final tablet product.

The Population Balance Equations (PBEs) are solved using the Method of Characteristics (MOC), which uses a moving grid. A grid size of 150 was used.

First, a scenario analysis was defined by using user-defined gas temperature and gas velocity input time series. These intended to mimick the dynamic local gas temperature and velocity the particle is exposed to when travelling along its path through the fluidized bed. Currently, these dynamic patterns are based on expert judgment, but in a later stage these should be derived from a Computational Fluid Dynamics (CFD) model. The objective here is to investigate the impact of fluctuating conditions on the characteristics of the distribution. It should be regarded as a proof of principle of using the calibrated and validated drying model under dynamic conditions of gas temperature and velocity.

A higher gas temperature increases the evaporation rate and as such the moisture content decreases faster. A constant gas temperature also benefits the drying rate. However, in both cases the standard deviation of the resulting distribution is larger compared to a standard case, meaning that some particles are considerably more dry compared to others which is undesired. An increase of 10°C for the gas temperature increases the standard deviation significantly over the entire time range. The increase of the standard deviation occurs during the filling period; when a high gas temperature is used, the moisture content will drop faster and the distribution will broaden more. A lower gas temperature slows down the fast drop in moisture content, and as such, the distribution will be narrower.

Focusing on the effect of the gas velocity, almost no difference can be detected in the mean and the standard deviation of the distribution of the moisture content when the gas velocity is varied. The gas velocity only has an influence on the first drying period, as the gas velocity is not involved in the equation for the second drying phase.

Second, a (global) sensitivity analysis (GSA) was performed. This entails the study of how the uncertainty in the output of a model (numerical or otherwise) can be apportioned to different sources of uncertainty in the model input, model structure or parameters [4]. A GSA provides useful information if one is to reduce the model output uncertainty as it points towards the source of the output uncertainty. Several tools to perform a GSA exist. Graphical GSA tools, such as Contribution to Sample Mean (CSM) and Contribution to Sample Variance (CSV) plots, are interesting since they only require a relatively low number of simulations.

Two-dimensional dotty plots are interesting to understand the influence of the input factors on the output. These plots showed that a small particle radius is important to allow a short drying time, and as such a low value for the mean of the distribution of the moisture content. Larger particle radii can be compensated for by a higher gas temperature at which the drying is performed. However, a low gas temperature governs a narrow distribution of the moisture content which clearly results in a trade-off between particle size and gas temperature. A trend can be detected for the combination of gas temperature and the filling time. When the filling time is large, a narrow distribution can be obtained by imposing a low gas temperature. In general a low value of the filling time is interesting to obtain a narrow distribution.

Based on the CSM the particle radius is the most influential parameter, followed by the gas temperature. The variance of the distribution is most sensitive towards the gas temperature, but also the filling time is influential on the variance.

Based on the GSA results, it could be concluded that the gas temperature is obviously the most sensitive parameter. The gas temperature is as such an important factor for process control and changing the process during operation. The gas temperature can easily be varied during operation based on real-time measurements. Particle radius is also important which can be fed back to the optimization of the granulation step. This clearly illustrates the necessity of an integrated approach when optimizing the coupled process.

[1] S.T.F.C. Mortier et al., Mechanistic modelling of fluidized bed drying processes of wet porous granules: A review, Eur. J. Pharm. Biopharm. 79 (2011) 205-225.

[2] S.T.F.C. Mortier et al., Mechanistic modelling of the drying behaviour of single pharmaceutical granules,

Eur. J. Pharm. Biopharm. 80 (3) (2012) 682-689.

[3] S.T.F.C. Mortier et al., Reduction of a single granule drying model: An essential step in preparation of a PBM with a continuous growth term, AIChE J. 59 (4) (2013) 1127-1138.

[4] A. Saltelli et al., Sensitivity Analysis in Practice: A Guide to Assessing Scientific Models (2004).

(674a) An Intensified Wet Media Milling Process for Faster Production of Dense, Stable Drug Suspensions With Sub-100 Nanometer Particles

Ravikumar, C.,
Afolabi, A., New Jersey Institute of Technology
Yaragudi, N. K. R., New Jersey Institute of Technology
Dave, R. N., New Jersey Institute of Technology
Bilgili, E., New Jersey Institute of Technology
Originally presented on: 11/7/2013 12:30:00 - 12:49:00

About 40% of drug substances identified through combinatorial screening programs in pharmaceutical industry are poorly water-soluble (Lipinski, 2002). Size reduction of drug crystals increases the specific surface area, which can improve the dissolution rate of drugs according to the Noyes–Whitney equation (Noyes et al., 1897). As a robust size reduction process, wet stirred media milling has found common use for the production of stable suspensions with drug nanoparticles, i.e., particles less than 1 μm (Bhakay et al., 2011; Bruno et al., 1996; Merisko-Liversidge et al., 2003). Typically, 100–500 nm particles were prepared by several hours of milling with relatively high energy consumption (Bose et al., 2012; Cerdeira et al., 2010; Knieke et al., 2013). In this study, we use a combined experimental and theoretical approach with the objective of achieving faster production of sub-100 nm BCS class II drug particles with reduced energy consumption. In the experiments, the effects of media (bead) size on the energy consumption and breakage kinetics of two model poorly water-soluble drugs, griseofulvin (GF) and naproxen (NPX) were studied. Then, this baseline process with the optimal bead size was intensified by increasing the stirrer tip speed, bead loading, and suspension flow rate systematically, as guided by a microhydrodynamic model of the process (Eskin et al., 2005). In all milling experiments, the suspensions were physically stabilized using hydroxypropyl methyl cellulose (neutral polymer) and sodium dodecyl sulfate (anionic surfactant), which were shown to have synergistic, electrosteric stabilization of fenofibrate, another poorly water-soluble drug (Knieke et al., 2013). Static light scattering, dynamic light scattering, scanning electron microscopy, and powder X-ray diffraction were used to characterize the milled drug particles.

Experimental investigation of the effects of bead size on GF particle size and mill energy consumption shows that smallest (100 micron) milling media (beads) studied with the baseline process were the most effective in producing sub-100 nm particles with lower energy consumption at the end of 6 h. Microhydrodynamic analysis of the bead–bead collisions using the model developed by Eskin et al. (2005) suggests that higher mill efficiency with the smaller beads is due to a significant increase in bead collision frequency and number of drug particle compression events despite a relatively small decrease in bead compression stress. A milling intensity factor, which is proportional to the energy dissipation rate during deformation of the drug particles (Afolabi et al., 2013), appears to capture the overall positive effects of the smaller beads on the observed faster breakage kinetics. To increase the breakage rate further, a higher milling intensity factor was targeted by increasing the mill tip speed, bead load, and feed flow rate, which is referred to as process intensification. The intensified process led to the production of sub-100 nm particles within 30 min. This resulted in a twelve times reduction in cycle time as compared to the baseline process with 6 h of milling, and a reduction in mill energy consumption up to 48%. Similarly, milling of NPX also led to sub-100 nm particles within 30 min. The use of small beads has enabled us to intensify the process, increase the breakage rate, and reduce the energy consumption without causing gross contamination of the drug product with worn-out beads. The intensified process methodology developed in this work presents an opportunity for formulators/engineers to design an intensified wet-milling process for production of sub-100 nm drug particles with reduced cycle time and acceptable media contamination and/or to optimize existing wet-milling processes.

References

Afolabi, A., Akinlabi, O., Bilgili, E., 2013. Impact of process parameters on the breakage kinetics of poorly water-soluble drugs during wet stirred media milling: a microhydrodynamic view, Eur. J. Pharm. Sci. (under review).

Bhakay, A., Merwade, M., Bilgili, E., Dave, R.N., 2011. Novel aspects of wet milling for the production of microsuspensions and nanosuspensions of poorly water soluble drugs. Drug Dev. Ind. Pharm. 37, 963–976.

Bose,S., Schenck,D., Ghosh, I., Hollywood,A., Maulit, E., Ruegger, C.,2012. Application of spray granulation for conversion of nanosuspensions in to dry powder form. Eur. J. Pharm. Sci. 47, 35-43.

Bruno, J.A., Doty, B.D., Gustow, E., Illig, K.J., Rajagopalan, N., Sarpotdar, P., 1996. Method of grinding pharmaceutical substances. US Pat. 5518187.

Cerdeira, A.M., Mazzotti, M., Gander, B., 2010. Miconazole nanosuspensions: Influence of formulation variables on particle size reduction and physical stability, Int. J. Pharm. 396, 210–218.

Eskin, D., Zhupanska, O., Hamey, R., Moudgil, B., Scarlett, B., 2005. Microhydrodynamic analysis of nanogrinding in stirred media mills. AlChE J. 51, 1346–1358.

Knieke, C., Azad, M.A., DavÃ©, R.N., Bilgili, E., 2013. A study of the physical stability of wet media-milled fenofibrate suspensions using dynamic equilibrium curves. CHERD, doi: 10.1016/j.cherd.2013.02.008

Lipinski, C.A., 2002. Poor aqueous solubility: an industry wide problem in drug discovery. American Pharm. Rev. 5, 82–85.

Merisko-Liversidge, E., Liversidge, G.G., Cooper, E.R., 2003. Nanosizing: a formulation approach for poorly-water soluble compounds. Eur. J. Pharm. Sci. 18, 113–120.

Noyes, A.A., Whitney, W.R., 1897. The rate of solution of solid substances in their own solutions. J. Am. Chem. Soc. 19, 930–934.

(674b) Particle Targeting Using Antigen-Antibody Interaction for Pharmaceutical Application

Tokarova, V., Institute of Chemical Technology, Prague
Pittermannova, A., Institute of Chemical Technology, Prague
Kral, V., Institute of Molecular Genetics
Stepanek, F., Institute of Chemical Technology, Prague
Originally presented on: 11/7/2013 12:49:00 - 13:08:00

This work is concerned with the surface modification of fluorescent silica nanoparticles by a monoclonal antibody IgG-M75 and the specific binding of such particles to surfaces coated by the PG domain of carbonic anhydrase IX (CA IX). CA IX is a trans-membrane protein specifically expressed in stomach but its overexpression is also associated with several types of carcinomas. The adhesion strength of antibody-bearing silica nanoparticles to antigen-bearing surfaces was investigated under laminar flow conditions in a microfluidic cell and compared to the adhesion of unmodified silica nanoparticles and nanoparticles coupled with a nonspecific antibody. Adhesion to HT-29 cancer cells (cell line derived from colorectal carcinoma) using flow cytometry was also investigated and compared with the cell line NIH 3T3 (sarcoma cell line) which does not displayed CA IX on its surface. The specific modification of nano- and microparticles by an antibody-like protein appears to be a feasible approach for the targeting of tumor cells.

(674c) Chitosan-Coated Core-Shell Microparticles for Synergistic Delivery of Vitamins and Anti-Cancer Drugs

Davoodi, P., National University of Singapore
Srinivasan, M. P., National University of Singapore
Wang, C. H., National University of Singapore
Originally presented on: 11/7/2013 13:08:00 - 13:27:00

As cancer takes its toll on human health and well-being, standard treatment techniques such as chemotherapy and radiotherapy often fall short of ideal solutions. In particular, adverse side effects due to excess dosage and collateral damage to healthy cells continue to pose challenges in cancer treatment. Controlled and targeted delivery of multiple drugs is required to address these issues, and one of the promising routes is the deployment of polymeric microparticles as encapsulants and carriers of the drugs.

In this work, multi-layered polymeric microparticles have been employed as encapsulants for a cocktail of remedial agents. In particular, we investigate co-delivery of ascorbic acid (vitamin C), AA, and two of the most commonly used anticancer drugs, paclitaxel (TAX) and doxorubicin (DOX), through the use of triple-layered microparticles. Microparticles of controlled particle size were fabricated through a combination of co-axial electrohydrodynamic atomization and emulsion-evaporation techniques. Poly ($_{D,L}$ lactide) and poly(lactic-co-glycolic acid), two biodegradable and biocompatible synthesized polymers, were used to encapsulate drugs within the core and inner-shell, respectively. Chitosan was used as the outer shell (third layer) to increase the hydrophilicity of the particles and to render the surface positively charged for better interaction with the negatively charged cell membrane. The triple-layered microparticles were characterized using SEM and laser scanning confocal microscopy. Zeta potential and FTIR experiments confirmed adsorption of chitosan at the surface of particles. The amount of protein adsorbed at the surface of the particles was measured using BCA protein Assey kit. Zeta potential measurements showed that the particles were stable in Phosphate buffered saline (PBS) for several days. In-vitro tests were conducted in order to optimize the release profiles and maximize therapeutic efficacy of the loaded agents.

The results demonstrated significant increase in the efficacy of the encapsulated drugs due to effective delivery of the drugs in an active form to cells.

(674d) Drug Polymer Constructs for Improved Powder Flow: Building Parametric Understanding and Manufacturability

Falco, N.,

Originally presented on: 11/7/2013 13:27:00 - 13:46:00

Drug substance solids often exhibit poor powder flow properties, which can pose handling or processing issues during pharmaceutical manufacturing. These issues are often approached by adding complexity in the form of additional excipients to modify flow or unit operations such as granulation or milling. A novel technique has been introduced for manufacturing constructs of active pharmaceutical ingredient (API) with low levels of polymer material to improve flow properties of drug substance solids. GSK APIs were selected for the study based on existing poor flow properties and opportunities to simplify complex formulation processes. Commercially-available polymer materials were selected based on similar crystallization solubility profiles to the API and potential for improving flow properties.

Initial studies at GSK had previously demonstrated the improved powder flow properties of drug polymer constructs and the viability of an antisolvent precipitation method of manufacture. In a reactor vessel typically used to prepare API, the final particle forming step was modified: low levels (less than 10% w/w) of polymer solution were added to the crystallization slurry; polymer precipitated onto the surface of the API; and small agglomerates of drug substance were formed. This change to the API crystallization has been demonstrated to deliver a drug substance which exhibits superior flow properties.

The investigation discussed extends earlier proof-of-concept work by building process knowledge, optimizing process parameters, and demonstrating scale independence. Parameters such as polymer addition rate, slurry agitation rate, and mixing during drying were studied. An analytical suite was developed to characterize and assess drug substance powders based on evidence of agglomeration, flow improvement, and changes in surface properties. Polarized light microscopy and in-situ crystallization techniques indicated agglomeration of the API upon addition of polymer. Flow analysis showed that API treated with polymer had increased density and flow properties. Surface characterization data further supported polymer deposition on API surface. Improved process understanding allowed for scale-up to a larger laboratory facility, demonstrating process robustness and manufacturability. Batches of drug-polymer construct have been successfully carried through to dosage form using a simplified manufacturing process.

(674e) Biodegradable Janus Particles for Drug Delivery: Bi-Compartmental Encapsulation of Two API of Disparate Solubility

Tomassone, M. S., Rutgers University
Winkler, J., Rutgers University
Garbuzenko, O., Rutgers University Ernest Mario School of Pharmacy
Minko, T., Rutgers, The State University of New Jersey
Originally presented on: 11/7/2013 13:46:00 - 14:05:00

Janus particles have unique chemical, optical, electrical, and physical properties that set them apart from their isotropic counterparts. Janus particles hold great promise as tiny colloidal carriers capable of meeting the demand for highly sophisticated drug delivery systems with multiple functionalities. The biphasic nature of Janus particles allows for temporally controlled release of multiple therapeutic and diagnostic agents in a specific order and with independent release rates. Biodegradable Janus particles were synthesized from a 2.5% w/v solution of PLGA/PCL (50/50 weight ratio) or PLGA/Precirol® (75/25 weight ratio) in dichloromethane (DCM) emulsified into an aqueous solution containing 0.3% w/v PVA and 0.1% w/v sodium dodecyl benzylsulfate (SDBS). Water-insoluble drugs are dissolved in the oil phase along with the polymers and lipids, while water-soluble drugs are adsorbed onto the particle surface post-synthesis. Emulsification and homogenization can be achieved using manual shear (low energy), a rotor stator homogenizer (moderate energy), a high pressure homogenizer (high energy), or a series of these. The higher the energy applied in the emulsification step, the smaller the emulsion droplet. After emulsification and homogenization, the solvent diffuses from the oil droplets into the aqueous phase and then evaporates out from the water/air interface. Solvent removal leads to precipitation of the immiscible polymer and/or lipid components into bicompartmental Janus particles. Biocompatibility and cell uptake of both polymer-polymer and polymer-lipid Janus particles have been studied *in vitro*. Fluorescent PLGA/Precirol® Janus nanoparticles were internalized by A549 human lung cancer cells after a 24 hour incubation period. The PLGA component was dyed with green-fluorescing FITC, while the Precirol® tail was dyed with red-fluorescing DiR. Superimposition of the images obtained from the green and red channels showed co-localization of the PLGA and Precirol® components inside the cells, indicating that the Janus particles remained intact during cell entry. Biodistribution of the PLGA/Precirol® Janus particles after inhalation was studied *in vivo* with mice. Two different sizes (155 nm and 450 nm) and methods of administration (inhalation and intravenous injection) were tested. As expected, it was found that the 450 nm particles administered via inhalation overwhelmingly localized to the lungs. The 155 nm particles delivered by inhalation were presumably exhaled after inspiration due to their small size. Both the 155 nm and 450 nm particles delivered intravenously and the 155 nm particles delivered through inhalation predominantly localized in the liver with minimal deposition in the lungs. To demonstrate co-encapsulation of two drugs with widely disparate solubility, PLGA/Precirol® Janus nanoparticles containing doxorubicin and curcumin were prepared for the treatment of lung cancer. Curcumin is a hydrophobic polyphenol with antioxidant and antitumor properties, and doxorubicin is a hydrophilic chemotherapeutic. These particles encapsulated curcumin in the polymer phase and doxorubicin in the lipid phase. Therapeutic efficacy of the PLGA/Precirol® Janus nanoparticles containing curcumin and doxorubicin was also fully demonstrated *in vitro* using A549 human lung cancer cells.

(674f) Stabilization of Solid Lipid Nanoparticle Suspensions

Yang, Y., University of Massachusetts Amherst
Corona, A. III, Procter & Gamble - Household Care
Schubert, B., Procter & Gamble - Household
Reeder, R., Procter & Gamble - Household Care Fabric and Home Care Innovation Center
Henson, M. A., University of Massachusetts Amherst
Originally presented on: 11/7/2013 14:05:00 - 14:24:00

Solid Lipid Nanoparticles (SLNs) have great potential as delivery systems for the encapsulation, protection and release of active lipophilic compounds (*e.g.,* drugs, nutraceuticals, antimicrobials, antioxidants and vitamins) in the pharmaceutical, agrochemical, food and personal care industries. SLNs are commonly prepared by making a hot oil-in-water nanoemulsion using high pressure homogenization followed by controlled cooling so that the fluid lipid droplets crystallize and produce solid nanoparticles ($r < 100$ nm). Compared to other colloidal delivery systems, SLNs offer several advantages for encapsulation of active components including improved physical stability, protection to chemical degradation and precise control over release rates. A major obstacle to the widespread industrial use of SLNs is their tendency to aggregate and form gels when stored at ambient temperatures. A widely accepted theory is that aggregation is driven by the lipid phase undergoing a polymorphic transformation from the thermodynamically unstable α-form to the stable β-form. An increase in surface area occurs as a spherical α-particle is transformed into a platelet-like or needle-like β-particle, causing a decrease in surfactant coverage on the particle surface. The newly formed surfaces contain lipophilic patches that result in hydrophobic attraction between particles.

Along with several other research groups, we have shown that SLN aggregation can be substantially reduced or even eliminated by mixing specific low-melting point oils with the molten lipid prior to homogenization. While the stabilization mechanism is not well understood, one theory is that the presence of the oil inhibits the large shape change that occurs for pure solidified lipid crystals, thereby reducing hydrophobic attraction between particles. Another theory is that shape change occurs but the oil coats the SLN surface and allows free movement of surfactant at the newly established oil-water interface to stabilize the particles. Currently, the design of stable SLNs must be performed by trial-and-error experimentation to identify a suitable oil, solid lipid and surfactant combination. To facilitate widespread industrial acceptance of SLNs, improved understanding of the stabilization mechanism and the development of systematic methods for designing stable systems are needed.

In this presentation, we report on the effects of different oils on the aggregation stability of certain lipids stabilized with nonionic surfactant in water. Aggregation dynamics were monitored by measuring particle size distributions with static light scattering. Polymorphic transformation behavior was examined with differential scanning calorimetry. We found that the oil type and concentration strongly affected SLN aggregation behavior. In the absence of oil, SLNs quickly aggregated and formed a gel. Increasing oil concentrations yielded increasingly stable dispersions while reducing SLN crystallization and melting temperatures. Furthermore, increasing oil concentrations increased the SLN polymorphic transformation rate such that negligible α-particle content was observed at oil concentrations that yielded stable dispersions. Cryogenic TEM images showed that increasing oil concentrations produced more spherical particles. Taken together, these results suggest that oil trapped within the growing SLN crystal matrix accelerates the polymorphic transformation but retards the large shape change normally associated with the transformation.

(674g) Dry (Solvent-less) Polymer Coating of Microparticles for Control Release Applications

Capece, M., New Jersey Institute of Technology
Dave, R. N., New Jersey Institute of Technology
Originally presented on: 11/7/2013 14:24:00 - 14:43:00

Coating is a necessary unit operation in the pharmaceutical industry for applications ranging from aesthetic improvement (e.g taste, appearance, texture) to the controlled or targeted release of active pharmaceutical ingredients (API) which is critical for the production of efficacious dosage forms. While solvent-based polymer coating methods are widely used, solvent-less or dry-polymer coating technologies have recently emerged as an attractive alternative to improve upon processing limitations and inefficiencies. This study introduces a solvent-less coating methodology by which microparticles can be coated with polymers without the use solvents, plasticizers, or heat treatments, noted as being major improvements upon solvent-based and current dry-polymer coating technologies. The method first creates an ordered mixture composed of a micronized polymer adhered to the surface of an API particle by virtue of van der Waals particle interactions resulting in a discrete polymer coating. Subsequent processing in a high intensity vibratory mixer induces mechanical impactions which deforms the discrete polymer into a continuous coating layer. As a practical example, ascorbic acid (50 – 500 μm) is coated with the polymer polyethylene wax (6.7 μm). Polymer coatings were able to prolong the dissolution time of ascorbic acid from seconds to hours depending on the coating thickness and processing conditions. Detailed characterization and modeling of the coating layer and the release mechanism was also performed. This novel dry-polymer-coating technique can reasonably be extended to a wide variety of applications in the pharmaceutical industry for control release applications.

(674h) Phase Behavior and Aggregation Stability of Nano- and Micro-Particles Interacting With Gelling Polymeric Matrices

Takhistov, P., Rutgers, The State University of New Jersey
Huynh, P., Rutgers University
Originally presented on: 11/7/2013 14:43:00 - 15:02:00

Development of effective delivery system for API requires a control of the structure and morphology of polymer matrix at the interface with API. The structure and morphology of such composites depend on many factors such as the method of formation, type of critical solution temperature, remoteness from the critical temperature, on the tendency of the film components to phase separation at the interface.

The introduction of nano- and micro-scale particles into polymer blends creates a large internal surface between the solid particles and the polymer, changing substantially the thermodynamic and mechanical properties of the polymer matrix Adsorption of polymers from solutions of their mixtures on solids plays an important role in the formation of API composite structures with evenly distributed i.e. non-aggregated particulate materials.

We present a comprehensive analysis of various types of aggregative behavior of API particulates embedded within the polymeric matrices. Examples include BCS Class II drugs interacting with cellulose ethers. The effect of the particles on the phase separation in forming polymeric networks may be also related to the selectivity of the component adsorption at the interface with a solid phase, leading to the redistribution of the system components between the bulk and the surface layers. We also present experimental data and theoretical models that predict transport properties and stability of API.

(688a) Aluminum Agglomeration Reduction in Composite Propellants Using Nanostructured Al/Hydrocarbon Particles

Sippel, T. R., Purdue University
Zhang, S., New Jersey Institute of Technology
Dreizin, E., New Jersey Institute of Technology
Son, S. F., Purdue University
Groven, L. J., Purdue University
Originally presented on: 11/7/2013 12:30:00 - 12:55:00

In recent years there has been significant interest in using nanoscale aluminum to alleviate product agglomeration in composite solid propellants. However, nAl still produces agglomerates several orders of magnitude larger than the nanoparticles and can result in unsatisfactory propellant strength, mixture viscosity, and poor aging characteristics. To alleviate some of these issues, micron sized aluminum (Al)/poly(tetrafluoroethylene) (PTFE) composite particles produced via high intensity milling have recently been considered. These particles consist of low levels of PTFE inclusion (10-30 wt.%), which result in altered ignition and agglomeration behavior when used in a composite propellant. However, their use can result in more toxic propellant combustion products (HF) and production of particles must be done carefully to prevent hazardous milling induced reactions.

Hydrocarbon inclusion in aluminum has also been shown to alter aluminum ignition behavior but has yet to be investigated in propellants. Unlike PTFE, the inclusion of hydrocarbons (e.g. low density polyethylene, LDPE) into aluminum through high intensity milling is safer as the mixture is less susceptible to milling induced reactions. Additionally, composite Al/hydrocarbon particles should not increase the toxicity of propellant combustion products.

In this work, 10 wt.% LDPE inclusion modified aluminum particles are used in a composite solid propellant to enhance aluminum combustion and reduce agglomeration. To this end, the effect of particle production method (cryomilled vs. high intensity milled) and resulting differences in particle shape, size, and morphology on propellant characteristics (density and viscosity), theoretical propellant performance, burning rate, metal particle ignition, agglomeration, and metal combustion efficiency are addressed.

(688b) Direct Observation of Effective Atomic Diffusion Distances In Zr/2Al Multilayers Due To Self-Propagating Reactions

Reeves, R. V., Sandia National Laboratories
Kotula, P. G.,
Adams, D. P., Sandia National Laboratories
Originally presented on: 11/7/2013 12:55:00 - 13:20:00

It is generally assumed that the diffusion distance for heterogeneous reactive materials is defined simply as half of the characteristic dimension of the constituent materials and that the heat release associated with reaction is finished when mass diffusion is complete. In a powder system, this assumption results in the diffusion distance being equivalent to the particle radius, and, in a lamellar system, the diffusion distance would be half of the thickness of a single layer of material. This assumption is important as it is used to estimate overall reaction times and heat release rates. In this study, actual diffusions distances of self-propagating synthesis reactions were investigated in sputter-deposited multilayers of Zr/2Al composition with a single Zr layer replaced by a Hf marker layer. Hafnium was selected as the marker material because Zr and Hf have very similar chemical behavior. This is due to lanthanide contraction, which causes Hf and Zr to have very similar electronegativities and atomic radii. Hf and Zr also react with Al to form intermetallic compounds with similar structures, including di-aluminide line compounds. These similarities between Zr and Hf allow the overall reaction to progress with little detrimental effect from the Hf inclusion. Multilayers containing a single marker layer were converted to product phases (nominally $ZrAl_2$/$HfAl_2$) in the self-propagating reaction mode. Both unreacted and reacted foils were then cross-sectioned and imaged in an aberration-corrected transmission electron microscope. Energy dispersive spectroscopy (EDS) was also performed to quantitatively identify the elemental distribution in both reacted and unreacted multilayer. EDS results clearly showed that the Hf atoms spread to over ten times the assumed diffusion distance. The results are disparate to the general assumption of diffusion distance and indicate that the atoms can remain mobile, even after complete mixing has occurred. A time scale of heat release as related to diffusion distance is also discussed.

Sandia National Laboratories is a multi-program laboratory managed and operated by Sandia Corporation, a wholly owned subsidiary of Lockheed Martin Corporation, for the U.S. Department of Energy's National Nuclear Security Administration under contract DE-AC04-94AL85000.

(688c) Effects of Various Solvents On Thermites

Vredenburgh, J.,
Meeks, K.,
Pantoya, M., Texas Tech University
Originally presented on: 11/7/2013 13:20:00 - 13:45:00

Developing thin film energetic materials using blade casting techniques has recently received attention for its advantages in deposition efficiency. This technique requires powdered thermite reactants to be combined with a solvent. Introducing solvents into the deposition processes prompts new questions about the influence of the solvent on combustion behavior and specifically flame speed. The objective of this study was to examine the effects of hexane, isopropanol and acetone on magnesium (Mg) + manganese oxide (MnO_2) and aluminum (Al) + MnO_2 on flame speeds. The powders were loaded into tubes and ignited by a hot wire. The flame speeds analyzed by a high-speed diagnostics show that the solvent used during mixing has an adverse effect on the flame speed of the powders. Results suggest that when the solvent evaporates, its residue remains on the particles and limits flame speed. The powders were baked for twenty-four hours at a temperature of 70°C to release solvent residue from the particles. The flame speeds of the baked powders were shown to increase significantly and powders mixed in isopropanol exhibited the highest increase.

(688d) Ignition of Fully-Dense Nanocomposite Thermite Powders By An Electric Spark

Williams, R., New Jersey Institute of Technology
Patel, J.,
Dreizin, E. L., New Jersey Institute of Technology
Originally presented on: 11/7/2013 13:45:00 - 14:10:00

Electro-Static Discharge (ESD) ignition of nanocomposite thermite powders prepared by Arrested Reactive Milling was investigated. Powders placed in a sample holder were subjected to an ESD produced by a discharging capacitor. The optical emission produced by the ignited powders was monitored using several photomultipliers equipped with interference bandpass filters. Time-resolved current and voltage traces of the ESD were also measured. Two ignition modes were detected: ignition of individual particles occurring during or immediately after the ESD, and powder ignition resulting in the combustion of a cloud of aerosolized particles, which was observed to begin after a substantial delay following the ESD pulse. Two materials studied in this work were nanocomposite powders $2Al{\cdot}3CuO$ and $8Al{\cdot}MoO_3$. For each thermite, the bulk composition remained fixed, while the mixing morphology was varied by preparing samples using different milling times. Experiments addressed the effects of material composition and morphology, energy input, powder layer thickness, and environment on the ignition mode observed and ignition delay for the case of powder ignition. In addition, effect of sample aging on their ignition behavior was considered. Finally, prepared thermites were blended with pure metal powders and ignition of the prepared blends was studied as well. When samples were placed on a conductive surface as a powder monolayer, ESD resulted in ignition of individual particles. Individual particle ignition was also observed in vacuum, for different powder layer thicknesses. For experiments in air and in argon, when the powder thickness layer varied from about 50 to 500 μm, a delayed powder ignition mode was observed. The ignition delays varied from 10 μs to several ms for different experimental conditions. For the cases of powder ignition, it was observed that $2Al{\cdot}3CuO$ powders had shorter ignition delays compared to $8Al{\cdot}MoO_3$. For both materials, ignition delays decreased for the powders prepared using longer milling times, in which the mixing between aluminum and oxide was achieved at a finer scale. The effect was especially pronounced for $8Al{\cdot}MoO_3$. For both materials, aged powders ignited after longer delays compared to the freshly prepared powders. Variation in the ESD energy and layer thickness (except for a monolayer) did not have a significant effect on the ignition behavior. Blending nanocomposite thermites with pure Al and Ti powders reduced their ESD ignition sensitivity significantly. When ignition was achieved, longer ignition delays were observed for the blended samples. Results of these experiments are expected to be useful in developing a model describing ESD-induced ignition in nanocomposite thermite powders and accounting for transfer of ESD energy to the powder as well as initiation of exothermic chemical reactions leading to ignition.

(688e) Heat Transfer Measurements and Modeling in Nano-Aluminum Combustion

Glumac, N., University of Illinois, Urbana-Champaign
Allen, D.,
Krier, H.,
Originally presented on: 11/7/2013 14:10:00 - 14:35:00

The mechanism of nano-aluminum combustion remains a topic of active debate. Classical combustion theory suggests that as the particle diameter approaches the nanoscale, gas-phase diffusion of oxidizer to the combustion front will no longer be the limiting process. Surface reaction or diffusion through a growing oxide barrier have been proposed as possible limiting processes. If a surface process limits the combustion rate, then it is often assumed that the concentration of the gas-phase oxidizer species at the particle surface rises to the freestream value. With the removal of the concentration gradient, the Reynolds analogy suggests that the thermal gradient will also be negligible, due to extremely rapid heat loss from the particles. Under these conditions, the particle temperature is not expected to rise above the ambient temperature significantly.

However, recent measurements by our group and others suggest that the particles do indeed exceed the ambient temperature under some conditions. We find that the degree of overshoot of the temperature above ambient is a sensitive indicator of the thermochemical processes in the vicinity of the particle. By monitoring both the particle temperature with pyrometery and the burning time with photometry in our heterogeneous shock tube, we generate observables that are very sensitive to the nature of the nano-aluminum combustion model.

For this work, we investigate particles of 18 to 120 nm in size at ambient temperatures of 1500 and 2000 K, in oxygen, carbon dioxide, and water vapor oxidizers. A very simple spherically symmetric model is used to interpret the results. The model assumes that some surface process limits the combustion rate, and uses a fully non-continuum approach to model the heat transfer by conduction and radiation. The two adjustable parameters in the model are a time-averaged oxidizer reaction probability and a time-averaged energy accommodation coefficient (EAC). The former constant directly gives the burning time, and the combination of the two constants provides the combustion temperature as a function of time.

The primary conclusion from this study is that temperature and burning times can only be consistent if the EAC of the particle at elevated temperature is of the order of 0.005, which is at least two orders of magnitude smaller than most studies assume. Values of the EAC in this range for burning metal particles were predicted by Igor Altman, and this work strongly supports his hypothesis. Time averaged sticking probabilities are found to be of the order of 0.001 to 0.002.

(688f) Influence of Organic Acid Coatings On Energetic Nanocomposite Combustion Performance

Kappagantula, K., Texas Tech University
Farley, C., Texas Tech University
Pantoya, M., Texas Tech University
Horn, J., Naval Surface Warfare Center
Originally presented on: 11/7/2013 14:35:00 - 15:00:00

Aluminum fuel particles, with and without self-assembled monolayers of two different organic acids: perfluoro tetradecanoic acid (PFTD) and perfluoro sebacic acid (PFS) were combined with molybdenum trioxide (MoO_3) to obtain three different energetic nanocomposites: Al-PFTD/MoO_3, Al-PFS/MoO_3 and Al/MoO_3. Combustion analysis using flame propagation tube and thermoequilibrium tests using DSC/TGA were performed for these nanocomposites. Flame speeds of Al-PFTD/MoO_3 was 86% higher than Al/MoO_3 whereas that of Al-PFS/MoO_3, almost half of Al/MoO_3. Results from the DSC showed that the apparent activation energy of the nanocomposites had an inverse relationship with the flame propagation velocity. This went on to show that the combustion behavior of nanocomposites could be influenced by coating them with organic acids. The Al–PFTD structure is more sterically hindered and exhibits lower bond dissociation energy. This chemistry promotes increased flame speeds. This study shows that surface functionalization can be used as an approach to control the reactivity of Al particles.

(734a) Ultrasmall Metal Nanoparticles With Molecular-Like Properties: Synthesis and Applications

Xie, J., National University of Singapore

Originally presented on: 11/7/2013 15:15:00 - 15:50:00

In very recent years, ultrasmall metal nanoclusters with core sizes smaller than 2 nm have emerged as a new direction in nanoparticle research due to their discrete and size dependent electronic structures and molecular-like properties, such as HOMO-LUMO transitions in optical absorptions, quantized charging, and strong luminescence. Synthesis of high quality metal nanoclusters in sufficiently large quantities is necessary for establishing reliable size-property relationships and exploring potential applications. This presentation will summarize our recent progress in the development of novel synthetic strategies for the production of monodisperse metal nanoclusters. The preparation of monodisperse metal nanoclusters is viewed as engineerable process where both the precursors (input) and their conversion chemistry (processing) may be rationally designed to achieve the desired outcome – monodisperse metal nanoclusters (output). We will demonstrate several efficient strategies for tailoring the precursor and the conversion process, and present our understanding of the processes involved. We will also highlight some application developments of metal nanoclusters in important sustainability topics including sensors for environment and human health, and biomedical applications.

(734b) Stability of Phosphonate-Functionalized Iron Nanoparticles

Greenlee, L. F., National Institute of Standards & Technology
Rentz, N. S., National Institute of Standards & Technology
Wilson, S. J., National Institute of Standards & Technology
Originally presented on: 11/7/2013 15:50:00 - 16:10:00

Iron nanoparticles are of interest in fields such as water treatment and alternative energy, due to their reactive properties and low cost. When combined with other metals, iron-metal nanoparticles can act as catalysts for a diverse set of reactions. In solution-based nanoparticle synthesis, organic stabilizers play a critical role in the properties of the nanoparticles, including size, morphology, composition, and colloidal stability. While polymers are often used in nanoparticle synthesis, chelator-type molecules can also be used as nanoparticle stabilizers. In this study, we characterize iron nanoparticles synthesized in water in the presence of a phosphonate chelator, amino tris(methylene phosphonic acid) (ATMP) for a range of molar ratios of ATMP to iron. An increase in the molar ratio from 0.05 to 0.8 decreases nanoparticle size from approximately 150 nm to less than 10 nm. Colloidal ATMP-stabilized iron nanoparticles are stable over a broad range of pH values, and zeta potential measurements were used to evaluate colloidal stability. Diffraction analysis indicates that ATMP-stabilized iron nanoparticles may have a nano-crystallite structure, with potential regions of amorphous iron. Characterization results of ATMP-stabilized iron nanoparticles are compared to results obtained previously for carboxymethyl cellulose (CMC)-stabilized iron nanoparticles. Quartz crystal microbalance measurements are used to evaluate the effect of surface functionalization on oxidation rate of the iron nanoparticles, and nanoparticle stability during storage is evaluated.

(734c) Magadiite Silylated With Sulfur-Functional Organosilanes: Investigation of Structure and Interlayer Accessibility

Mao, Y., University of South Carolina
Li, S., University of South Carolina
zur Loye, H. C., University of South Carolina
Ploehn, H. J., University of South Carolina
Originally presented on: 11/7/2013 16:10:00 - 16:30:00

Sodium magadiite (Na-MGD) is a prospective active filler for elastomer applications. In this work, Na-MGD was functionalized with sulfur-containing silane coupling agents, either bifunctional (bis (triethoxysilylpropyl) polysulfide, Si69) or monofunctional (3-mercaptopropyltriethoxysilane, MPTES) using cetyltrimethylammonium bromide modified MGD (CTA-MGD) as the intermediate. During grafting, the organosilanes displace CTA^+ cations from CTA-MGD surface, leading to a decrease in the interlayer spacing as measured by XRD; however, some CTA^+ cations remain in the MGD interlayer. The amounts of organosilane molecules grafted on MGD surface indicate that MPTES leads to a higher grafting density than Si69. Having functionalized the MGD interlayer surfaces, the accessibility of the sulfur-functional interlayer was evaluated using squalene, a model compound for natural rubber. Various sulfur-functional MGD materials (CTA-MGD, Si69-MGD, and MPTES-MGD) were reacted with pure liquid squalene. FTIR results show that new Si-O-C bonds are formed after reaction of squalene with sulfur-functional MGDs, but not with CTA-MGD. TGA and elemental analysis were used to quantify the amount CTA+, silane, and squalene in the products. XRD results demonstrate that squalene can access the interlayer sulfur only when the silane graft density is relatively low. This work shows the value of using squalene as a model compound for characterizing the structure and properties of sulfur-functional MGD prior to preparation of MGD-based elastomer composites.

(734d) Polymer-Gold Composite Particles Via Pickering Emulsion Polymerization: Synthesis, Characterization and Application

Zhang, M.,
Dai, L., Arizona State University
Phelan, P.,
Originally presented on: 11/7/2013 16:30:00 - 16:50:00

A one-step Pickering emulsion polymerization route was successfully applied to synthesize core-shell polystyrene-gold composite particles. Interestingly, in addition to core-shell structured polystyrene-gold composite particles, we found asymmetric polystyrene-gold nanocomposite particles are also produced, where a single gold nanoparticle is attaching on each polystyrene substrate. It is found these asymmetric particles are formed via "seeded-growth" mechanism. The core-shell and asymmetric polystyrene-gold composite particles proves to be efficient catalysts as they successfully catalyzed Rhodamine B reduction reaction. Further, PNIPAAM were successfully incorporated to produce "smart" polystyrene-PNIPAAM-gold composite particles. Without the aid of cross linker, the synthesized particles have up to 72% size decrease percentage. The polystyrene-PNIPAAM-gold asymmetric particles are found to have tunable optical properties due to thermo-responsiveness of PNIPAAM. It is found that absorption/scattering intensity could be adjusted significantly and the refractive index of substrates of the asymmetric particles influence their SPR properties. These tunable optical properties show great potential for biomedical and optical applications.

(734e) Synthesis and Characterization of Tin Selenide Nanocrystals Using Air-Stable Precursors

Qi, Y., University of Massachusetts
Mountziaris, T. J., University of Massachusetts
Originally presented on: 11/7/2013 16:50:00 - 17:10:00

Near-infrared (NIR) absorbing semiconductor nanocrystals, also known as quantum dots (QDs), have unique properties, such as size tunable bandgap, and potential applications in photothermal therapy, photovoltaics, and infrared absorbing glass coatings or interlayers. SnSe QDs are attractive as NIR materials, because they do not contain a heavy metal, such as Cd, Hg or Pb. The synthesis of SnSe colloidal QDs from organometallic precursors that are highly toxic and pyrophoric has been reported in the literature (e.g., Franzman et al., *J. Am. Chem. Soc.*, 2010, 132 (12), pp 4060–4061; Baumgardner et al., *J. Am. Chem. Soc.*, 2010, 132 (28), pp 9519–9521). Our study focuses on the development of a new synthesis method for SnSe QDs that employs an air-stable tin(II) chloride-oleylamine (OLA) complex and selenium powder dissolved in trioctylphosphine (TOP) as precursors. SnSe QDs with average diameter between 1.7 and 2.5 nm were synthesized by injecting the precursors in a hot coordinating solvent consisting of OLA and TOP. The effects of reaction conditions on the growth kinetics and final morphology of the QDs were investigated. A high OLA concentration in the growth solvent was found to be necessary for activating the reaction forming SnSe nuclei. On the other hand, TOP was found to hinder SnSe nucleation and, at increased concentrations, to yield larger particles. The relative amounts of precursors, OLA and TOP in the growth mixture control the growth kinetics and the aggregation of the QDs. The evolution of QD size and morphology as function of reaction time for different relative amounts of precursors, OLA and TOP has been studied using TEM. The formation of aggregates has been studied as function of reaction conditions and methods of controlling their size and shape have been identified. Growth conditions that provide precise control over the final particle size and prevent particle aggregation have been identified. The ultimate objective of this study is the development of a practical method for manufacturing SnSe QDs using readily available, air-stable precursors.

(734f) Asymmetric Functionalization of Shape-Anisotropic Polymer Nanoparticles

Guignard, F., University of Fribourg
Lattuada, M., ETH Zurich
Originally presented on: 11/7/2013 17:10:00 - 17:30:00

In the research in the field of nanoparticle in the last years, much attention has been devoted to the synthesis of anisotropic nanoparticles, also called Janus Nanoparticles. Anisotropicity may be provided in different ways, usually either in terms of non-homogeneous surface functionalization, or as compartmentalized internal structure. These special features render the nanoparticles much more interesting than their isotropic counterparts, especially regarding their self-assembling properties.

In this work, a two-step emulsion based polymerization has been applied to synthesize shape-anisotropic dumbbell nanoparticles, which size and morphology can be tuned by changing the process parameters. Narrowly distributed surfactant-free polystyrene nanoparticles are used as seeds. They are coated with a random copolymer of styrene and 3-trimethyloxyxilyl propyl acrylate (MPS), bringing silane groups on the surface of the seeds. The particles are then swollen with a monomer solution before a second polymerization. Due to the hydrophilic shell on the surface of the nanoparticles, the newly formed polymer is bulging out of the seeds, giving birth to dumbbell-like nanoparticles. The resulting nanoparticles are not only anisotropic in shape, but also in term of surface chemistry. The first bulb is bearing silane groups coming from the MPS, while the second is only made of polystyrene. This specific feature can be used for further asymmetric functionalize the dumbbells. Silane chemistry can be applied, which will result in the functionalization of the first hemisphere only. Superparamagnetic iron oxide nanoparticles (SPIONS) can be coated by surfactant having silane group, offering the possibility to attach SPIONS only to one bulge. The same approach has been used for platinum nanocrystals, leading to particles that can be used as Janus nanomotors in H_2O_2 solution.

(737a) Scaling-Up the Gas-Phase Synthesis of Metallic Nanoparticles By Means of Electrical Discharges

Kruis, E., University Duisburg-Essen
Stein, M., University of Duisburg - Essen
Hontanon, E., University of Duisburg
Originally presented on: 11/7/2013 15:15:00 - 15:50:00

Scientific journals and patent data banks are continuously filled with reports on new synthesis methods for nanoparticles. Although the therein-reported particle properties may be useful to test specific applications, it is unlikely that these synthesis methods are viable for scaling-up to industrial dimensions or that the desired particle properties found on a smaller scale would be conserved after scaling up. The preservation of the designed nanostructures and nanosystems with novel or predefined properties when the process is implemented at a commercially viable scale is unlikely without significant interdisciplinary cooperation between scientists, engineers and end-users.

The BUONAPART-E project, a project funded by the EU, involving 21 partners from universities as well as industry, aims to demonstrate that a physical nanoparticle synthesis process can be economically scaled-up. BUONAPART-E investigates **B**etter **U**pscaling and **O**ptimization of **Nanopart**icle and Nanostructure Production by Means of **E**lectrical Discharges. Experimental results as well as literature data indicate that an energy efficiency of 100 kWh/kg has been reached for a single unit. The challenge addressed in BUONAPART-E, which can only be met with new knowledge of the hitherto unknown fundamental mechanisms taking place, is to obtain an increase in the production rate, while retaining energy efficiency. The process allows for the synthesis of different materials using the same production platform. The basic evaporation unit (called hereafter the Optimal Single Unit or 'OSU') is a set of electrodes. A large number of these units can be placed in a single housing, contributing to the cost-effectiveness of the process. The use of many single production units in parallel, which can be thoroughly optimized and tested on a lab scale for a given material, ensures that a highly-effective scale-up of the synthesis process in terms of cost and energy consumption is possible. This strategy is analogous to supercomputers based on low-cost graphic cards from the consumer market. Further equipment, such as pumps, power supply to the OSUs and the particle collection unit, can be scaled-up as single units leading to additional cost benefits.

The final decision about the implementation of a new industrial production line is taken on the basis of a thorough economic analysis. Initial decisions concerning raw materials/reactants for synthesis as well as choice of the unit operations with the necessary capital investment and power requirements are vital for the economic viability of a process. Taking into account the high price of chemical precursors for nanomaterials as well as the costs of solvents and related operations such as purification and disposal, simple cost estimations lead us to believe that our choice of physical synthesis based on an electrical discharge from solid elements in inert gases is economically well-founded.

The process is designed in an environmentally friendly way as a closed loop system that avoids the use of hazardous precursors, solvents and stabilizers. The main input is electric power for the electric discharges and the pumps for gas recycling. The closed-loop cycle necessitates higher quality flanges than normally used in industry. This results in an inherently much lower possibility of workplace exposure to nanoparticles. The proposed process does not call for complicated and expensive safety measures, as quasi-atmospheric inert gas is used as transport medium and hazardous chemicals are avoided. All these factors make this process a clear example of green and sustainable synthesis process. The motivation behind the choice of spark discharge as a synthesis platform stems from the experience of BUONAPART-E consortium that this technology is cost-efficient to implement and easy to use, while still allowing for rapid process optimization. On the other hand, increased production rate of a single unit would lead, however, to engineering problems for this unit, such as an inabilities to dilute and rapidly cool to conserve small sizes, turbulence leading to broad size distributions, a lack of design and scaling-up rules, and scaling-up processes that are both lengthy and cost-intensive. As the basic unit is just a pair of electrodes and a large number of these pairs can be placed in a single housing, it is evident that a cost-effective scaling-up strategy include the use of multiple units in parallel.

In BUONAPART-E will be demonstrated the feasibility of simultaneous operation of a large number of electrode pairs. The OSU is optimized with special emphasis given to energy efficiency. The resources of the project will be focused on the development of a technology driven by requirements for energy efficiency and allowing a no-risk, highly optimized scale-up, which fulfill industrial demands. The feasibility of a scale-up will be assessed in the project, along with criteria dealing with health, security and environment.

In this talk, we will show the characteristics of the various discharge regimes (spark, glow discharge, thermal arc) as related to metal nanoparticle production. Of special interest are the production rates as well as specific electricity consumption.

This work was supported by the European Union´s Seventh Framework Program (EU FP7) under Grant Agreement No. 280765 (BUONAPART-E).

(737b) Adjustment and Online Determination of Primary Particle Size in Transferred Arc Synthesis

Stein, M., University of Duisburg - Essen
Kiesler, D.,
Kruis, E., University Duisburg-Essen
Originally presented on: 11/7/2013 15:50:00 - 16:15:00

Metallic nanomaterials are used in a large number of applications. Their production however still challenges scientists and engineers all over the world, since most processes are not capable of scaling-up the production rate of metal nanoparticles. Some are inefficient in terms of energy or material consumption or too expensive to scale-up, others produce micron sized particles instead of nanoparticles, when the production rate is increased. A transferred arc process has been found suitable for scaling-up the production rate of metal nanoparticles (Stein et al. 2013). However, this process as well tends to fluctuate in terms of particle size, due to changing process parameters (e.g. electrode gap), when the production rate is increased. Traditional characterization devices for particle size (XRD, BET) are not capable to determine those fluctuations as they present a mean particle size of the collected powder. Therefore a novel characterization system is set up to determine the production rate, specific electricity consumption, mobility size and an estimated primary particle size directly online at the process.

A transferred arc process is optimized to increase the production rate (> 2 g/h) and thereby minimize the specific energy consumption (< 500 kWh/kg) of copper nanoparticles. The optimization has been done by an adjustment of the experimental setup to reduce particle losses and give a better size control of particle size. A carrier gas composition of argon and nitrogen is thereby found to be the most relevant parameter for particle size control. Copper particles between 15 – 90 nm can be adjusted. An online measurement system containing a Tapered Element Oscillating Microbalance (TEOM), an Electrical Low Pressure Impactor (ELPI) and a Scanning Mobility Particle Sizer (SMPS) is used to characterize the synthesized particles. The parallel measurement of ELPI and SMPS allows determining the effective density and fractal dimension of the formed agglomerates (Virtanen et al. 2004), giving the opportunity to calculate the primary particle size in the agglomerates online. The effective density is thereby the fitting parameter between the number size distributions based on mobility (SMPS) and aerodynamic diameter (ELPI). An equation is formed, with the knowledge of the BET diameter, which allows calculating the fractal dimension directly from the fitted effective density. Since the effective density and fractal dimension is needed to calculate the primary particle size from the mobility diameter (Sorensen 2011), it is possible to calculate the primary particle size online.

The research leading to these results has received funding from the European Union's Seventh Framework Program under grant agreement N° 280765 (BUONAPART-E).

Sorensen C M (2011) The Mobility of Fractal Aggregates: A Review. Aerosol Science and Technology 45:765–779.

Stein M, Kiesler D, Kruis F E (2013) Effect of carrier gas composition on transferred arc metal nanoparticle synthesis. J Nanopart Res 15:1400

Virtanen A, Ristimäki J, Keskinen J (2004) Method for measuring effective density and fractal dimension of aerosol agglomerates. Aerosol Sci. Technol 38:437–446

(737c) Aerosol Assisted Synthesis of Gold Nanoparticles

Lähde, A., University of Eastern Finland

Koshevoy, I., University of Eastern Finland

Karhunen, T., University of Eastern Finland

Torvela, T., University of Eastern Finland

Pakkanen, T. A., University of Joensuu

Jokiniemi, J., VTT Fine Particles

Originally presented on: 11/7/2013 16:15:00 - 16:40:00

Gold nanoparticles (AuNPs) also called gold colloids are the most stable metal nanoparticles [Daniel and Astruct 2004]. AuNPs are known to have some unique properties including the surface plasmon resonance (SPR), which arises from the collective oscillation of conduction electrons and, dominates the optical spectra of metallic nanoparticles (NPs) [de la Garza et al. 2010]. This makes them attractive for many potential applications in the fields of biotechnology (e.g. biosensors, cell and tissue engineering, open MR imaging), energy (e.g. high-K capacitors, solar cells) and information technology (e.g. fiber-optic communications systems, printable electronics) [Wu et al. 2008]. However, the properties and applicability of AuNPs are strongly dependent on the particles size, interparticle distance, nature of the protecting organic shell, and shape of the nanoparticles.

There are a number of methods used for the preparation of AuNPs including sol-gel and other solvent based methods, layer-by-layer self-assembly, and aerosol assisted methods [Daniel and Astruct 2004, Majumdar et al. 1996]. In particular, aerosol assisted methods like spray pyrolysis are promising as they can produce submicron-sized particles with high elemental and phase purity. However, the preparation of AuNPs with aerosol methods typically require high pressures, temperatures and/or the use of strong acids like HCl and HNO_3 [de la Garza et al. 2010]. Furthermore, the particle size, concentration, chemical stability and degree of aggregation need to be controlled, since these properties strongly affect the optical, electrical, chemical and biological properties of the gold nanoparticles [Amendola and Meneghetti 2009].

In this paper we present an atmospheric pressure aerosol assisted synthesis of nanocolloidal gold. The aim of the study was to prepare plasmomic nanoparticles with well determined size, shape, morphology and composition using a single-stage, continuous aerosol synthesis route. Homoleptic cluster of $(AuC_2R)_{10}$ (R=2,6-dimethyl-4-heptanol) was synthesized [Koshevoy et al. 2012] and used as the precursor for the studies. The particle formation is based on the thermal transformation of the precursor in a high-temperature reactor (T=200 ? 800 °C). A systematic investigation of the influence of process parameters on powder characteristics, including particle and crystallite size, and surface properties, has been carried out.

The particle properties including particle size, shape and crystallinity could be altered by changing the reactor conditions. Phase-pure gold nanoparticles with the crystallite size between 2.4 and 4.2 nm were obtained depending on the process conditions. The result is very different from typical spray pyrolysis product where only one particle is formed from one starting droplet as here many nanosized Au particles are obtained from one droplet. A dramatic change in the color from bright yellow to very dark red of the collected powder was observed as the temperature increased (Figure 1 inserts). In addition, a clear transformation of droplet like particles formed at 200 °C to solid gold nanoclusters obtained at 800 °C was observed. At 200 °C spherical liquid like particles were observed indicating that the particles were formed mainly by droplet drying (Figure

1 A-B). At the intermediate temperature, 400 °C, spherical gold particle clusters were formed by thermal reactions taking place within the droplet (Figure 1 C-D). Finally, irregular agglomerates of gold nanoparticles with the primary size below 5 nm were obtained at 800 °C (Figure 1 E-F). No significant sintering of the primary Au nanoclusters was observed even at the temperatures above 400 °C. This can be partly explained by the steric hindrance of the side chains. In addition, at 800 °C a surface layer was observed around the particles in TEM images. Based on the Raman measurements the surface layer consisted of carbonaceous material originating from the thermal decomposition of the precursor compound.

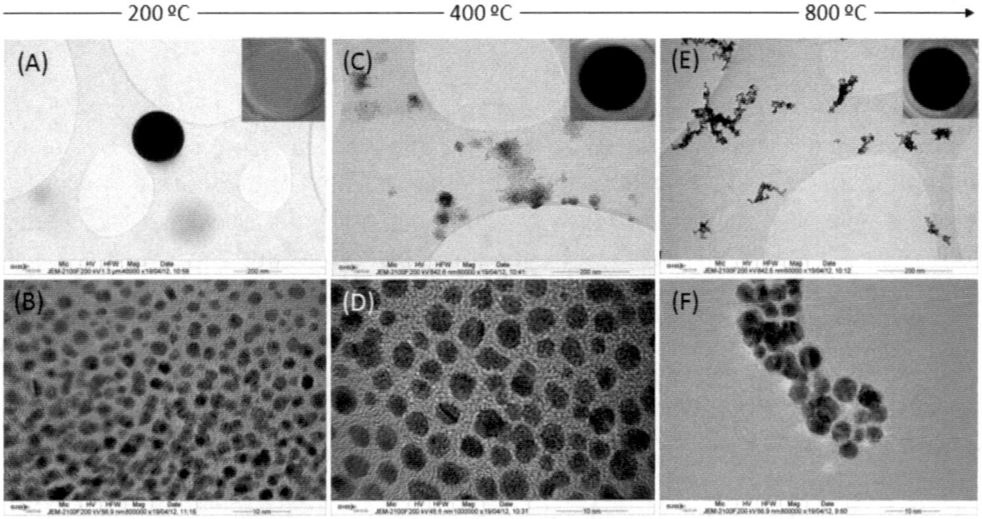

Figure 1. TEM images of AuNPs prepared at (A-B) T=200 °C, (C-D) T=400 °C and (E-F) 800 °C. The particles collected on the filter are shown as inserts.

Acknowledgements

This work was supported by the strategic funding of the University of Eastern Finland under the NAMBER spearhead project.

References

Amendola, V. Meneghetti, M. (2009) *J. Phys.Chem.* 113, 4277.

Daniel, M.-C., Astruct, D. (2004) *Chem. Rev.* 104, 293.

de la Garza, M., Hernández, T., Colás, R., Gómez, I. (2010) *Mater. Sci. Eng. B* 174, 9.

Koshevoy, I.O., Chang, Y.-C., Karttunen, A.J., Selivanov, S.I., Jänis, J., Haukka, M., Pakkanen, T., Tunik, S.P., Chou, P.-T. (2012). *Inorg.Chem.* 51, 7392.

Majumdar, D., Kodas, T.T, Glicksman, H.D. (1996) *Adv. Mater.* 8, 1020.

Wu, D., Xu, X., Liu, X. (2008) *J. Chem. Phys.* 129, 074711.

(737e) Optimization of Flame Reactors for Pilot-Scale Nanoparticle Manufacture

Grohn, A. J., Swiss Federal Institute of Technology (ETH Zurich)
Pratsinis, S. E., Swiss Federal Institute of Technology (ETH Zurich)
Wegner, K., ETH Zurich
Originally presented on: 11/7/2013 17:05:00 - 17:30:00

A present challenge with flame spray synthesis of nanoparticles is the translation of current laboratory-scale reactors into an industrial production environment. To achieve this goal the production rate has to be increased while maintaining constant nanoparticle characteristics in a safe and economic process. Flame spray reactor operating costs consist mainly of the employed precursor and process gases (Wegner et al., 2011). The goal here is to minimize the utilization of hydrocarbon solvents and to replace oxygen dispersion gas with air, yielding a more environmentally friendly process with better energy utilization.

The effect of preheating air and precursor streams on the spray combustion as well as product particle properties was investigated for the example of zirconia nanoparticle synthesis at rates up to 2 kg/h. Therefore, the entrained air was preheated electrically or via recovery of the process heat and the up to 100 L/min of dispersion gas were gradually exchanged by air. While a non-preheated spray flame operated with increasing levels of air did produce carbon-contaminated products or was blown out completely, the preheated system yielded stable flames and cleaner products. This is attributed to enhanced combustion kinetics and expanded flammability limits. Employing a flame enclosure allowed to confine the combustion-generated heat, further stabilized the process and allowed to preheat part of the entrainment air. Addition of steam to the flame aimed at counteracting soot formation (Dryer, 1977) but instead affected the specific surface area of product powders.

Overall, the process parameter window for stable combustion and nanoparticle formation could be broadened, process costs reduced and product quality improved by pre-conditioning of reactant streams and optimizing reactor geometry.

References:

Dryer, F. (1977). Water addition to practical combustion systems—concepts and applications. Symposium (international) on combustion, Elsevier.

Wegner, K., B. Schimmöller, B. Thiebaut, C. Fernandez and T. Rao (2011), KONA 29: 251-265

(749a) Metal-Based Reactive Materials With Biocidal Reaction Products

Abraham, A., New Jersey Institute of Technology

Schoenitz, M., New Jersey Institute of Technology

Dreizin, E. L., New Jersey Institute of Technology

Originally presented on: 11/7/2013 15:15:00 - 15:37:00

Reactive materials with biocidal combustion products capable of inactivating aerosolized microorganisms have been developed recently. The focus was on additives to conventional energetic formulations involving halogens, known to have strong biocidal properties. Materials developed to date can be classified into three main groups: metal-based fuels, oxidizers, and thermites (metal fuel and oxidizer compositions) with at least one component producing biocidal combustion products. In addition to generating the desired combustion products, reactive material components must be stable and compatible with common binders to endure conventional processing routine. Furthermore, insensitivity to spark, impact, friction, and other common ignition stimuli is also desired for material handling.

Our work has focused on incorporating halogens or halogen containing compounds in aluminum. Powders containing up to 20 wt % of iodine with compositions of Al-I_2 [1], Al-B-I_2 [1], and Al-CHI_3 have been developed and tested. In this talk, efforts aimed at developing new metal-based components with a substantially increased halogen content will be discussed.

The main challenge is to stabilize the halogen in a reactive metal matrix, so that the material can be handled and formulated as a regular metal fuel, e.g., aluminum powder, commonly added in energetic formulations. The approach taken in this study is based on mechanical alloying (or ball milling) starting metal powder and a halogen-containing material to produce a stable composite powder. Previous work showed that cryo-milling was necessary to stabilize iodine in aluminum matrix. Milling at room temperature could be used to stabilize iodine in a binary Al-B composite. However, the amount of iodine that could be stabilized could not exceed 20 wt %. Thus, new material compositions need to be explored to produce materials with greater halogen concentrations.

To select most promising halogen-containing starting components, properties of different metal halides were surveyed ranking them based on the maximum enthalpy, DH, required to convert the halide into the most stable metal oxide and oxidized halogen. This enthalpy is taken per mole of halogen. The results are shown in Fig. 1 where the estimated reaction enthalpy is plotted vs. weight % of the respective halide that should be added to aluminum in order to prepare a reactive material with 40 wt. % of halogen. The 40 wt % benchmark doubling the presently achieved content of iodine in the Al-based powders serves as a target halogen content for the new reactive materials. In Fig. 1, the most desired compositions would have the greatest negative reaction enthalpy and the lowest wt. % of metal halide (allowing their encapsulation into aluminum matrix); thus, they would be represented by symbols located at the lower left portions of the plots, shown separately for chlorine, bromine and iodine. It is clear that the iodine-containing materials are most favorable energetically. Among all materials considered, only a few could be selected taking into account their toxicity, stability, and cost. In particular, TiI_4 is selected for initial experiments. Because SiI_4 is not readily available commercially, ball-milling elemental iodine with Si and Al is also of interest. In addition, a chlorine-containing starting material, $NbCl_5$ was selected to evaluate the effect of chlorine as a biocidal fuel additive. Due to the interest in stabilizing these metal-halides in the aluminum matrix, composite particles may need to be coated with polymers such as paraffin wax and/or Teflon.

Preparation, characterization, and testing of several new halogen-containing Al-based fuels will be discussed. Both stability and oxidation mechanisms of the prepared materials are assessed using thermal gravimetry performed, respectively in inert and oxidizing gas environments. Materials are further characterized using x-ray diffraction and electron microscopy. Ignition of the prepared powders is studied by coating them onto an electrically heated filament. Combustion dynamics of the prepared powders is tested by feeding them into a hydrocarbon flame and measuring intensities and durations of the emission pulses produced by particles.

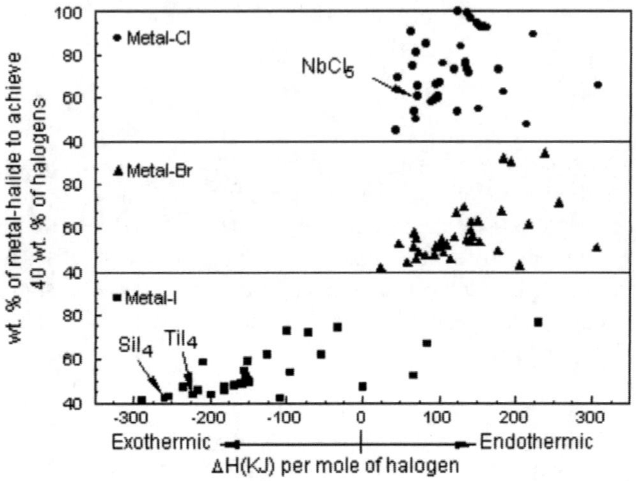

Figure 1: Survey of metal-halide oxidation enthalpies

Reference

[1] E.L. Dreizin et al., *Int. J. of Energ. Mater. Chem. Propul.,* 10, p 297 (2011) ADDIN EN.REFLIST

(749b) Nanoscale Aluminum, Silicon, and Aluminum/Silicon Mixture Formation For Use In Pyrotechnic Applications

Morris, L. A., US Army, RDECOM-ARDEC
Doorenbos, Z., Innovative Materials and Processes, LLC
Puszynski, J., South Dakota School of Mines and Technology
Kapoor, D., US Army RDECOM-ARDEC
Originally presented on: 11/7/2013 15:37:00 - 15:59:00

The incorporation of nanoscale materials into pyrotechnic, energetic, and explosive formulations has fostered the development of a variety of novel tunable materials with controllable combustion characteristics. The addition of metallic nanoparticles to energetic formulations has been shown to modify burn rates and reaction propagation in a variety of pyrotechnic formulations. Specifically, high surface area nanoscale metals with large combustion enthalpies can be used as fuels to enhance energetic reactions. Silicon and aluminum are both highly combustible additives of interest to the Department of Defense. While nano-scale silicon and aluminum are both readily combustible in their unpassivated state, it is important to note that their ignition and oxidation behaviors are different. Silicon has a higher ignition temperature than aluminum, and it is believed silicon it is less susceptible to oxidation. Therefore, combining these two fuels into a single heterogeneous nanoscale composite in variaous stoichiometric ratios is advantageous for tuning properties. In order to maximize the chemical reactivity of these fuels premature oxidation during nanoscale synthesis must be prevented.

The top-down synthesis of heterogeneous aluminum-silicon nanoscale composites was undertaken using inert gas high energy milling and cryogenic milling. The effect of milling parameters (e.g. time, intensity, temperature, and use of a process control agent) on the specific surface area, morphology, intermixedness, and reactive content of these composites will be discussed. Finally, the oxidative properties of these composites will also be presented.

(749c) Synthesis and Performance of Nano-Processed Terephthalic Acid Smoke Obscurants

Morris, L. A., US Army, RDECOM-ARDEC
Hall, M., US Army, ARDEC

Originally presented on: 11/7/2013 15:59:00 - 16:21:00

Terephthalic acid (TA) smoke obscurant was developed by the U.S. Army as an environmentally benign replacement to hexachloroethane and white phosphorus based smoke for use in the M-83 grenade and M-8 smoke pot. Although TA smoke mix is a safer pyrotechnic smoke with respect to inhalation, it is a less effective visual obscurant than the legacy mixes. Visual obscurants are deployed to conceal ground movements from an enemy line of sight. It was hypothesized that reducing the particle size of the TA smoke mix will cause a more complete burn and thus create a thicker visual obscurant cloud. Here, the particle size of terephthalic acid smoke mix was reduced into the nano-regime through high energy milling. The effect of various processing parameters is presented; the nano-processed smoke mixes are characterized for particle size, morphology, and sensitivity. The smoke mixes were pressed into large-scale grenade configurations. Large-scale chamber and live-fire testing was performed, and the results will be discussed and compared to legacy obscurants.

(749d) Using Quenched Glass Particles to Recover Thermal Histories of Pyrotechnical Events

Vummidi Lakshman, S., New Jersey Institute of Technology
Schoenitz, M., New Jersey Institute of Technology
Dreizin, E. L., New Jersey Institute of Technology
Originally presented on: 11/7/2013 16:21:00 - 16:43:00

A typical pyrotechnical event generates environment temperatures exceeding 2000 K, and lasts from a few microseconds to seconds. To obtain the thermal histories of these pyrotechnical events, traditional temperature measuring techniques are either too fragile or sluggish to describe the event with good accuracy.

This study explores the possibility to use glass particles as thermosensors. The structures of glasses are strongly dependent on the quench rate during glass formation. In addition, silicate and borate glasses have unique Raman signatures corresponding to the connectedness of their respective network forming units [SiO_4] and [BO_3]. These signatures therefore change depending on quench rate. This effect is well understood and quantified considering cooling times that measure in minutes to years.

Glass particles that are injected into the combustion environment will experience a cooling rate that depends on both, the environment temperature as it cools down, and the particle size. On this time scale, cooling takes place in less than a second. Recovering particles after the event, analyzing the Raman signature on a per-particle basis, and correlating this signature with the fictive temperature of the glass will allow to determine the cooling rate of the combustion environment in the range of the observed fictive temperatures. Different glasses can potentially be used to extend the temperature range from which cooling rate information can be recovered.

An experimental setup has been built where aerosolized silicate and borate glass particles are heated to temperatures in excess of their melting points (max. 1600 K). The particles are then propelled through a ~100 μm orifice into an adjustable low-pressure environment, where they cool at rates corresponding to the pressure and their size, generally in the range 10^3 - 10^4 K/s. Particles are collected on a filter and analyzed by Raman microscopy. The extent to which the Raman signature and therefore the fictive temperature is sensitive to cooling rate changes on this time scale will be quantified.

(749e) On Mechanism of Splitting Water By Its Reaction With Aluminum and Magnesium

Nie, H., NJIT
Schoenitz, M., New Jersey Institute of Technology
Dreizin, E. L., New Jersey Institute of Technology
Originally presented on: 11/7/2013 16:43:00 - 17:05:00

Reactions of metals with water are of interest for hydrogen generation for a wide range of applications, from in-situ operated fuel cells to propulsion of underwater vehicles. In our previous work, the mechanism of water split reaction with aluminum was studied by monitoring the heat generation with micro-calorimetry and weight loss in thermo-gravimetric measurements for liquid water and steam, respectively. In both experiments, two spherical aluminum powders were used with different but overlapping particle size distributions. The experimental results were interpreted taking into account actual particle size distributions of aluminum powders and assuming a location for the reaction interface at either the outer surface of the oxidizing particles or the inner metal-oxide interface. Matching the oxidation dynamics for particles of the same size and belonging to powders with different size distributions allows one to determine the location of the reaction interface. For liquid water, results suggest that the reaction occurs at the surface of the shrinking aluminum core, but for reactions of aluminum with steam at elevated temperatures, the results indicate that the reaction is most likely occurring at the outer surface of the growing alumina shell. It is further observed that the alumina shell becomes fragile. When the shell break once it grew thick enough, a new shell begins to grow around the fresh aluminum surface. Here, these conclusions are used to develop a simplified, diffusion-limited reaction model considering different species diffusing for oxidation of aluminum in liquid water and in steam. Reaction kinetics is quantified using previous thermo-analytical measurements and using isoconversion techniques. In addition, the oxidation behavior of spherical magnesium powders in water is investigated using the same approach as was used previously for aluminum. Two measurements for the reaction of spherical magnesium with liquid water and steam are performed with heat flow calorimetry and thermo-gravimetry, respectively. The results are interpreted considering actual particle size distributions. Once the mechanisms of metal-water reactions are elucidated, modified materials, including aluminum/magnesium alloys can be developed enabling one to optimize the hydrogen generation by the water split reaction.

(749f) Production of Submicron-Sized RDX Particles Using a Nucleation Agent

Kim, J. W., Sogang university
Lee, S. Y.,
Koo, K. K., Sogang University
Originally presented on: 11/7/2013 17:05:00 - 17:27:00

In order to prevent munitions related accidents, the insensitive high energetic materials to unplanned stimuli without decreases in military operational capability have been extensively investigated. One way to reduce sensitivity is submicronization of high energetic particles because critical temperature to ignite increases significantly by decreasing the size of internal defects. In general, when 30 ¢ªC saturation solution of hexahydro-1,3,5-trinitro-1,3,5-triazine (RDX)/acetone is quenched into water of 0 ¢ªC, vigorous nucleation is immediately induced. On the other hand, with addition of small amount of polyvinylpyrrolidone (PVP, 0.01 wt%), it was found that nucleation of RDX did not occur because PVP acted as nucleation retarder. However, ultrasonic irradiation to this high supersaturated solution showed that extremely high nucleation rate was lead and submicron-sized RDX particles with very uniform crystal size distribution (average length of about 700 nm) were obtained with no agglomeration. Morphology of submicron-sized RDX particles were tetragonal pyramid consisting of eight {111} side faces.

(749g) The Potential Challenges and Benefits of Utilizing Novel, Nanostructured and Nanoscale Materials in DoD Applications (INVITED)

Redner, P., US Army, RDECOM-ARDEC
Oyler, K., US Army RDECOM-ARDEC
Csernica, C., US Army, RDECOM-ARDEC
Mehta, N., US Army RDECOM-ARDEC
Kapoor, D., US Army RDECOM-ARDEC
Puszynski, J., South Dakota School of Mines and Technology
Originally presented on: 11/7/2013 17:27:00 - 17:49:00

Nanoscale and nanostructured materials offer unprecedented opportunities to improve on legacy energetic materials, and to achieve performance objectives that these materials cannot currently meet. There has been great success by the US Army Research, Development and Engineering Command, Armament Research, Development and Engineering Center (RDECOM-ARDEC) and other organizations in formulating material systems that will provide many potential benefits to the Warfighter. These benefits are derived from the ability to "tune" these formulations by modifying the constituent fuels and oxidizers via particle size, surface area, fuel to oxidizer ratio, etc. Burn rates, light outputs, sensitivity, added effects after detonation, and other key parameters can be adjusted according to system requirements. However cost, reproducibility, availability and scalability all pose major challenges to the adoption of these materials by Department of Defense customers. This paper will discuss past and recent successes, current technological and economic roadblocks, and the various paths towards integration in multiple weapons systems.

(752a) Controlled Synthesis of Nanoparticles Using a Coaxial Turbulent Jet Mixer

Lim, J. M., Massachusetts Institute of Technology
Gilson, L. M.,
Chopra, S.,
Langer, R. S., Massachusetts Institute of Technology
Farokhzad, O. C., Brigham and Women's Hospital - Harvard Medical School
Karnik, R., MIT
Originally presented on: 11/7/2013 15:15:00 - 15:35:00

Nanoparticles have shown great promise for the diagnosis and therapy of major diseases such as cancer, cardiovascular, and infectious diseases. Use of microfluidics for synthesis of these nanoparticles has been of great interest due to controllability and reproducibility in their physicochemical properties. However, there are intrinsic limitations in conventional polydimethylsiloxane (PDMS) microfluidic systems for the synthesis of nanoparticles. First, only a handful of organic solvents are compatible with conventional PDMS microfluidic systems, which hinders PDMS microfluidic systems from serving as versatile platforms for synthesis of various types of nanoparticles. Second, productivity of microfluidic systems is lower than that of batch reactors due to low flow rates, making it challenging to produce large quantities of nanoparticles required for *in vivo* or clinical studies. Here, we report a novel coaxial turbulent jet mixer capable of synthesizing various types of functional nanoparticles with high production rates suitable for *in vivo* studies and clinical trials, while maintaining the controllability and reproducibility of PDMS microfluidic systems.

In the coaxial turbulent jet mixer, the inner flow stream containing raw materials is mixed with outer stream of non-solvent by turbulent jet flow for generation of uniform nanoparticles by rapid solvent exchange method called nanoprecipitation. The mixing of the inner and outer streams was visualized using phenolphthalein, a pH indicator. Two dimensionless parameters, flow velocity ratio and Reynolds number (Re), were used to classify the mixing behavior into flow regimes corresponding to laminar, transition, and turbulence regimes. Operating in the turbulence regime, the mixing time was tunable in the range of 1 - 40 ms by changing the Re. Nanoparticles obtained using the coaxial turbulent jet mixer at high Re (rapid mixing) were more homogeneous and smaller than those synthesized by bulk mixing, because the mixing time scale is more controllable and shorter than the characteristic aggregation time scale. Since the coaxial turbulent jet mixer is compatible with various organic solvents, it is versatile system where various types of nanoparticles including poly(lactide-co-glycolide)-b-polyethyleneglycol (PLGA-PEG) nanoparticles, iron oxide nanoparticles, lipid vesicles, and polystyrene nanoparticles can be made. In each case, the nanoparticle size decreased as mixing time decreased (*i.e.*, as Re increased). Various functional agents including anticancer drug, insulin, and fluorescent dye could be loaded in the nanoparticles for diagnostic and therapeutic applications during the nanoprecipitation. The coaxial turbulent jet mixer developed in these studies can be used to make functional nanoparticles with high production rates and reproducible manner suitable for clinical studies and mass production.

(752b) Hierarchically Porous Microparticles VIA Spray Drying

Waldron, K., Monash University
Wu, W. D., Monash University
Wu, Z., Monash University
Liu, W., Monash University
Selomulya, C., Monash University
Zhao, D., Fudan University
Chen, X. D., Monash University
Originally presented on: 11/7/2013 15:35:00 - 15:55:00

The versatility of mesoporous silica with tailored pore size and functionality renders it ideal for a wide range of applications such as bio-adsorption and as catalyst substrates. A rapid synthesis method combining evaporation induced self-assembly and surfactant templating via spray-drying can overcome drawbacks associated with traditional synthesis routes, namely long processing time, non-scalable batch operations, and that most powders are sub-micron polydisperse particles. A micro-fluidic jet spray-dryer, developed at Monash University, was used to produce monodisperse microparticles with mesoporous characteristics. We demonstrated that hierarchical structures could be generated through the addition of co-templates in the precursors. Microparticles produced with pluronic block copolymer (F127) with the addition of Eudragit RS 30D ®, a polymer-colloidal dispersion of ~ 120 nm nanoparticles as a co-template, produced microparticles with very open, porous structure with macropores of the ~ 120 nm, along with ~9 nm mesopores. These particles exhibited highly ordered hexagonal mesostructures in addition to the macropores created by the removal of the polymer, with surface areas up to 460 m^2/g and pore volumes up to 0.68 cm^3/g. Through the variation in weight ratio of the co-templates and the silica different hierarchical structures can be obtained. These macro-meso-structures can be used for a range of applications such as bioseparation and bioadsorption in dynamic systems and for immobilisation of bio-catalysts.

(752c) Nano Encapsulation of Active Ingredients

Lakshminarayanan, V.,
Meesters, G., DSM
Mendes, E.,
Originally presented on: 11/7/2013 15:55:00 - 16:15:00

In this work the production of polymeric micelles is studied. Their size and size distributions are measured. Also the stability of the micelles will be measured and discussed. Secondly the polymeric micelles are loaded with an active ingredient. The loading of these polymeric micelles with the active is measured by HPLC methods. Next to this several other techniques will be discussed to elucidate in what state, e.g. crystaline or amorphous, the active is present in the polymeric micel. Also the loaction of the acive in the polymeric micel will be discussed. Techniques like TEM and Cryo TEM are used for this.As a final step, the quantification of the active is carried out to realize the complete characterization of the system. With complete information of the encapsulation system, it is possible to control the and optimize the variables of the system to better tune the nano-system towards the application for which it is developed

(752d) Effect of Nanosmearing On the Rate of Nanostructure Formation in Individual Grains and Tablet Manufacturing

Kick, C., Western Michigan University
Ari-Gur, P., Western Michigan University
Pingali, K., Western Michigan University
Originally presented on: 11/7/2013 16:15:00 - 16:35:00

The influence of mixing order, flowing agent formulation, and compaction on tablet hardness, hydrophobicity, coating uniformity, extent of coating, micoporosity, mesoporosity, and nanoporosity was studied to understand how nanostructures affect bio-availability of the active pharmaceutical ingredient (API). Multiple pharmaceutical formulations of both lubricated and un-lubricated blends were prepared and mixed in a v-blender under fixed shear and strain conditions to characterize the formation of microstructures and nanostructures on the API. The changes in powder properties and surface topography throughout the various blends can be attributed to these formations of nanostructures, which was the focus of our study. These nanostructures were characterized using gas absorption and Scanning Electron Microscopy (SEM) with Electron Dispersive Spectroscopy (EDS) mapping. While the uniformity and extent of coverage on the API particle interface controls the powder flowability, only a fraction of the MgSt lubricant used in the flowing agent formulation coats the API surface and forms the nanostructure. The powder flowability of the blends was quantified in a gravitational displacement rheometer (GDR), which showed that the flow index did improve when the extent of the coating area was increased. However, the extent of coating for the Cab-O-Sil was non-uniformly distributed on the API and much weaker than MgSt, which can be attributed to the interaction between the coating and the API particle surface. When tablets were produced on a rotary tablet press, mercury intrusion, gas absorption, and SEM imaging analysis software were used to characterize the size and distribution of the nanostructures formed. The best flow properties in a blend were exhibited with a formulation consisting of 1% MgSt and 0.5% Cab-O-Sil. The distribution or presence of Cab-O-Sil was seen to enhance the distribution and smearing of MgSt on the particle surface only when the Cab-O-Sil concentration did not exceed 0.5% of the formulation. The variations in uniformity, extent of coating, and characteristic nanostructure formation in the powders are key parameters that influence tablet performance and bioavailability, which has wide applications for bio-medicine manufacturing.

(752e) Cell Free Production of Flagellin As a Vaccine Adjuvant and Its Ordered Display On Virus-Like Particles

Lu, Y., Stanford University
Swartz, J. R., Stanford University
Originally presented on: 11/7/2013 16:35:00 - 16:55:00

Bacterial flagellin has been explored as a potential vaccine adjuvant for enhancing immune responses. In this report, we describe *Escherichia coli*-based cell-free protein synthesis (CFPS) as a method to rapidly produce soluble phase 1 flagellin (FliC) protein from *Salmonella typhimurium*. Especially when flagellin is urgently needed for new and highly potent vaccines at short notice, for example, to combat pandemic influenza threats, CFPS can provide such a means for rapid, cost-effective and high-level production of protein products in a few hours. The synthesized flagellin product had high affinity for the Toll-Like Receptor 5 (hTLR5) (EC50 = 2.4 ± 1.4 pM); approximately 10 to 50 fold higher than previously reported. Observations from protein purification, SDS-PAGE, alanine scanning and multiple mutations indicated that the C-terminal D0 domain of flagellin has disordered tertiary structure, making it susceptible to proteolytic degradation. This proteolysis could be reduced by protease inhibitors, the use of protease-deficient cell extracts, or deletion of the flagellin D0 domain.

We hypothesized that the TLR5 receptor activation could be enhanced if flagellin were presented in an ordered array on the surface of a virus-like particle (VLP). We incorporated a non-natural amino acid displaying an alkyne moiety into flagellin using the CFPS system and attached flagellin to hepatitis B core virus-like particles (VLPs) using bioorthogonal azide-alkyne cycloaddition reactions. The ordered and oriented VLP display of flagellin increased its specific TLR5 stimulation activity by approximately 10-fold.

In this study, we set out to produce more potent vaccine designs. Unexpectedly, we also gained an important insight into our innate immune systems. Cell-free protein synthesis overcame serious proteolytic challenges, and the intact flagellin then exhibited unexpectedly high affinity to the toll-like receptor, TLR5. It stimulated cellular responses at 2 pM (10^{-12}M) concentrations, at least a thousand fold lower than for most biological interactions. When attached to VLPs, the flagellin was even more effective. When bacteria invade our bodies, they find an extremely hospitable environment. In order to overcome this threat, our immune systems must respond very rapidly and very strongly. Our measurements indicate that immune cells can detect incredibly low concentrations of flagellin so they can quickly mobilize to win the battle.

(752f) Polycarboxybetaine Esters for Gene Delivery: Molecular Tuning, Controlled Degradation and Vaccine Applications

Sinclair, A.,

Bai, T., University of Washington

Zhang, L., University of Washington

Jiang, S., University of Washington

Nowinski, A.,

Originally presented on: 11/7/2013 16:55:00 - 17:15:00

The safe and controlled delivery of genes to targeted cells promises exceptional advancements in clinical disease treatment, next-generation vaccines, and tissue engineering. [1-3] An ideal polymeric vector would condense nucleic acids into stable polyplexes that can successfully navigate the extracellular environment and pass through the cell membrane, as well as protect genes from degradation inside and outside the target cells and assist with escape from the endosome or trafficking vesicle. We have developed a gene delivery platform based on polycarboxybetaines (PCBs) modified with degradable esters, with the goal of optimizing each function without detriment to the others.

When the anionic carboxylate groups in zwitterionic PCB side chains are esterified, the polymer is rendered cationic and thus able to bind and condense nucleic acids. [4,5] A tertiary amine can be introduced in place of the quaternary ammonium to make the polymer pH-responsive, buffering pH changes during polyplex trafficking and providing a mechanism via the "proton sponge" for endosomal or phagosomal escape. Another key function of this platform is the hydrolytic conversion of cationic PCB esters to zwitterionic PCB. The rate of this charge shifting can be easily modified by the incorporation of different ester leaving groups and different distances between the ester and the tertiary amine groups. This appends a "smart" characteristic to this polymer; DNA is first strongly complexed by cationic PCB-esters, but is efficiently released when ester degradation reveals the anionic carboxylate groups, rendering PCB zwitterionic and nontoxic.

In our recent study [6], the side chain length was varied to include either one or two alkyl spacers denoted between the amine and the carboxylate/ester groups. Additionally, each chain length was synthesized with both an ethyl ester and a tert-butyl ester hydrolytic group. These molecular modifications result in monomers and resulting polymers with varying proton buffering capacities and hydrolytic profiles. An alkyl spacer length of one brings the pKa of each amine into the optimal range for endosomal buffering, and the hydrolytic ester can be easily modified to speed or slow charge switching. An optimized polymer is able to deliver plasmid DNA an order of magnitude more effectively than polyethyleneimine, without demonstrating any cytotoxicity.

We have also developed a novel photolabile o-nitrobenzyl ester of polycarboxybetaine (PCB-NBE) to give this platform a UV- sensitive "switch" for active degradation control, to study how the charge neutralization caused by ester degradation directly catalyzes DNA release from a polyplex. Photoinitiated ester degradation precipitates the rapid release of over 70% of complexed DNA from PCB-NBE polyplexes. This molecule is promising for triggered release of nucleic acid therapies in vitro.

Extending the functionality of this polymer class, we have applied a similar platform to the passively targeted delivery of DNA vaccines with polycarboxybetaine micro-"gels". [7] These hydrogel microparticles target the phagocytosis pathway of immunostimulating cell types to delivery nucleic acid vaccines adsorbed to their surface. Tertiary amines and hydrolytic esters confer similar benefits of attraction, buffering, and release as they do in nanosized polyplexes. A model vaccine delivered to macrophage cells has resulted in over 10x higher expression than that delivered from a cationic CTAB microparticle control. This platform is also nontoxic and easily tunable.

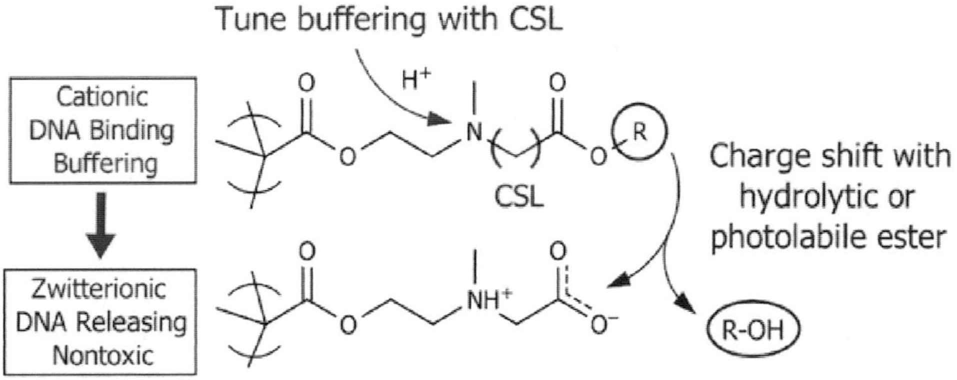

References:

[1] Sheridan, C. *Nat. Biotechnol.* **2011**, 29, 121-128.

[2] Ferraro, B.; Morrow, M. P.; Hutnick, N. A.; Shin, T. H.; Lucke, C. E.; Weiner, D. B. *Clin. Infect. Dis.* ***2011**, 53, 296-302.*

[3] Vallet-Regi, M.; Ruiz-Hernandez, E.; Gonzalez, B.; Baeza, A. J. *Biomater. Tissue Eng.* ***2011**, 1, 6-29.*

[4] Zhang, Z.; Cheng, G.; Carr, L. R.; Vaisocherová, H.; Chen, S.; Jiang, S. *Biomaterials* **2008**, 29, 4719-4725.

[5] Carr, L. R.; Jiang, S. *Biomaterials* **2010**, 31, 4186-4193.

[6] Sinclair, A.; Bai, T.; Carr, L. R.; Ella-Menye, J-R.; Zhang, L.; Jiang, S. *Biomacromolecules* **2013** (DOI: 10.1021/bm400221r)

[7] Zhang, L.; Sinclair, A.; Cao, Z.; Ella-Menye, J-R.; Xu, X.; Carr, L. R.; Pun, S. H.; Jiang, S. *Small* **2013** (DOI: 10.1002/smll.201202727)

(752g) Hollow Stimuli-Responsive Magnetic Silica Microparticles for Controlled Release and MRI Imaging

Sarvasova, N., Institute of Chemical Technology
Stepanek, F., Institute of Chemical Technology, Prague
Originally presented on: 11/7/2013 17:15:00 - 17:35:00

The possibility to control the release of active substances represents one of the most important attributes of targeted drug delivery (TDD). Here we are presenting one of the possible means for such delivery, which could be provided by microscopic, synthetically created particles from composite material based on silica, thermoresponsive polymer poly-N-isopropylacrylamide (PNIPAM, LCST 32°C) and magnetite nanoparticles. All of the materials mentioned above provide desired microparticles with unique characteristics such as endurance and hardness of silica, the ability of thermoresponsive polymers to change its volume in dependence on temperature and the possibility of induction heating as a result of the presence of magnetite nanoparticles. By combining these attributes, the final microparticles could act as a vessel for TDD able to transport the delivered substance into its destination and then release it there as a reaction to an external stimulus being the application of alternating magnetic field.

The synthesis of such microparticles consists of several steps. Firstly, it is essential to prepare macroporous hollow silica microparticles with incorporated magnetite nanoparticles using a solid template method. This is achieved by following a chain of reactions starting with the production of solid polystyrene (PS) template in the form of monodisperse spheres with relatively uniform size in the range of 1 mm. Then the silica/PS/Fe_3O_4 composite microparticles are obtained using modified Stober method based on the reactions of condensation and hydrolysis. During this process, Fe_3O_4 nanoparticles are gradually absorbed into the growing silica seeds, thus creating continuous composite silica/ Fe_3O_4 network on the surface of the PS templates. The last step, constituting of the PS core removal, is easily done by dissolving in toluene and leads to the creation of hollow silica/Fe_3O_4 microparticles with pores of the approximate size near 100 nm. In order to proceed to the final product being silica/PNIPAM/Fe_3O_4 hybrid microparticles, the already existing silica/Fe_3O_4 microparticles have to be functionalized by introducing the double bond onto its surface utilizing 3-(trimethoxysilyl)propyl methacrylate (3-MOP), allowing PNIPAM to be subsequently attached onto the microparticles as a result of radical copolymerization of 3-MOP silica with NIPAM. The as prepared hybrid microparticles were characterized in terms of its composition and morphology. Additionally, also the possibility of radiofrequency heating and release process of a model substance (Nile Red) were investigated. Since it would be highly beneficial to be able to observe the behavior of microparticles containing magnetite in 3D surroundings, the method of MRI was used.

(752h) Iodinated Nanoparticles for Diagnostic Medical Imaging Applications

Tang, C., North Carolina State University
Chacko, A. M.,
Mikitsh, J.,
Alavi, A.,
Prud'homme, R. K., Princeton University
Originally presented on: 11/7/2013 17:35:00 - 17:55:00

Non-invasive medical imaging techniques such as X-ray-computed tomography (CT) and positron emission tomography (PET) are powerful tools in diagnostic medicine. In these applications, contrast agents are needed to enhance the diagnostic quality of the images. Ideally, contrast agents would be biocompatible and contain high payloads of contrast agent thus nanoparticle formulations of radiopaque or radioactive materials offer opportunities for improvements in diagnostic imaging. In this work, we explore the use of Flash NanoPrecipitation (FNP) as a platform technology to produce such nanoparticle formulations of iodine (radiopaque) and its radioactive isotopes. In FNP, amphiphilic block copolymers direct self-assembly of biocompatible nanoparticles with high concentrations of encapsulated components. The precipitation process is controlled by carefully tuning the time scales of micromixing, self-assembly, and nucleation and growth. Using this method, various routes to iodinated nanoparticles have been explored. In one approach, an iodinated polymer can be encapsulated within the nanoparticles during the FNP process. Alternatively, the FNP can be used to encapsulate polyphenolic compounds and the phenolic moieties within the nanoparticles can be iodinated. Comparison using TEM and elemental analysis of these two approaches using sodium iodide with sodium hypochlorite as a model chemistry for iodination will be presented. Radioiodination of various size nanoparticles (80 to 300 nm) for PET imaging applications will also be discussed.

(758a) Development and Implementation of a Process to Control Off-Gassing

Lee, A. G., Bristol-Myers Squibb
Lo, E. T., Bristol-Myers Squibb Company
Natalie, K., Bristol-Myers Squibb Co,
Vu, T., Bristol-Myers Squibb Company
Rossano, L., Bristol-Myers Squibb Company
Originally presented on: 11/7/2013 15:15:00 - 15:35:00

The use of the *tert*-butyloxycarbonyl (BOC) moiety is a highly effective protection strategy for amines in the synthesis of small molecule active pharmaceutical ingredients. While evaluating these chemistries, the evolution rate of carbon dioxide and other volatile byproducts from these reactions is a major consideration for scale up. In our development of a BOC protection reaction to produce a pharmaceutical intermediate, the physical property of a solid intermediate severely hindered mixing in the vessel. As the reaction progressed, changes to the solid's physical property resulted in rapid uncontrolled off-gassing. We discuss the development of a process using acid and base charge equivalents, charge rate and pH to address the solid intermediate formation and gas evolution. We scaled up the process in the kilo plant to demonstrate a range of process parameters which yielded good control over gas evolution while maintaining acceptable quality and yield.

(758b) Chemical Engineering Strategies to Directly Deliver Process Properties Across Development and Manufacturing Scales

Glennon, B., UCD
Carr, M., APC Ltd
Whelan, J., University College Dublin
Barrett, M., University College Dublin
Originally presented on: 11/7/2013 15:35:00 - 15:55:00

In recent years there has been a growing necessity to reduce the time, cost and risk associated to the development and scale-up of pharmaceutical processes. The increasing pressures to direct deliver critical quality attributes (CQAs) and process performance criterion (cycle times, capacities etc) across manufacturing and supply scales compounding this challenge. As a result of this process engineering strategies, to enable streamlined, reliable and robust scale-up, has become increasing important.

Within this presentation novel process engineering strategies and technologies, underpinned by process modelling, will be presented that enable the delivery of process characteristics (i.e. particle size, shapes, PSDs, flowability, reaction purities, process kinetics etc). Applications of advanced process control, process simulation, kinetic assessments, particle engineering strategies, processing technologies (batch and continuous) will be discussed via numerous development and scale-up case studies.

(758c) The Chemical Engineer Strikes Back â€“ A New Hope for Commercial Process Development

Rothstein, S.,

Originally presented on: 11/7/2013 15:55:00 - 16:15:00

The process challenges of optimizing and scaling up the reaction to make 2,6-diaminopyrazine-1-oxide (DAPO), a precursor to a highly promising insensitive energetic material, will be discussed. The original process featured a precisely controlled dual addition of a liquid base and a solid reagent to limit the formation of more than 30% impurity and minimize starting material decomposition. These conditions limited this labor intensive reaction to the laboratory scale and produced DAPO with inconsistent yield and purity. Our understanding of the chemical reactivity led to improved reaction conditions featuring liquid doses rather than a solid dose, thus allowing for the rate of the reaction to be controlled by automated dosing. Using a kinetic model built with DynoChem® software, a robust process was optimized to improve the yield of DAPO and significantly reduce impurity formation. The reaction was scaled up using the optimized conditions and provides a potential commercial process for DAPO.

(758d) Reaction Engineering for a Sulfonamide Intermediate

Marek, J. C., AbbVie, Inc.
Zhao, Y., AbbVie, Inc.
Hill, D. R., AbbVie, Inc.
Originally presented on: 11/7/2013 16:15:00 - 16:35:00

Reaction engineering of a sulfonamide synthesis from early through mid and late stage development is discussed, with emphasis on consideration of reaction pathways. Reaction of sulfonyl chloride with ammonia in acetone solvent eliminates the use of aqueous ammonium hydroxide at extreme cryogenic temperatures (-60 °C), or for solid ammonium carbamate to achieve the amidation. The order of addition in the semi-batch amidation reaction was chosen to minimize use of excess equivalents of amidation reagent. In the revised process, the ammonia / acetone solution is charged into a solution of sulfonyl chloride, maintained at sub-ambient temperature (NMT -10 °C). Hence, both temperature and reagent equivalents are used to control aniline sulfonamide impurity levels. Formerly, the sulfonyl chloride was added to an excess of ammonium carbamate reagent at sub-ambient temperatures, and was promptly quenched to avoid over-reaction. The revised process affords robustness in design space including less potential for over reaction to aniline impurity.

(758e) Scale-Up Oscillatory Helical-Baffled Reactor for Multiphase Reactions

Phan, A. N., Newcastle University
Harvey, A. P., Newcastle University
Originally presented on: 11/7/2013 16:35:00 - 16:55:00

Scale-up oscillatory helical-baffled reactor for multiphase reactions

Anh N. Phan and Adam P. Harvey

School of Chemical Engineering & Advanced Materials (CEAM), Newcastle University, NE1 7RU, UK

Corresponding to anh.phan@ncl.ac.uk

Mesoscale oscillatory helical-baffled reactor (MOHB), a helical wire embedded inside a smooth tube without modification to the tube, was constructed and characterised at very low net flow rates (net flow Reynolds numbers, $Re_n \leq 10$) over a wide range of oscillation conditions. The fluid mixing is controlled by the oscillatory flow and is thus independent of net flow. The flow becomes chaotic at high oscillatory Reynolds numbers and the secondary flow (swirling) becomes active, which imposes an axial velocity component. The coupled effect of the vortex flow and the swirling flow in the presence of helical baffles augments the significant radial mixing and greatly broadens the operating/design window of these reactors. Plug flow performance can be achieved at a ratio of oscillatory Reynolds number to net flow Reynolds number of approximately 10-250 and a helical turn to tube diameter ratio in a range of 0.2-0.6. The highest degree of plug flow was achieved at Strouhal number $0.2 < St \leq 0.1$ and $Re_o = 400\text{-}700$ with the number of tanks in series equal to the number of helical turns. At a scale of 5mm inner diameter, the MOHB is considered to be effective and robust in scale-down production of high added value products, decreasing the reagents requirements and waste.

Is there a correlation between this mesoscale and a conventional scale? We scaled the mesoscale reactor up to 13 times and 130 times in volume by maintaining geometrical parameters (opened cross-sectional area and helical turn to tube diameter ratio) and oscillatory conditions.Â It was found that the same mixing behaviour in terms of residence time distribution profiles of tracers was obtained in the large scales as in the mesoscale.

Keywords: scale-up, helical-baffled reactor, oscillatory flow, plug flow

(758f) Process Scale-Up From Drying of a Single Drop to Production Scale Spray Dryer Using Levitated Single Droplet Analysis (DKA) and Computation Fluid Dynamics (CFD)

Walter, M., GEA Process Engineering Inc.
Originally presented on: 11/7/2013 16:55:00 - 17:15:00

Process Scale-Up from Drying of a Single Drop to Production Scale Spray Dryer Using Levitated Single Droplet Analysis (DKA) and Computation Fluid Dynamics (CFD)

Since its recent introduction, the single droplet analyzer (DKA - Digital Kinetics Analyzer) and Computational Fluid Dynamics (CFD) has allowed optimization of spray dryer design.Â Prior to utilizing DKA combined with CFD, one would have to theorize how a powder would behave during drying under various conditions and design the spray dryer accordingly.Â Â New Computational Fluid Dynamics modeling power combined with a novel means to examine in detail how a single drop of feed dries has allowed spray dryers to be more efficient, safer, smaller, of higher capacity, and with fewer operational headaches than in the past.

Â

In addition to improvements in spray dryer design, DKA allows us to examine how individual particles dry and how different drying parameters affect the dried product.Â This allows the spray dryer to produce a ? custom particle? through adjustment of spray drying parameters determined by DKA.Â Â For example, some might want a strong particle not easily damaged during transport; others might require a lightweight powder that dissolves easily.Â Use of DKA can determine if a given feed will dry into hollow spheres, shriveled particles, ?doughnut? shapes, etc. while providing insight into design of drying equipment and process parameters for attaining a certain type of particle.

(758g) An Integrated Management System to Optimize Process Operations Within the Pharmaceutical Industry

Drayton, R. A., THS, Inc
Gloor, G. J., Imperial College London
Grills, J.,
Originally presented on: 11/7/2013 17:15:00 - 17:35:00

One of the major challenges in developing a pharmaceutical process is the need to supply material of a quality representative to the final product, and at the same time, develop a safe and scalable process that can evolve into the final manufacturing process. In order to overcome this challenge, requires careful integration of process chemistry development, analytical techniques and process engineering of the chemical synthetic route. Historically, process safety has been regarded as a manufacturing concern, leaving the collection and analysis of process safety information, such as thermal stability and thermo-chemistry to the latter stages of process development. Since, other constraints such as product impurity was developed earlier in the process, this introduced the need for engineering solutions to process safety problems later in the process. This could have been avoided by straightforward changes in process chemistry, had these problems been identified and determined earlier in development. In addition, the process safety information that is obtained from physical experiment to address the process safety problems, can also provide valuable insight into, *inter alia*, process robustness and optimum operating conditions. Thus, the opportunity to leverage this information is lost by separating process safety from process chemistry and regarding it as an independent compliance function.

It is proposed here, to integrate the activities generally associated with process safety, such as thermal stability testing, reaction engineering and safety management into the development process for pharmaceuticals and fine chemicals. This will enable the data generated by the process safety function to be part of the dossier of process information that can be utilized by chemists and engineers in developing the "operating envelope". In addition, companies can avoid the delay and expense due to process changes that result from safety recommendations at latter stages in the process, and which conflicts with the process declarations in regulatory filings. This paper presents a novel integrated management system that can optimize process operations within the pharmaceutical industry.

(817a) Numerical Simulation Investigation On Particle Loading Process of Nano-Fiber Filter

Lu, P., Institute of Process Engineering, Chinese Academy of Sciences
Hua, C., Chinese Academy of Sciences
Bai, F., Institute of Process Engineering, Chinese Academy of Sciences
Tong, R., University of Science and Technology Beijing
Zhang, X., Institute of Process Engineering, Chinese Academy of Sciences
Originally presented on: 11/8/2013 12:30:00 - 12:55:00

For worsening air pollution□fibrous filter is respected to have higher performance to capture suspended particle in industrial production and life area, so nano-fiber is considered to design next generation filter because of low pressure drop and high particle capture efficiency. In this project, theoretical analysis, numerical simulation and experimental method were used to deeply investigate effect of slip-boundary of nano-fiber on flow filed and particle loading process in single fiber model and nano-fibrous filter model respectively. The Fluent software with User-Defined Function was used for simulation. The Euler – Lagrange method was used to solve flow filed and particles trajectory, particle interface was captured by voxel method and slip-boundary condition was realized by permeable fiber assumption. The particles deposition on single nano-fiber with different mechanism (diffusion, inertia and interception) was compared with normal fiber. In addition, a mathematic model was generated for nano-fibrous filter to describe effect of dust holding capacity (DHC) on subsequent pressure drop and particle capture efficiency. The predicted DHC data of nano-fibrous filter agreed well with experimental value of hand sheet. This research will provide available theoretical analysis and numerical results for design of next generation nano-fiber filter.

(817b) Development of Carbon Nanofibers By Electrospinning for Aerosol Filtration

Chattopadhyay, S.,
Rutledge, G. C., Massachusetts Institute of Technology
Originally presented on: 11/8/2013 12:55:00 - 13:20:00

Among various air filtration technologies, nanofibrous media are of interest, since theoretical calculations predict that these materials have high efficiencies to remove sub-micrometer aerosol particles with relatively lower energy consumption. Conventional fibrous filters made of nanofibers exhibit a local minimum of fractional collection efficiency, which is called the most penetrating particle size (MPPS). This is usually observed for particles 0.1-0.5 µm in diameter. Increasing filtration efficiency in this critical size range could be targeted by altering fiber composition, size and morphology. A method called "electrospinning" to prepare fibers with sub-micrometer diameters has received considerable attention in aerosol filtration research due to the reduction in filter mass required to obtain nanoparticle filtration efficiencies comparable to currently available commercial filters. Recently, electrospun carbon fibers prepared by carbonization of precursors like polyacrylonitrile, polycarbosilane, polyvinylidene fluoride and cellulose acetate fibers have been reported for various applications. During carbonization, the fiber structure remained unaffected but the mean diameter decreased by a factor of about 3-4. Additionally, properties like high electrical conductivity, wettability and specific surface area observed in such studies may offer other advantages for air filtration performance.

To this end, we report the development of continuous cellulose acetate and carbon fibers with diameters in the range of 0.1-1 µm and large surface areas (300-3000 m^2/g) as high efficiency particle filtration media. Important filter properties such as aerosol removal efficiency at low pressure drop are studied vis-à-vis electrospinning and carbonization parameters. For the current study, engineered nanoparticles comprising monodisperse oil droplets of di-2-ethyl hexyl sebacate (0.09 to 0.35 µm in physical diameter), are generated using a condensation monodisperse aerosol generator (CMAG). This aerosolizer generates a high number concentration (10^7 counts/cc) of spherical and charge-neutral particles by condensing oil vapor onto sodium chloride nuclei, based on the *Sinclair-LaMer* principle. The size distribution, number concentration, relative humidity and temperature of the aerosol were customized in an aerosol diluter before exposure to the test filter, which is housed in a standard 13-mm aerosol filter holder. The face velocity is regulated (1-100 cm/s) by a sump operating downstream of the filter assembly, and the pressure drop across the filter is monitored using a pressure transducer. The real time number concentrations of aerosol are measured using a scanning mobility particle sizer (SMPS), which measures the electrical mobility of aerosol in the size range 0.02-0.8 µm. The ratio of particle concentration, with and without the filter material and the geometric mean diameter of the particles that pass through the filter material are defined as 'particle penetration' and MPPS, respectively. Initial results indicate that the diameter (0.1-1 µm) and cross-section shape (ribbon and cylindrical) of electrospun cellulose acetate fibers can be controlled through selection of polymer solution and electrospinning parameters. The measured filtration efficiency for fibers having ribbon and cylindrical cross sections were found to be comparable, but the filters of cylindrical fibers were easier to handle and compact. The percentage retention by these cylindrical fiber filters was altered when solutions having different concentrations of cellulose acetate were used. Solutions with lower polymer concentration (15 and 17 wt %), which are anticipated to generate smaller diameter fibers, had 1.3 times higher percentage retention than counterparts formed from solutions of 19 wt % polymer. Likewise, the MPPS of filters decreased (from 90 nm to 77 nm) with decreasing solution concentration. In general, this seems to corroborate the understanding that smaller fiber diameter increases the filtration efficiency and decreases the MPPS. However, the anticipated decrease in fiber diameter was not evident through scanning electron micrographs, which showed that the fibers vary from 0.1-0.5 µm in diameter. Studies are underway to investigate other properties like surface area and solidity that could affect filtration properties. In addition, the filtration capabilities of carbonized polymeric electrospun will be reported for their possible application in high efficiency aerosol filters.

(817c) Modeling Investigation On Distillation Process Optimization of Benzene-Clohexene-Cyclohexane System By Using Aspen Plus

Tong, R., Institute of Process Engineering, Chinese Academy of Sciences
Mi, W., University of Science and Technology Beijing
Lu, P., Institute of Process Engineering, Chinese Academy of Sciences
Bai, F., Institute of Process Engineering, Chinese Academy of Sciences
Hua, C., Chinese Academy of Sciences
Originally presented on: 11/8/2013 13:20:00 - 13:45:00

In this work, Aspen plus software was used to model and optimize distillation process of Benzene(BZ)-Clohexene(HE)-Cyclohexane(HA) system with Dimethylacetamide □DMAC□as extractant. The thermodynamic model used NTRL. The effect of important operation parameter (theoretical plate number, feed position, solvent ratio, reflux ratio, and extractant feed position and temperature) on products separation efficiency (recovery percentage of BZ, mass percentage of BZ, HE and HA) was modeled and investigated in BZ separation tower and recovery tower respectively. The modeling results indicated when distillation process was operated in following condition, the separation efficiency was best and energy consumption was the lowest.

1) In Benzene separation tower, the theoretical plates was 36, feed position of raw material on the 12nd theoretical plates, solvent ratio 3.18, and reflux ratio 2.4; extractant feed position was on the 2^{nd}theoretical plates and feed temperature was 64℃.

2) In recovery tower, the best operation condition was 26 theoretical plates, feed position 10nd and reflux ratio 1.1.

(817d) A Mass Transfer Model of Absorption of Carbon Dioxide in a Bubble Column Reactor

Cheng, L., University of Cincinnati
Li, T., University of cincinnati
Keener, T. C., University of Cincinnati
Lee, J. Y., University of Cincinnati
Originally presented on: 11/8/2013 13:45:00 - 14:10:00

A bubble column reactor was tested and evaluated for carbon dioxide removal from flue gases by using magnesium hydroxide slurry. An absorption model was developed for system analysis and CO_2 removal data analysis. The overall mass transfer coefficients for CO_2-H_2O and CO_2-$Mg(OH)_2$-H_2O systems have been calculated. The chemistry of the system, temperature effects, and changes of equilibrium constants also have been discussed.

AUTHOR INDEX

AUTHOR INDEX

AUTHOR INDEX

AUTHOR INDEX

AUTHOR INDEX

AUTHOR INDEX

AUTHOR INDEX

AUTHOR INDEX

AUTHOR INDEX

American Institute of Chemical Engineers
3 Park Avenue
New York, New York 10016-5991

CURRAN ASSOCIATES INC
proceedings.com